Seismic Exploration of Hydrocarbons in Heterogeneous Reservoirs

New Theories, Methods, and Applications

Seismic Exploration of Hydrocarbons in Heterogeneous Reservoirs

New Theories, Methods, and Applications

Jing Ba
Department of Earth and Atmospheric Sciences
University of Houston, Houston, Texas, US

Qizhen Du
School of Geosciences, China University of Petroleum
(East China), Qingdao, China

José M. Carcione
Istituto Nazionale di Oceanografia e di Geofisica Sperimentale,
Trieste, Italy

Houzhu (James) Zhang
Geophysical Technology Department, ConocoPhillips,
Houston, Texas, US

Tobias M. Müller
Energy Flagship, CSIRO, Perth, Australia

ELSEVIER

AMSTERDAM • BOSTON • HEIDELBERG • LONDON
NEW YORK • OXFORD • PARIS • SAN DIEGO
SAN FRANCISCO • SINGAPORE • SYDNEY • TOKYO

Elsevier
Radarweg 29, PO Box 211, 1000 AE Amsterdam, Netherlands
The Boulevard, Langford Lane, Kidlington, Oxford OX5 1GB, UK
225 Wyman Street, Waltham, MA 02451, USA

Notices

Knowledge and best practice in this field are constantly changing. As new research and experience broaden our understanding, changes in research methods, professional practices, or medical treatment may become necessary.

Practitioners and researchers must always rely on their own experience and knowledge in evaluating and using any information, methods, compounds, or experiments described herein. In using such information or methods they should be mindful of their own safety and the safety of others, including parties for whom they have a professional responsibility.

To the fullest extent of the law, neither the Publisher nor the authors, contributors, or editors, assume any liability for any injury and/or damage to persons or property as a matter of products liability, negligence or otherwise, or from any use or operation of any methods, products, instructions, or ideas contained in the material herein.

British Library Cataloguing-in-Publication Data
A catalogue record for this book is available from the British Library

Library of Congress Cataloging-in-Publication Data
A catalog record for this book is available from the Library of Congress

ISBN: 978-0-12-420151-4

Typeset by Thomson Digital

Contents

5. Characteristics of Seismic Wave Propagation in Viscoelastic Anisotropic Fractured Reservoirs

Xianzheng Zhao, Xuming Bai, Qizhen Du

6. Reverse-Time Migration: Principles, Practical Issues, and Recent Developments

Houzhu (James) Zhang

List of Contributors

Jing Ba, Department of Earth and Atmospheric Sciences, University of Houston, Houston, Texas, US

Xuming Bai, Bureau of Geophysical Prospecting INC., China National Petroleum Corporation, Zhuozhou, Hebei, China

José M. Carcione, Istituto Nazionale di Oceanografia e di Geofisica Sperimentale, Trieste, Italy

Eva Caspari, Applied and Environmental Geophysics Group, University of Lausanne, Lausanne, Switzerland

Qizhen Du, School of Geosciences, China University of Petroleum (East China), Qingdao, China

Gang Fang, Qingdao Institute of Marine Geology, Qingdao, China

Xufei Gong, School of Geosciences, China University of Petroleum (East China), Qingdao, China

Yuqian Guo, Institute of Geomechanics, Chinese Academy of Geological Sciences, Xicheng, Beijing, China

Boris Gurevich, Energy Flagship, CSIRO, Perth, Australia; Department of Exploration Geophysics, Curtin University, Perth, Australia

Xinzhen He, Amu Darya Petroleum Company Ltd., CNPC, Beijing, China

Maxim Lebedev, Department of Exploration Geophysics, Curtin University, Perth, Australia

Jinsong Li, Research Institute of Petroleum Exploration and Development, PetroChina, Beijing, China

Sofia Lopes, Department of Exploration Geophysics, Curtin University, Perth, Australia

Tobias M. Müller, Energy Flagship, CSIRO, Perth, Australia

Qiaomu Qi, Department of Exploration Geophysics, Curtin University, Perth, Australia

J. Germán Rubino, Applied and Environmental Geophysics Group, University of Lausanne, Lausanne, Switzerland

Weitao Sun, Zhou Pei-Yuan Center for Applied Mathematics, Tsinghua University, Haidian, Beijing, China

Danilo Velis, Facultad de Ciencias Astronómicas y Geofísicas, Universidad Nacional de La Plata, and CONICET, Argentina

Xinfei Yan, Research Institute of Petroleum Exploration and Development, PetroChina, Beijing, China

Hao Yu, Research Institute of Petroleum Exploration and Development, PetroChina, Beijing, China

Zhenyu Yuan, School of Geosciences, China University of Petroleum (East China), Qingdao, Shandong, China

Houzhu (James) Zhang, Geophysical Technology Department, ConocoPhillips, Houston, Texas, US

Lin Zhang, School of Geosciences, China University of Petroleum (East China), Qingdao, Shandong, China

Mingqiang Zhang, School of Geosciences, China University of Petroleum (East China), Qingdao, China

Xingyang Zhang, Research Institute of Petroleum Exploration and Development, PetroChina, Beijing, China

Xianzheng Zhao, Huabei Oilfield Branch, China National Petroleum Corp., Renqiu, Hebei, China

Chapter 1

Introduction

Jing Ba[1], Qizhen Du[2], José M. Carcione[3], Houzhu (James) Zhang[4], Tobias M. Müller[5]

[1]*Department of Earth and Atmospheric Sciences, University of Houston, Houston, Texas, US;*
[2]*School of Geosciences, China University of Petroleum (East China), Qingdao, China;*
[3]*Istituto Nazionale di Oceanografia e di Geofisica Sperimentale, Trieste, Italy;*
[4]*Geophysical Technology Department, ConocoPhillips, Houston, Texas, US;*
[5]*Energy Flagship, CSIRO, Perth, Australia*

Chapter Outline

Over the past decade, the world's petroleum and natural gas consumption exceeded four times the expected oil/gas reserves so that the supply from oil companies could not meet the demands of human activities. The difference between proven reserves and increasing demands is still growing and the remaining recoverable reserves could run out in 40 years according to the latest oil/gas production rate.

The global demands of hydrocarbons have been driven by the increase of world's population and, particularly, the industrialization of the Asian countries. After several decades of exploration in large structural hydrocarbon reservoirs, high technology has been developed to attack more complex stratigraphic hydrocarbon traps, and recently, unconventional oil/gas resources, such as shale gas (oil), tight gas (oil), gas hydrates, heavy oil, and oil sands, which cannot be effectively explored and extracted using traditional exploration and production techniques.

Seismic Exploration of Hydrocarbons in Heterogeneous Reservoirs. http://dx.doi.org/10.1016/B978-0-12-420151-4.00001-5

1.1 CHALLENGES IN HYDROCARBON SEISMIC EXPLORATION

With the depletion of conventional reservoirs, oil companies are facing new technological challenges to deal with deep exploration, highly dipping structures, strong lateral variations, and complex reservoir rocks showing anisotropy and attenuation induced by fractures and layering. These challenges require new rock-physics theories and processing methods to be developed, such as new rheological equations and modeling and imaging algorithms.

1.1.1 Seismic Attenuation

Wave-induced local fluid flow is the main physical mechanism to explain the observed seismic attenuation and velocity dispersion in fluid-saturated rocks (Müller et al., 2010). Dispersion and attenuation are dependent on several factors such as fluid saturation and distribution (patch size and shape) and pore structure (e.g., Cadoret et al., 1998). Although the global-flow attenuation was predicted by Biot (Biot, 1962), a theory of mesoscopic flow was not established until the 1970s (White, 1975). In fact, the mesoscopic-loss mechanism is believed to correctly describe the attenuation effects at seismic frequencies (Ba et al., 2008, 2011; Carcione, 2014). Seismic wave signatures affected by this mechanism, are inherently frequency dependent. Dispersion and attenuation depend on several factors, such as partial saturation, variations in pore structures and matrix compressibility. Therefore, the study of dispersion and attenuation characteristics can yield to useful information about the rock properties including porosity, permeability, and fluid saturation.

1.1.2 Seismic Anisotropy

Fine layering is a geological system contributing to the formation of sedimentary basins and is the cause of seismic effective anisotropy. Moreover, fine layers may be composed of permeable rocks, such as sandstones, and have fractures and aligned cracks that contain hydrocarbons under certain pressure and temperature conditions. They can be described by equivalent transversely isotropic or/and lower-symmetry media, which occurs when the wavelength is long compared with the dimensions of the layers and the separation between fractures. This leads to varying velocity and attenuation of seismic waves with offset and azimuth. Therefore, the study of the wave response of these systems is essential to interpret seismic data. Backus obtained the five average-elasticity constants when the single layers are lossless and transversely isotropic with the symmetry axis perpendicular to the layering plane. The presence of viscous fluids induces attenuation- or Q-anisotropy of seismic waves with related dispersion effects. Carcione (1992) generalized Backus averaging to the anelastic case, obtaining the first model for Q-anisotropy. Other effective media can usually be used in describing fractured systems (e.g., Hudson, 1980). Schoenberg (1980) introduced a suitable interface condition to model fractures, the so-called linear-slip model, based on the discontinuity of the displacement field across the interface.

1.1.3 Reverse-Time Migration and Wavefield-Propagation Operators

Seismic imaging is the primary approach to estimate the subsurface structures for hydrocarbon exploration. Today, reverse-time migration (RTM; McMechan, 1983) is one of the most advanced imaging methods largely due to the rapid development of computer hardware and high-performance computing. With the use of full-wave modeling for wavefield extrapolation, RTM has no dip limitations, can image complicated geological features, and is able to handle arbitrary rheologies, from isotropic to VTI (transversely isotropic with a vertical symmetry axis), TTI (transversely isotropic with a tilted symmetry axis), and orthorhombic media. True-amplitude migration, which computes reflection-coefficient maps from the recorded data, provides a solid physical basis for the quantitative analysis of amplitude variations with offset and angle. Propagation operators are essential for true-amplitude migration. True-amplitude migration have been proposed to compensate the effects of geometric spreading in one-way wave equation migration (Zhang et al., 2003) or the effects of transmission losses in reverse-time migration (Deng and McMechan, 2007; Deng and McMechan, 2008). High-frequency asymptotics is a powerful tool providing a relatively simple approximation, that is, the WKBJ (Wentzel–Kramer–Brillouin–Jeffreys) solution of the wave equation, where geometric spreading and transmission losses can be compensated. Based on this tool, Docherty (1991) and Zhang et al. (2003) developed the true-amplitude migration/inversion theory (Bleistein, 1987).

1.1.4 Rock-Physics Modeling and Quantitative Seismic Interpretation

Complex fluid distributions and pore structures generate seismic fields that have to be interpreted with proper rock-physics models. The classical approach has been developed by Gassmann (1951), which relates the bulk modulus to the fluid, grain, and frame properties of fully saturated porous media. However, Gassmann theory is not suitable for applications to partially saturated media, due to the assumption of a single (effective) fluid saturating the pore space. Ba et al. (2011) introduced the Biot–Rayleigh (BR) theory for wave propagation in inhomogeneous rocks, deriving a set of suitable wave-propagation equations on the basis of physically measurable properties to meet the needs of practical applications. The equations are based on the classical Biot theory and describe propagation in porous rocks with mesoscopic heterogeneities. The BR theory has successfully been extended and applied to the detection of gas in sandstone reservoirs by Ba et al. (2012). One of the main issues in rock-physics modeling and fluid detection with seismic waves is the difficulty to adapt the theories to solve practical problems. Ba et al. (2013) attempted to achieve this scope using the BR equations and a simple processing workflow to detect hydrocarbons in limestone reservoirs by means of rock-physics templates. In the presence of strong lateral heterogeneities, a more effective workflow has to be developed, for instance, combining rock physics and the method of logging (well)-constrained seismic inversion.

1.2 MAIN CONTENTS OF THE BOOK.

This book addresses four relevant issues of seismic exploration of hydrocarbons in heterogeneous reservoirs, as illustrated in the following sections.

1.2.1 Wave-Propagation Theories and Experiments

Chapters 2 and 3 focus on new developments of wave-propagation theories and laboratory experiments in reservoir rocks. Chapter 2 illustrates the Biot–Rayleigh theory for wave motion in a heterogeneous fluid-saturated rock, regarding inclusions embedded in a host medium. The theory describes the cases of a double-porosity solid saturated with a single fluid and a single-porosity medium saturated with two immiscible fluids. Examples show how fluid type, saturation, and porosity affect the seismic wave properties and how the theory compares to experimental data obtained for sandstones and carbonates. The effect of fluid substitution is particularly analyzed based on the ultrasonic experimental data performed on sandstones containing oil. Chapter 3 outlines recent advances in the interpretation of controlled laboratory experiments in view of a more quantitative understanding of velocity- and attenuation-saturation relations. Herein the interaction between elastic waves and fluid patches is interpreted in terms of wave-induced pressure diffusion. This mechanism assumes that elastic waves can induce an oscillatory relative motion between the solid and fluid phases that leads to dissipation and thus wave attenuation. For a review of this mechanism, we refer to Müller et al. (2010). On the basis of the works by Lopes et al. (2014) and Qi et al. (2014), new aspects of wave-induced pressure diffusion are discussed in this chapter. The issue of extrapolating laboratory-scale data to field-scale observations is addressed with numerical simulations, as in the CO_2 geo-sequestration example, where it is shown that centimeter-scale fluid patches affect the characteristics of seismic waves. Core flooding laboratory experiments and reservoir-scale injection/production operations differ by orders of magnitude in length scale. The quantitative interpretation of velocity- and attenuation-saturation relations at different scales and frequency bands offers new insight for future research.

1.2.2 Seismic Modeling in Anisotropic Rocks

Chapter 4 presents the seismic properties of fine layers and fractures in a solid background and the generalization to the more realistic poroelastic case. Wave propagation through fractures, cracks, and any partially bonded interface is described with suitable boundary conditions, consider first the scattering coefficients of a single fracture and then obtaining the wave characteristics of a dense set of fractures in solid and porous media. Measurable quantities such as the phase and energy (ray) velocities and the attenuation and quality factors are obtained from the dispersion equation and energy considerations in terms of the medium properties and propagation directions. Finally, quasistatic numerical harmonic experiments are described to compute the stiffness components of an equivalent medium (layered or fractured) in heterogeneous media. Chapter 5 describes algorithms for simulating wave propagation in fractured reservoirs including effective anisotropy and viscoelasticity, on the basis of the Schoenberg linear-slip model (Schoenberg, 1980) and the Kelvin–Voigt mechanical element. The relationships between the stiffness moduli and the microscopic fracture properties are based on the Hudson theory. The 3D seismic modeling algorithm is based on a rotated staggered grid finite-difference (RSGFD) scheme (Saenger et al., 2000).

1.2.3 Developments in Reverse-Time Migration and Wave Operator

Chapter 6 provides an overview of the principles, practical issues, and recent developments of RTM, focusing on practical aspects rather than mathematical developments. Imaging in anisotropic media (Thomsen, 1986) is one of milestones in the development of RTM and had a significant impact on imaging the subsurface in the Gulf of Mexico. This topic is fully covered in Section 6.3. Regarding geological environments with strong velocity contrasts such as subsalt areas in the Gulf of Mexico and West Africa, illumination is an important issue (Section 6.5.4). It is shown how to improve the illumination with real-data examples. RTM is closely related to the developments of other technologies such as seismic acquisition. Recently, broadband and long offset data have shown their values in providing good images and models in subsalt areas (Mothi and Kumar, 2014). These new developments are reported in Section 6.6. Chapter 7 reviews the wave operators used in the forward and backward propagation of RTM. Then, a true-amplitude migration formula is derived for variable density and velocity media. The chapter analyses the issues of compensating the geometric spreading and transmission losses under the high-frequency approximation. Finally numerical examples are provided to verify the theoretical derivations and the compensation method.

1.2.4 Quantitative Hydrocarbon Seismic Detection

Chapter 8 illustrates a case history to estimate porosity and hydrocarbon saturation in a heterogeneous carbonate reservoir, and a workflow to apply multiscale rock-physics templates to seismic data, where CT scans are performed on partially saturated rocks to determine the fluid distribution. A rock-physics method is applied to limestone gas reservoirs located on the right bank block of Amu Darya River, whereas the template is calibrated with laboratory, well-log, and seismic data. The estimation of rock porosity and fluid saturation is performed by applying the rock-physics template to prestack inversion results. The reservoir pore structures at each well are investigated by thin-section analyses and seismic data calibrations. Finally, the main results are compared to actual well-test production reports.

REFERENCES

Ba, J., Carcione, J.M., Cao, H., Du, Q.Z., Yuan, Z.Y., Lu, M.H., 2012. Velocity dispersion and attenuation of P waves in partially-saturated rocks: wave propagation equations in double-porosity medium. Chinese J. Geophys. 55 (1), 219–231 (in Chinese).

Ba, J., Cao, H., Carcione, J.M., Tang, G., Yan, X.F., Sun, W.T., Nie, J.X., 2013. Multiscale rock-physics templates for gas detection in carbonate reservoirs. J. Appl. Geophys. 93, 77–82.

Ba, J., Carcione, J.M., Nie, J.X., 2011. Biot-Rayleigh theory of wave propagation in double-porosity media. J. Geophys. Res. 116, B06202.

Ba, J., Nie, J.X., Cao, H., Yang, H.Z., 2008. Mesoscopic fluid flow simulation in double-porosity rocks. Geophys. Res. Lett. 35, L04303.

Biot, M.A., 1962. Mechanics of deformation and acoustic propagation in porous media. J. Appl. Phys. 33, 1482–1498.

Bleistein, N., 1987. On the imaging of reflectors in the earth. Geophysics 52, 931–942.

Cadoret, T., Mavko, G., Zinszner, B., 1998. Fluid distribution effect on sonic attenuation in partially saturated limestones. Geophysics 63, 154–160.

Carcione, J.M., 1992. Anisotropic Q and velocity dispersion of finely layered media. Geophys. Prosp. 40, 761–783.

Carcione, J.M., 2014. Wave fields in real media: wave propagation in anisotropic, anelastic, and porous and electromagnetic media, third ed. Elsevier, Amsterdam.

Deng, F., McMechan, G.A., 2007. True-amplitude prestack depth migration. Geophysics 72, S155–S166.

Deng, F., McMechan, G.A., 2008. Viscoelastic true-amplitude prestack reverse-time depth migration. Geophysics 73, S143–S155.

Docherty, P., 1991. A brief comparison of some Kirchhoff integral formulas for migration and inversion. Geophysics 56, 1164–1169.

Gassmann, F., 1951. Über die Elastizität poröser Medien: Vier. der Natur. Gesellschaft in Zürich 96, 1–23.

Hudson, J., 1980. Overall properties of a cracked solid. Mathematical Proceedings of the Cambridge Philosophical Society. 88, 371–384.

Lopes, S., Lebedev, M., Müller, T.M., Clennell, M.B., Gurevich, B., 2014. Forced imbibitions into a limestone: measuring P-wave velocity and water saturation dependence on injection rate. Geophys. Prospect. 62, 1126–1142.

McMechan, G.A., 1983. Migration by extrapolation of time-dependent boundary values. Geophys. Prospect. 31, 413–420.

Mothi, S., and Kumar, R., 2014. Detecting and estimating anisotropy errors using full waveform inversion and ray-basedtomography: a case study using long-offset acquisition in the Gulf of Mexico: 84th Annual International Conference & Exhibition, Society of Exploration Geophysicists.

Müller, T.M., Gurevich, B., Lebedev, M., 2010. Seismic wave attenuation and dispersion resulting from wave-induced flow in porous rocks – a review. Geophysics 75, 75A147–75A164.

Qi, Q., Müller, T.M., Gurevich, B., Lopes, S.C., Lebedev, M., Caspari, E., 2014. Quantifying effect of capillarity on dispersion and attenuation in patchy-saturated rocks. Geophysics 79, WB35–WB50.

Saenger, E.H., Gold, N., Shapiro, S.A., 2000. Modeling the propagation of elastic waves using a modified finite-difference grid. Wave Motion 31 (1), 77–92.

Schoenberg, M., 1980. Elastic wave behavior across linear slip interfaces. J. Acous. Soc. Am. 68 (5), 1516–1521.

Thomsen, L., 1986. Weak elastic anisotropy. Geophysics 51, 1954–1966.

White, J.E., 1975. Computed seismic speeds and attenuation in rocks with partial gas saturation. Geophysics 40 (2), 224–232.

Zhang, Y., Zhang, G., Bleistein, N., 2003. True amplitude wave equation migration arising from true amplitude one-way wave equations. Inverse Prob. 19, 1113–1134.

Chapter 2

Wave Propagation and Attenuation in Heterogeneous Reservoir Rocks

Jing Ba[1], Zhenyu Yuan[2], José M. Carcione[3], Yuqian Guo[4], Lin Zhang[2], Weitao Sun[5]

[1]*Department of Earth and Atmospheric Sciences, University of Houston, Houston, Texas, US;*
[2]*School of Geosciences, China University of Petroleum (East China), Qingdao, Shandong, China;*
[3]*Istituto Nazionale di Oceanografia e di Geofisica Sperimentale, Trieste, Italy;*
[4]*Institute of Geomechanics, Chinese Academy of Geological Sciences, Xicheng, Beijing, China;*
[5]*Zhou Pei-Yuan Center for Applied Mathematics, Tsinghua University, Haidian, Beijing, China*

Chapter Outline

2.1 INTRODUCTION

The mechanism of local fluid flow explains the high attenuation and dissipation of low-frequency seismic waves in fluid-saturated reservoir rocks. When acoustic waves propagate through small-scale heterogeneities, pressure gradients are induced between regions of dissimilar properties. The mesoscopic-scale length of the heterogeneity is intended to be larger than the grain size, but much smaller than the wavelength of pulse. If the matrix compressibility varies significantly,

Seismic Exploration of Hydrocarbons in Heterogeneous Reservoirs. http://dx.doi.org/10.1016/B978-0-12-420151-4.00002-7

diffusion of the pore fluid between different regions constitutes a mechanism that is important at seismic frequencies.

An attempt to introduce squirt-flow effects has been presented by Dvorkin et al. (1995) in a microscopic scale, in which a force applied to the area of contact between two grains produces a displacement of the surrounding fluid in and out of this area. White (1975) and White et al. (1975) were the first to introduce the mesoscopic loss mechanisms based on poroelasticity. Gas pockets in a water-saturated porous medium and porous layers alternately saturated with water and gas are considered, respectively. These are the so-called "patchy saturation" models. The mesoscopic-loss theory has been further studied by Shapiro and Müller (1999), Johnson (2001), Müller and Gurevich (2004), and Pride et al. (2004).

Johnson (2001) developed a model for patches of arbitrary shape, which has two geometrical parameters: the specific surface area and the size of the patches. Murphy (1982) and Knight and Nolen-Hoeksema (1990) have observed the patchy saturation effects on acoustic properties. Cadoret et al. (1995) investigated the relevant phenomenon in the laboratory at the frequency range 1–500 kHz, and the results showed that two different saturation methods result in different fluid distributions and produce two dissimilar values of velocity for the same saturation degree.

Generalization of Biot's theory to a composite or a double-porosity medium is generally based on Biot's poromechanical approach by using Lagrange's equations (Biot, 1962). Berryman and Wang (2000) derived phenomenological equations for the poroelastic behavior of a double-porosity medium to account for storage porosity and fracture porosity in oil/gas reservoirs. Another Biot-type poroelasticity theory describes the mesoscopic loss generated by lithological patches having different degrees of consolidation (Pride et al., 2004). Ba et al. (2008a) and Ba et al. (2008b) have solved the double-porosity governing equations and simulated the wavefields by using the pseudo-spectral method. Liu et al. (2009) solved equivalent poroviscoacoustic equations by approximating the mesoscopic complex moduli in the frequency domain using Zener mechanical models.

In this chapter, we develop a new double-porosity theory, namely the Biot–Rayleigh theory, based on Biot's theory of poroelasticity and a generalization of Rayleigh's theory of fluid collapse to the porous case. We further generalize it to the case of a single-porosity solid saturated with two immiscible fluids. By analyzing the influence of the local fluid-flow velocity fields in the inclusion, the poroelasticity theories are reformulated when the fluid kinetic energy of the inclusions cannot be neglected (it is the case in heterogeneous oil/water reservoirs). Comparisons are performed between the theoretical predictions and experimental measurements and fluid substitution effects in actual rocks are discussed.

2.2 BIOT–RAYLEIGH THEORY OF WAVE PROPAGATION IN HETEROGENEOUS POROUS MEDIA

For a double-porosity medium as shown in Figure 2.1, wave propagation induces local fluid flow (LFF) between the inclusions and the host medium due to

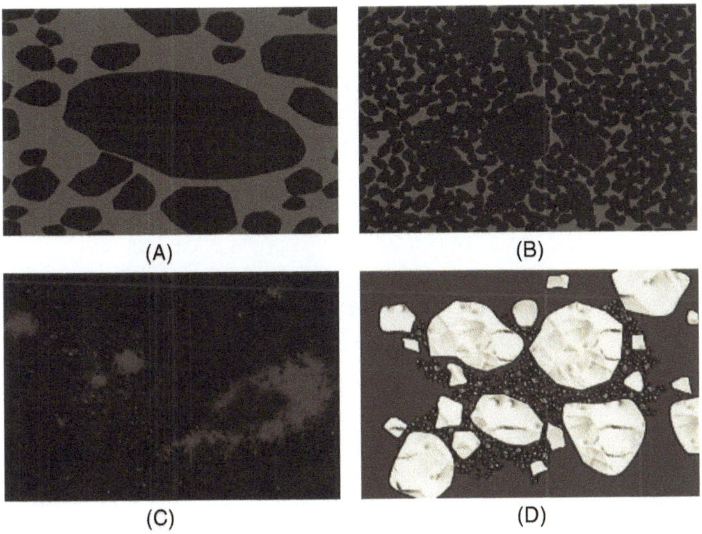

FIGURE 2.1 Synoptic diagram showing four types of double-porosity media in nature. (A) Flat throats connected to stiff pores at a microscopic scale. (B) Patches of small grains embedded in a matrix formed of large grains. (C) Strongly dissolved dolomite, in which powder crystals are present in the pores forming a second matrix. (D) Partially melted ice matrix, whose pores are filled with a mixture of loosely contacted micro-ice crystals and water.

the different compressibilities. For the purpose of describing the wave-induced LFF, we assume that LFF occurs between soft and stiff pores. We then propose a double-porosity approach by using a generalization of Rayleigh's theory of liquid collapse of a spherical cavity in the framework of Biot's poromechanics (Biot, 1962).

In order to describe quantitatively the dynamical process of wave-induced LFF in heterogeneous fluid/solid composites, the dynamical governing equations are derived from first principles. First, the kinetic energy expression is derived by considering a spherical inclusion imbedded in a uniform porous host medium. Then, the kinetic energy and strain potential energy functions are combined with Lagrange equations, and three dynamical equations are obtained describing wave propagation. We derive the relevant stiffness coefficients by performing "gedanken experiments." Then, analyses of plane-wave solutions are conducted to obtain the phase velocity and quality factor as a function of frequency.

We consider the inclusion model shown in Figure 2.2, with the following assumptions: (1) The inclusions are spherical, homogeneous and of the same size; (2) the boundary conditions between the inclusions and the host medium are open; (3) the radius of the inclusion is much smaller than the wavelength; and (4) the inclusion volume ratio is low, so the interaction between inclusions can be neglected. Assume that R_0 is the radius of the (spherical) inclusion at

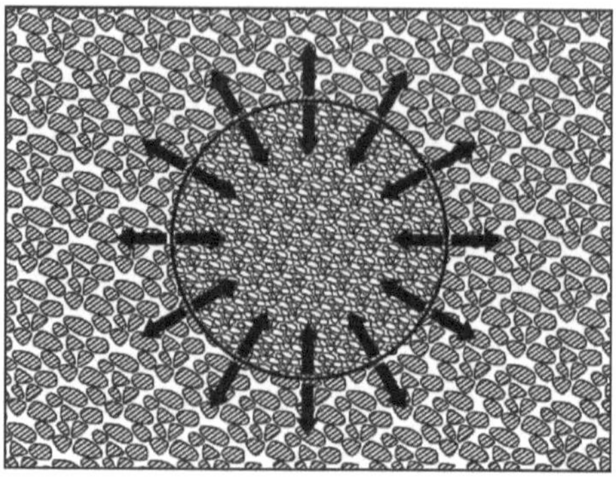

FIGURE 2.2 Double-porosity model containing spherical inclusions. The circle indicates the boundary of the inclusion.

rest, and R is the dynamical radius of the fluid sphere after the LFF process (see Figure 2.2), which is a function of time. The inclusions and host medium are assumed to be completely connected (Deresiewicz and Skalak, 1963). In actual rocks, there may be partial connection or no connection between different components.

The time variable is denoted by t and the spatial variables by x_i, $i = 1, 2, 3$. The displacement vector of the matrix is $\mathbf{u} = (u_1, u_2, u_3)^T$ and the components of the average fluid displacement vector in the two porosity systems (host medium and inclusions) are denoted as $\mathbf{U}^{(m)} = (U_1{}^{(m)}, U_2{}^{(m)}, U_3{}^{(m)})^T$, $m = 1, 2$. In a double-porosity medium, the difference between the fluid displacement in the inclusion and host medium is much larger than that of the solid phase, due to the fluid's lower bulk modulus. In this context, one solid displacement vector and two fluid displacement vectors are considered. The corresponding strain components are

$$\begin{aligned} e_{ij} &= \frac{1}{2}(\partial_j u_i + \partial_i u_j), \\ \xi_{ij}^m &= \frac{1}{2}(\partial_j U_i^{(m)} + \partial_i U_j^{(m)}), \end{aligned} \tag{2.1}$$

and the dilatations are

$$\begin{aligned} e &= e_{ii} = \partial_i u_i, \\ \xi_m &= \xi_{ii}^{(m)} = \partial_i U_i^{(m)}, \end{aligned} \tag{2.2}$$

where ∂_i denotes a partial derivative with respect to the spatial variable x_i and the Einstein summation of repeated indices is assumed. The strain energy in a double-porosity medium is a function of five independent variables,

$$W = W(I_1, I_2, I_3, \xi_1, \xi_2), \tag{2.3}$$

where

$$I_1 = e_{ii},$$
$$I_2 = \begin{vmatrix} e_{11} & e_{12} \\ e_{12} & e_{22} \end{vmatrix} + \begin{vmatrix} e_{22} & e_{23} \\ e_{23} & e_{33} \end{vmatrix} + \begin{vmatrix} e_{33} & e_{13} \\ e_{13} & e_{11} \end{vmatrix}.$$
$$I_3 = \begin{vmatrix} e_{11} & e_{12} & e_{13} \\ e_{12} & e_{22} & e_{23} \\ e_{13} & e_{23} & e_{33} \end{vmatrix}. \tag{2.4}$$

We consider a Taylor expansion of W and neglect all terms whose orders are higher than two,

$$2W = (A+2N)I_1^2 - 4NI_2 + 2Q_1 I_1 \xi_1 + R_1 \xi_1^2 + 2Q_2 I_1 \xi_2 + R_2 \xi_2^2, \tag{2.5}$$

where A, N, Q_1, R_1, Q_2, and R_2 are stiffnesses.

A scalar ζ is then used to denote the fluid strain's increment in the LFF process generated by the wave. An internal vibration process accompanies wave propagation, which causes rock relaxation. The deformation of the pore fluid in the two types of pores is rebalanced before potential energy is transferred to kinetic energy. We have

$$2W = (A+2N)I_1^2 - 4NI_2 + 2Q_1 I_1(\xi_1 + \phi_2\zeta) + R_1(\xi_1 + \phi_2\zeta)^2,$$
$$+2Q_2 I_1(\xi_2 - \phi_1\zeta) + R_2(\xi_2 - \phi_1\zeta)^2, \tag{2.6}$$

where ϕ_m are porosities.

Then, the LFF interaction is represented by the fluid variation ζ. Fluid flow from $m = 1$ to $m = 2$ implies the fluid variation $-\phi_1\zeta$. The opposite flow induces the variation $\phi_2\zeta$ in system 1, such that the conservation of fluid mass is satisfied, that is, $\phi_1(\phi_2\zeta) + \phi_2(-\phi_1\zeta) = 0$.

Based on the model of Figure 2.2, the effective fluid volume variation in system 1 is expressed as $\phi_1\zeta \approx 1 - V_0/V$, where V_0 and V are the volumes of the fluid sphere corresponding to the radius of inclusion R_0 and the dynamical radius of the fluid sphere R, respectively. Then

$$\zeta = \frac{1}{\phi_1}\left(1 - \frac{R_0^3}{R^3}\right). \tag{2.7}$$

The quantities v_m, $m = 1, 2$ are used to denote the volume fraction of the m porous media present in an averaging volume. The total porosity is given by

$$\phi = \phi_1 + \phi_2, \tag{2.8}$$

where

$$\phi_m = v_m \phi_{m0}, \tag{2.9}$$

and ϕ_{m0} is the porosity of each phase.

Based on Biot's theory (Biot, 1962), the kinetic energy is a function of the fluid-particle velocity relative to the solid phase. Biot's kinetic energy T is generalized to the double-porosity case as

$$2T = \rho_0 \sum_i \dot{u}_i^2 + \rho_f \sum_m \int_{\Omega_m} \sum_i (\dot{u}_i + g_i^{(m)} + l_i^{(m)})^2 d\Omega_m, \tag{2.10}$$

where $\mathbf{g}^{(m)} = (g_1, g_2, g_3)^{\mathrm{T}}$ and $\boldsymbol{l}^{(m)} = (l_1, l_2, l_3)^{\mathrm{T}}$ denote the relative microvelocity fields of the fluid in each phase, associated with the global and local flows, respectively, Ω_m is the pore volume, and

$$\rho_0 = (1 - \phi)\rho_s, \tag{2.11}$$

with ρ_s and ρ_f the solid and fluid densities, respectively. The symbols Σ_i and Σ_m denote summation from 1 to 3 corresponding to the Cartesian coordinates and summation from 1 to 2 corresponding to the two porosity phases, and a dot above a variable indicates time differentiation.

The kinetic energy in terms of vectors $\mathbf{U}^{(m)}$ is

$$2T = \rho_{00} \sum_i \dot{u}_i^2 + 2\sum_m \rho_{0m} \sum_i \dot{u}_i \dot{U}_i^{(m)} + \sum_m \rho_{mm} \sum_i (\dot{U}_i^{(m)})^2 + 2T_L, \tag{2.12}$$

where

$$\begin{aligned} \rho_{00} &= \rho - \sum_m \phi_m (2\rho_f - \bar{\rho}_m \phi_m), \\ \rho_{mm} &= \bar{\rho}_m \phi_m^2, \\ \rho_{0m} &= \rho_m - \rho_{mm}, \end{aligned} \tag{2.13}$$

and T_L is the LFF kinetic energy density which is related to compressional deformations.

In analyzing the kinetic energy of the LFF process, the relation between the fluid sphere's dynamical radius and fluid increment can be established. Let R be the dynamical radius of the fluid sphere, $\dot{R}$ the particle velocity at the boundary, $r > R$, and $\dot{r}$ the particle velocity outside the inclusion, such that these particle velocities correspond to the radial component of vector $\mathbf{U}$. Since this study deals with a fluid-saturated porous solid, Rayleigh's theory is generalized to the case of a porous medium.

First, the boundary condition at the surface of the inclusion is $\phi_{10}\dot{r} = \phi_{20}\dot{R}$. Then, the Rayleigh equation that expresses the conservation of fluid (between the host medium and the inclusion) can be generalized as

$$4\pi R^2 \dot{R} \phi_{20} = 4\pi r^2 \dot{r} \phi_{10}, \tag{2.14}$$

which satisfies the boundary condition at $r = R$. By following Rayleigh (1917), the kinetic energy outside the inclusion is expressed as

$$T_1 = \frac{1}{2} \phi_{10} \rho_f \int_R^{\infty} 4\pi r^2 \dot{r}^2 \, dr, \tag{2.15}$$

where T_1 gives the kinetic energy of a single inclusion. Substituting eq. (2.14) gives

$$T_1 = 2\pi\rho_f R^3 \dot{R}^2 (\phi_{20}^2 / \phi_{10}). \tag{2.16}$$

The volume ratio of the inclusion is $v_2 \approx (4/3)\pi R_3 N_0$, where N_0 is the number of inclusions per unit volume of composite. Using eq. (2.9),

$$R^3 = \frac{3\phi_2}{4\pi\phi_{20}N_0}. \tag{2.17}$$

Using eqs (2.7) and (2.17) and multiplying (2.16) by N_0, we get

$$\dot{R} \approx \frac{1}{3}\phi_1 R_0 \dot{\zeta}, \tag{2.18}$$

which is a relation between the dynamical radius and fluid increment. The kinetic energy density for all the inclusions can then be derived as

$$T_L = \frac{1}{6}\left(\frac{\phi_1^2\phi_2\phi_{20}}{\phi_{10}}\right)\rho_f R_0^2 \dot{\zeta}^2. \tag{2.19}$$

The volume ratio of inclusions is assumed to be low and the distance between two inclusions large enough, so that the interaction between them can be neglected. Wave scattering in a solid matrix and LFF can happen simultaneously in a fluid-saturated double-porosity system. However, the scattering effect is much weaker than the fluid's swell-shrinking motion at the frequency bands under consideration. In this work, scattering is neglected.

Similar to the calculation of the kinetic energy T_L (see eq. (2.15)), the dissipation function of the LFF process is

$$D_L = \frac{1}{2}\phi_{10}^2\left(\frac{\eta}{\kappa_1}\right)\int_R^\infty 4\pi r^2 \dot{r}^2 dr = 2\pi\phi_{20}^2\left(\frac{\eta}{\kappa_1}\right)R^3\dot{R}^2, \tag{2.20}$$

where eq. (2.14) have been used, and η is the viscosity and κ_1 is the permeability of the host medium. D_L is divided by the volume of the inclusion to obtain the dissipation function per unit volume. Finally, using eqs (2.17) and (2.18), we have

$$D_L = \frac{3}{2}\frac{\eta\phi_2\phi_{20}}{\kappa_1}\dot{R}^2 = \frac{1}{6}\left(\frac{\eta}{\kappa_1}\right)\phi_1^2\phi_2\phi_{20}R_0^2\dot{\zeta}^2. \tag{2.21}$$

Now, we discuss the determination of the stiffness and density coefficients. Johnson (1986) performed three "gedanken" experiments and derived the explicit expressions for the four Biot elastic constants of a single-porosity medium.

Here, similar experiments are discussed for a double-porosity medium. First, the stress–strain relations are derived for the solid and fluid phases as

$$\tau_{ij} = \frac{\partial W}{\partial e_{ij}}, \qquad \tau^m = \frac{\partial W}{\partial \xi_m}, m = 1, 2. \tag{2.22}$$

Neglecting the LFF process, we obtain

$$\tau_{ij} = \left(Ae + \sum_m Q_m \xi_m \right) \delta_{ij} + 2Ne_{ij}, \qquad \tau^{(m)} = Q_m e + R_m \xi_m, \tag{2.23}$$

where δ_{ij} is Kronecker's delta.

By performing idealized experiments (Biot and Willis, 1957; Johnson, 1986), the expressions of the stiffness coefficients are

$$\begin{aligned} A &= (1-\phi)K_s - \frac{2}{3}N - \frac{K_s}{K_f}(Q_1 + Q_2), \\ Q_1 &= \frac{\beta \phi_1 K_s}{\beta + \gamma}, \quad Q_2 = \frac{\phi_2 K_s}{1 + \gamma}, \\ R_1 &= \frac{\phi_1 K_f}{\beta / \gamma + 1}, \quad R_2 = \frac{\phi_2 K_f}{1/\gamma + 1}, \\ \gamma &= \frac{K_s}{K_f} \left(\frac{\beta \phi_1 + \phi_2}{1 - \phi - K_b / K_s} \right). \end{aligned} \tag{2.24}$$

where β denotes compressibility ratio between the two components.

On the other hand, the calculations of the density coefficients follow Biot's theory (Biot, 1956), which are

$$(1-\phi)\rho_s = \rho_{00} + \sum_m \rho_{0m}, \qquad \phi_m \rho_f = \rho_{0m} + \rho_{mm}, m = 1, 2. \tag{2.25}$$

In terms of the tortuosity τ_m, we have

$$\rho_{mm} = v_m \tau_m \phi_{m0} \rho_f = \tau_m \phi_m \rho_f, \tag{2.26}$$

such that

$$\bar{\rho}_m = \frac{\tau_m \rho_f}{\phi_m}. \tag{2.27}$$

Following Berryman (1979),

$$\tau_m = \frac{1}{2}\left(1 + \frac{1}{\phi_{m0}}\right). \tag{2.28}$$

Then, the density coefficients are

$$\begin{aligned} 2\rho_{00} &= 2(1-\phi)\rho_s - (\phi - 1)\rho_f, \\ 2\rho_{0m} &= (\phi_m - v_m)\rho_f, \\ 2\rho_{mm} &= (\phi_m + v_m)\rho_f. \end{aligned} \tag{2.29}$$

Furthermore, the equation of motion can be obtained from Hamilton's principle, where the Lagrangian density of a conservative system is defined as

$$L = T - W. \tag{2.30}$$

Substituting the kinetic energy expressions, the potential energy expressions and the dissipation functions into Lagrange equation, the equations of motion in of double-porosity medium are derived. Lagrange's equation can be written as

$$\partial_t\left(\frac{\partial L}{\partial \dot{u}}\right) + \partial_t\left[\frac{\partial L}{\partial(\partial_j u)}\right] - \frac{\partial L}{\partial u} + \frac{\partial D}{\partial \dot{u}} = 0, \tag{2.31}$$

where u represents one of the components of vectors $\mathbf{u}$ and $\mathbf{U}^{(m)}$, $m = 1, 2$, taken here as generalized coordinates. The remaining equation regards the variable ζ,

$$\partial\left(\frac{\partial L}{\partial \dot{\zeta}}\right) - \frac{\partial L}{\partial \zeta} + \frac{\partial D}{\partial \dot{\zeta}} = 0. \tag{2.32}$$

The following equations are then derived:

$$\begin{aligned} & N\nabla^2\mathbf{u} + (\mathrm{A}+\mathrm{N})\nabla e + Q_1\nabla(\xi_1 + \phi_2\zeta) + Q_2\nabla(\xi_2 - \phi_1\zeta) \\ & \quad = \rho_{00}\ddot{\mathbf{u}} + \sum_m\left[\rho_{0m}\ddot{\mathbf{U}}^{(m)} + b_m(\dot{\mathbf{u}} - \dot{\mathbf{U}}^{(m)})\right], \\ & Q_1\nabla e + R_1\nabla(\xi_1 + \phi_2\zeta) = \rho_{01}\ddot{\mathbf{u}} + \rho_{11}\ddot{\mathbf{U}}^{(1)} - b_1(\dot{\mathbf{u}} - \dot{\mathbf{U}}^{(1)}), \\ & Q_2\nabla e + R_2\nabla(\xi_2 - \phi_1\zeta) = \rho_{02}\ddot{\mathbf{u}} + \rho_{22}\ddot{\mathbf{U}}^{(1)} - b_2(\dot{\mathbf{u}} - \dot{\mathbf{U}}^{(2)}), \\ & \phi_2[Q_1 e + R_1(\xi_1 + \phi_2\zeta)] - \phi_1[Q_2 e + R_2(\xi_2 - \phi_1\zeta)] \\ & \quad = \frac{1}{3}R_0^2\phi_1^2\phi_2\phi_{20}\left(\frac{\rho_f}{\phi_{10}}\ddot{\zeta} + \frac{\eta}{\kappa_1}\dot{\zeta}\right). \end{aligned} \tag{2.33}$$

This system of coupled equations has 10 unknowns (u_i, $U_i^{(m)}$, $i = 1, 2, 3$; $m = 1, 2$, and ζ), describing wave dispersion and attenuation due to the LFF process. All these coefficients can be quantitatively estimated on the basis of measurable quantities.

Plane-wave analysis is performed to study the propagation phenomena in a double-porosity medium. Homogeneous body waves are assumed; that is, the propagation and attenuation directions coincide. By substituting the plane-wave solution, a cubic equation for P waves and one solution for S waves are obtained.

The complex velocity is defined as

$$v = \frac{\omega}{k}, \tag{2.34}$$

such that

$$v_p = \left[\mathrm{Re}\left(\frac{1}{v}\right)\right]^{-1} \tag{2.35}$$

is the phase velocity,

$$\alpha = -\omega\,\mathrm{Im}\left(\frac{1}{v}\right), \tag{2.36}$$

the attenuation factor, and

$$Q = \frac{\pi f}{\alpha v_p} = \frac{\mathrm{Re}(v)}{2\,\mathrm{Im}(v)} \tag{2.37}$$

is the quality factor, which is defined as the total energy (potential plus kinetic) divided by the dissipated energy, where $f = \omega/(2\pi)$ is the frequency.

In order to illustrate the results of the present theory, the same rock properties of Pride and Berryman (2003) are used to obtain the phase velocities and quality factors. The phase velocity of the fast P waves (P1) as a function of frequency is shown in Figure 2.3. P1 velocities increase with frequency in the seismic band. This low-frequency velocity dispersion is induced by mesoscopic

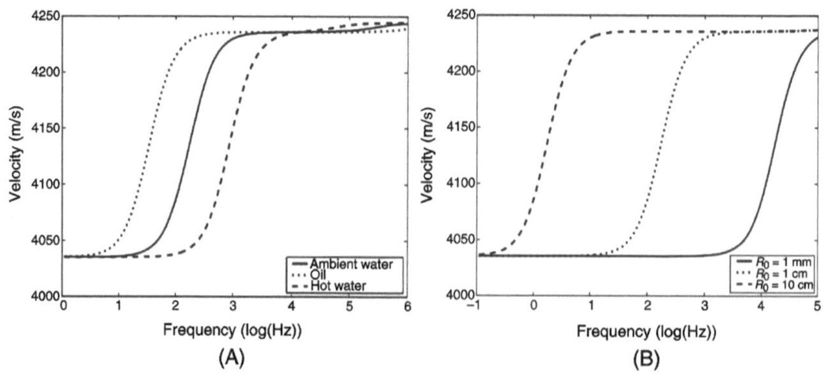

FIGURE 2.3 P-wave phase velocities as a function of frequency for (A) different fluids and $R_0 = 1$ cm and (B) different radii of the inclusions and ambient water.

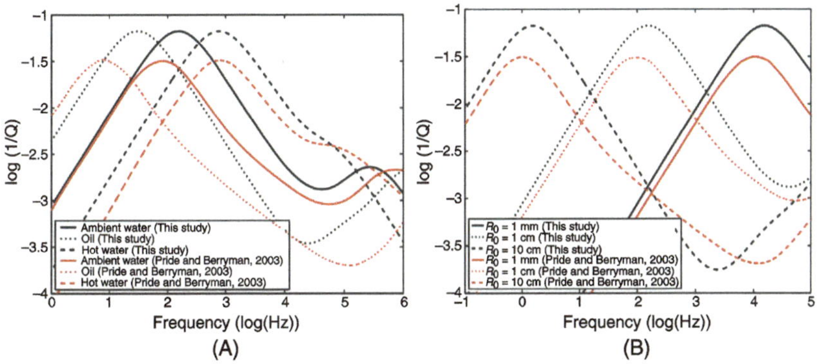

FIGURE 2.4 P-wave dissipation factors as a function of frequency for (A) different fluids and $R_0 = 1$ cm and (B) different radii of the inclusions and ambient water.

LFF. Figure 2.4 displays the dissipation factor as a function of frequency for different fluids (Figure 2.4A) and different inclusion radii (Figure 2.4B), where the results predicted by Pride and Berryman (2003) are also shown. The smaller relaxation peak at the high frequencies represents the loss mechanism caused by the Biot mechanism (Biot, 1962). The two theories show a similar behavior. The mesoscopic peaks are located in the 1–1000 Hz range and the Q factor ranges from 15 to 30, which are typical values for reservoir rocks. The LFF peaks move toward the low frequencies with increasing fluid viscosity and larger inclusion sizes.

The velocity and attenuation of two slow P waves (P2 and P3) are also estimated. The wave velocities corresponding to the slow P waves and different fluids and inclusion radii are shown in Figure 2.5A and B, respectively. Figure 2.6 shows the corresponding dissipation factors. The P2 and P3 waves are diffusive modes at low frequencies, which are respectively related to the fluid in the two

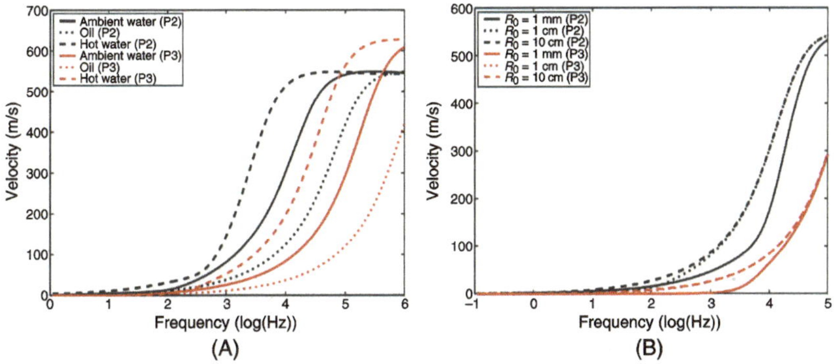

FIGURE 2.5 P2- and P3-wave phase velocities (black and red lines) as a function of frequency for (A) different fluids and $R_0 = 1$ cm and (B) different radii and ambient water.

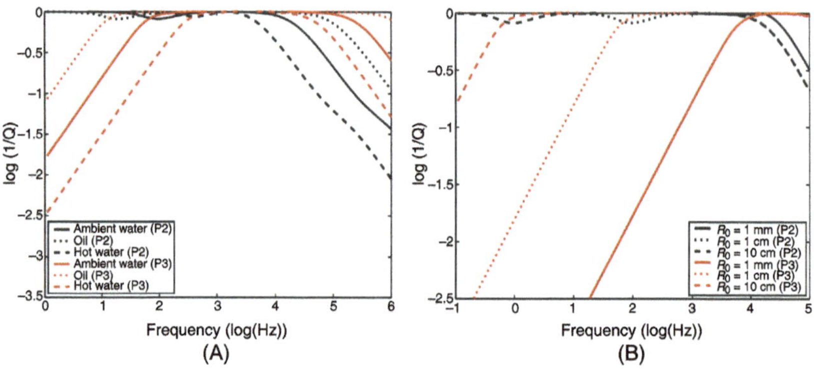

FIGURE 2.6 P2- and P3-wave dissipation factors (black and red lines) as a function of frequency for (A) different fluids and $R_0 = 1$ cm and (B) different radii and ambient water.

components. The P2 and P3 velocities are approximately zero at low frequencies and increase with frequency. The P2 inverse quality factors (dissipation factors) are around unity in the zero frequency limit and decrease with frequency. The slow P-wave diffusion is caused by viscous friction between the solid matrix and the pore fluid (Biot, 1956). The P3 dissipation factors increase with frequency in the seismic band. In the ultrasonic band, the P3 inverse quality factors decrease with frequency.

The properties of the S waves are given in Figure 2.7. In this case, the curves are independent of the radii of the inclusions. The attenuation peaks move to higher frequencies for increasing fluid viscosity, which agrees with the characteristics of Biot's friction mechanism. The attenuation is caused by the relative motion between the solid skeleton and the pore fluid.

In this work, a double-porosity theory is presented to describe wave propagation and mesoscopic attenuation in isotropic poroelastic media. The medium consists of spherical inclusions embedded in a host medium. The fluid flow

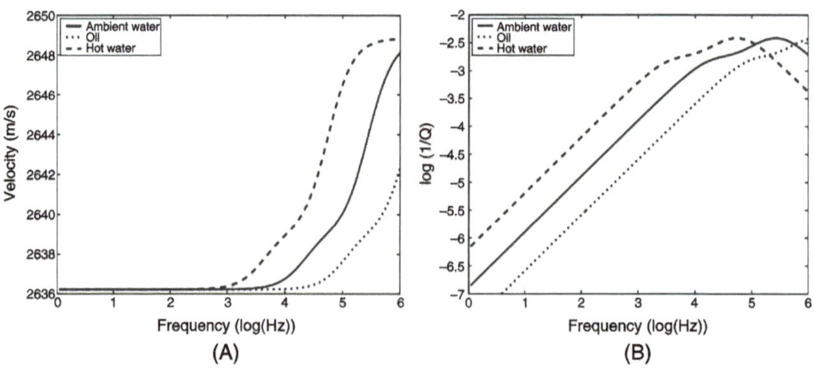

FIGURE 2.7 (A) S-wave phase velocities and (B) dissipation factors versus frequency for different fluids.

induced by the wave motion is described by Rayleigh's theory of liquid collapse of a spherical cavity. The elastic coefficients are theoretically determined by "gedanken" experiments and can be estimated from measurable quantities. In the numerical example for a sandstone, the theory predicts the usual fast P and S waves and two slow P waves. The local fluid flow causes significant fast P-wave dispersion and attenuation in the 1–1000 Hz frequency band, for inclusion sizes between 1 mm and 10 cm. The friction force between the solid matrix and the pore fluid in the wave-propagation direction has caused the diffusion of the two types of slow P waves.

2.3 BIOT–RAYLEIGH THEORY OF WAVE PROPAGATION IN PATCHY-SATURATED RESERVOIR ROCKS

The Biot–Rayleigh (BR) theory consists of spherical inclusions embedded in a homogeneous host medium, which can be applied in two cases: a single-porosity solid saturated with two immiscible fluids and a double-porosity solid saturated with a single type of fluid. In this section, the BR theory (Ba et al., 2011) is adapted to the case of patchy saturated rocks, where the gas-saturated porous solid is an inclusion in a water-saturated background. The contrasts in density, viscosity and bulk modulus between the gas- and water-saturated zones constitute mesoscopic heterogeneities. Therefore, the patchy-saturated rock can be considered as a special case of a double-porosity medium with regions of different compliances.

For a single-porosity solid saturated with two immiscible fluids (for instance, a rock partially saturated with gas and water), the inclusions are gas pockets (or water pockets) whereas the host medium is saturated with water (or gas), and the inclusions and host medium share the same skeleton in the BR theory. It is assumed that the pores between the inclusion and the host medium are open and the inclusions do not interact.

The derivation of the wave-propagation equations for the patchy-saturated BR theory is similar to that of the original BR theory. The strain energy density is the same, but $m=1,\ 2$ now represent the displacements of the host medium and inclusion respectively (the two fluid phases of a single-porosity matrix saturated with two fluids), and ϕ_1 and ϕ_2 are the absolute porosities of the host medium and inclusion (absolute porosities for the two fluid phases in the partially saturated medium).

The kinetic energy density T_L related to a compression deformation is (Ba et al., 2011):

$$T_L = \frac{1}{6}\left(\frac{\phi_1^{\,2}\phi_2\phi_{20}}{\phi_{10}}\right)\rho_{f_1} R_0^2 \dot{\zeta}^2. \tag{2.38}$$

In the above formula, $\phi_m = v_m \phi_{m0} = v_m \phi_0$ $(m=1,2)$ since the inclusions and host medium are assumed to share the same pore structures. Furthermore, they have the same permeability κ.

The local fluid-flow dissipation function is

$$D_L = \frac{1}{6}\left(\frac{\eta_1}{\kappa}\right)\phi_1^2\phi_2\phi_{20}R_0^2\dot{\zeta}^2 \tag{2.39}$$

where η_m ($m = 1, 2$) are the fluid viscosities. The Lagrangian equation for the current system is

$$\frac{\partial}{\partial t}\left(\frac{\partial L}{\partial \dot{q}_j}\right) + \sum_{k=1}^{3}\frac{\partial}{\partial q_k}\left[\frac{\partial L}{\partial q_{j,k}}\right] + \frac{\partial(D + D_L)}{\partial \dot{q}_j} = \frac{\partial L}{\partial q_j}, \tag{2.40}$$

where $L = T - W$ and q_j are generalized coordinates representing the components of $\left(u_i, U_i^{(m)}, \zeta\right)$ ($m = 1,2;\ i = 1,2,3$).

The above Lagrangian formula leads to the following dynamical equations

$$\begin{aligned} &N\nabla^2\mathbf{u} + (A+N)\nabla I_1 + Q_1\nabla(\xi_1 + \phi_2\zeta) + Q_2\nabla(\xi_2 - \phi_1\zeta) \\ &= \rho_{00}\ddot{\mathbf{u}} + \sum_{m=1}^{3}\left[\rho_{0m}\ddot{\mathbf{U}}^{(m)} + b_m\left(\dot{\mathbf{u}} - \dot{\mathbf{U}}^{(m)}\right)\right] \end{aligned} \tag{2.41}$$

$$Q_1\nabla I_1 + R_1\nabla(\xi_1 + \phi_2\zeta) = \rho_{01}\ddot{\mathbf{u}} + \rho_{11}\ddot{\mathbf{U}}^{(1)} + b_1\left(\dot{\mathbf{U}}^{(1)} - \dot{\mathbf{u}}\right) \tag{2.42}$$

$$Q_2\nabla I_1 + R_2\nabla(\xi_2 - \phi_1\zeta) = \rho_{02}\ddot{\mathbf{u}} + \rho_{22}\ddot{\mathbf{U}}^{(2)} + b_2\left(\dot{\mathbf{U}}^{(2)} - \dot{\mathbf{u}}\right) \tag{2.43}$$

$$\begin{aligned} &\phi_2 Q_1 I_1 + \phi_2 R_1(\xi_1 + \phi_2\zeta) - \phi_1 Q_2 I_1 - \phi_1 R_2(\xi_2 - \phi_1\zeta) \\ &= \left(\frac{\rho_f}{\phi_0}\ddot{\zeta} + \frac{\eta_1}{\kappa}\dot{\zeta}\right)\frac{\phi_1^2\phi_2}{3}R_0^2\phi_0 \end{aligned} \tag{2.44}$$

There are 10 unknowns in these coupled equations. The transport property κ and the fluid viscosity η_m differ from those of the original BR theory.

The stiffness coefficients can be derived by means of idealized experiments. Following the calculations proposed by Biot (1956), the expressions for the density coefficients are

$$\begin{aligned} \rho_{mm} &= \frac{\rho_{f_m}}{2}\phi_m\left(1 + \frac{1}{\phi_{m0}}\right) \\ \rho_{0m} &= \frac{\rho_{f_m}}{2}\phi_m\left(1 - \frac{1}{\phi_{m0}}\right) \\ \rho_{00} &= \rho_0 - \frac{1}{2}\sum_{m=1}^{2}\rho_{f_m}\phi_m\left(1 - \frac{1}{\phi_{m0}}\right) \end{aligned} \tag{2.45}$$

Next, the patchy-saturated BR model is tested with some experiments for two different rocks, namely, Casino Otway basin sandstone (Lebedev et al., 2009)

TABLE 2.1 Rock Properties

Parameters	Values
ϕ	0.167
K_s (GPa)	37
ρ_s (kg/m^3)	2650
ρ_0 (kg/m^3)	2200
V_p (m/s)	2450
V_s (m/s)	1840
κ(D)	7.26×10^{-3}
K_{water} (GPa)	2.25
ρ_{water} (kg/m^3)	1000
η_{water} (Pa·s)	0.001
K_{gas} (GPa)	0.050
ρ_{gas} (kg/m^3)	1200
η_{gas} (Pa·s)	1.00×10^{-7}

and high-porosity Estaillades limestone (Cadoret et al., 1995). The predicted P-wave velocities are shown as a function of frequency and water saturation.

Lebedev et al. (2009) measured the ultrasonic velocities as a function of water saturation for the Casino Otway basin sandstone from the quasi-static-saturation and dynamical-saturation experiments. The P-wave dispersion and attenuation are calculated with different models at a given frequency (1 MHz) as in the experiments. The curves depend on saturation, frequency, and water patch size. The rock parameters from Lebedev et al. (2009) used in the calculation are listed in Table 2.1.

The results present a consistent increasing trend bounded by the Biot–Gassmann–Hill (BGH) and Biot–Gassmann–Wood (BGW) limits (Figure 2.8). However, even at the same frequency, porosity, and saturation, the P-wave dispersion changes as the patch size changes. As in White (1975) and Dutta and Odé (1979a, 1979b), the gas-filled regions are spherical and all inclusions have the same radius R_a. The cube is approximated by a sphere of radius R_b which has the same volume, so that the gas saturation S_{gas} in the spherical unit cell is the same as that in the original cubic cell, that is, $S_{gas} = (R_a/R_b)^3$. When we fix R_b, the gas saturation is determined by a varying R_a.

The diffusion length in the experiments is of the order of 0.1 mm (Lebedev et al., 2009). When we use a comparable patch size $R_b = 0.2$ mm (fitting parameter), the experimental data compares well to the BR theory. The mesoscopic fluid flow due to the heterogeneities is responsible for the significant acoustic attenuation. For a partially saturated porous medium having only

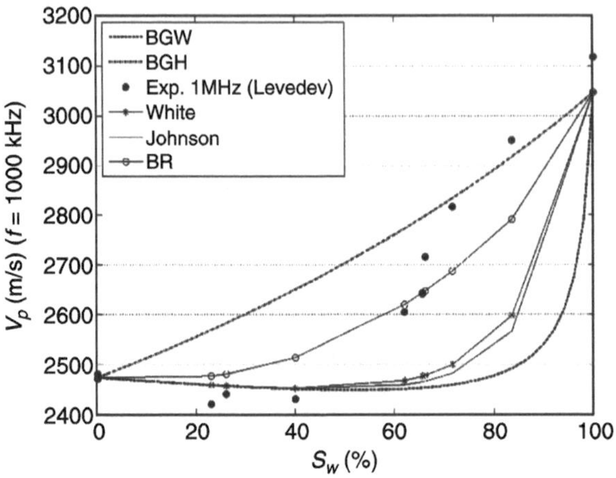

FIGURE 2.8 Comparison between the measured and theoretical P-wave velocities in the Casino Otway basin sandstone. The experimental data, BGW and BGH curves are also shown. Patch size is 0.2 mm.

heterogeneities due to the saturating fluids, the distribution of fluid pressures is governed by the diffusion equation with a diffusion length λ_d. When λ_d is much smaller than the patch size, the BGH bound is expected. On the contrary, if λ_d is larger than the patch size, the velocities follow the BGW prediction. The transition between these two bounds occurs when the scale of λ_d is comparable to the patch size.

In another example, Cadoret et al. (1995) measured elastic-wave velocities in limestones at three different frequency ranges with varying water saturations. Eight different quarry limestones covering a wide range of porosities and permeabilities were chosen for the sonic and ultrasonic experiments. The P-wave velocities of Estaillades limestone with a porosity of approximately 30% and a permeability of 225 mD have been measured using large resonant bars at approximately 1 kHz, whereas the conventional pulse transmission technique was used at 500 kHz. The samples were saturated with the drainage and imbibition techniques (Cadoret et al., 1995).

The rock porosity and permeability that are used in the calculation (Table 2.2) are obtained from Cadoret et al. (1995). The mineral properties are taken from Mavko et al. (2003). The bulk modulus K_b and shear modulus N are estimated from the measured P- and S-wave velocities of dry rocks (Cadoret et al., 1995). The same permeability is assumed for phases 1 and 2. With these parameters, we calculated $V_p(S_w)$ at 1 kHz (Figure 2.9) and 500 kHz (Figure 2.10). The inclusion size is 1.05 mm.

As shown in Figure 2.9, the experimental data (see Cadoret et al. (1995)) and theoretical velocities are bounded by the BGW and BGH limits. In addition, the velocities are close to the low-frequency velocity limit (BGW). There is a sudden decrease in velocity when water saturation decreases from 100% to

TABLE 2.2 Properties of the Estaillades Limestone

Property	Value
ϕ	0.30
K_s (GPa)	76.8
ρ_s (kg/m^3)	2710
K_b (GPa)	7.45
N (GPa)	6.09
κ (D)	0.225
K_{water} (GPa)	2.25
ρ_{water} (kg/m^3)	997
η_{water} (Pa·s)	0.001
K_{gas} (GPa)	1.44×10^{-5}
ρ_{gas} (kg/m^3)	1240
η_{gas} (Pa·s)	1.0×10^{-5}

95%. As gas saturation increases, the velocity gradually increases. Beyond 60% water saturation, the BR model provides the best match.

Figure 2.10 shows that the experimental data are distributed close to the high-frequency limit (BGH). At low S_w, the theoretical predictions are close to

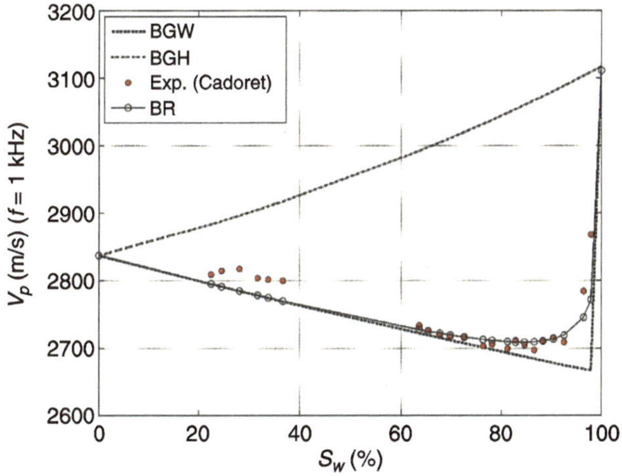

FIGURE 2.9 Comparison between the measured and theoretical P-wave velocities corresponding to Estaillades limestone as a function of gas–water saturation. The porosity is 0.30 and the frequency is 1 kHz.

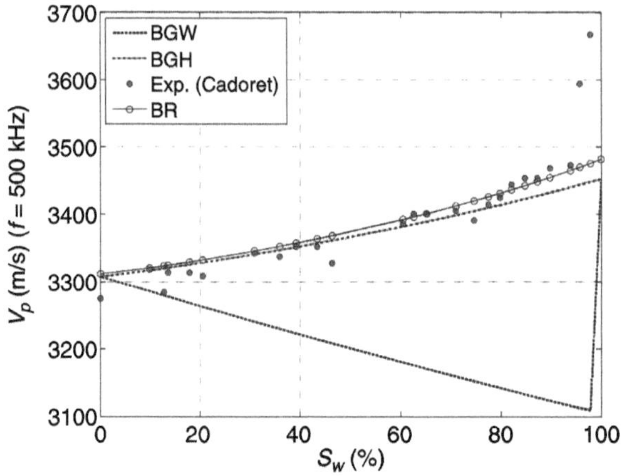

FIGURE 2.10 Comparison between the measured and theoretical P-wave velocities for Estaillades limestone as a function of gas–water saturation. The porosity is 0.30 and frequency is 500 kHz.

the experimental measurements. When S_w approaches 100%, the experimental velocity strongly increases. The two limits and the BR model underestimate the P-wave velocity in the case of full-water saturation.

The comparisons show that for the sandstone, the velocity prediction agrees well with the data, particularly at water saturations between 30% and 70%, and for the limestone with a porosity of 0.3, the BR model matches the velocities at both the sonic and ultrasonic ranges.

2.4 WAVE PROPAGATION IN PARTIALLY SATURATED ROCKS: NUMERICAL EXAMPLES

Substantial research is available regarding wave attenuation and dispersion in porous media induced by inhomogeneities at different scales (e.g., Gassmann, 1951; Biot, 1956; Biot, 1962; White, 1975; Dutta and Odé, 1979a, 1979b). Inhomogeneities constitute the spatial variations of solid properties, permeability, fluid mobility, fluid distribution, and so forth. Numerical modeling is helpful for providing further understanding on the relationships between wave dispersion/attenuation features and their controlling factors. In this section, we compare the results obtained with the White (1975), Dutta and Odé (1979a, 1979b), Johnson (2001) and BR (Ba et al., 2011) models. The rock-matrix and pore-fluid properties given in Table 2.3 are taken from Johnson (1986), White (1975), Carcione (2007) and Dutta and Odé (1979b). We consider sandstones saturated with oil, gas, and brine.

2.4.1 Influence of Fluid Composition

Figures 2.11–2.14 show the P-wave velocity and attenuation as a function of frequency for different fluid compositions (the BGW and BGH bounds reported).

TABLE 2.3 Rock Properties Used in Numerical Examples

Parameter	Values	Parameter	Values	Parameter	Values	Parameter	Values
ϕ	0.3	K_{brine} (GPa)	2.25	ρ_{gas} (kg/m^3)	0.1×10^3	ρ_{oil} (kg/m^3)	0.7×10^3
K_2 (GPa)	37	η_{brine} (Pa·s)	0.003	ρ_{brine} (kg/m^3)	0.1×10^3	K_{water} (GPa)	2.4
K (GPa)	48	K_{gas} (GPa)	0.022	K_{oil} (GPa)	0.7	η_{water} (Pa·s)	0.001
μ (GPa)	5.7	η_{gas} (Pa·s)	1.5	η_{oil} (Pa·s)	0.001	ρ_{water} (kg/m^3)	1.0×10^3
κ (m^2)	10^{-12}	ρ_2(kg/m^3)	2.65×10^3				

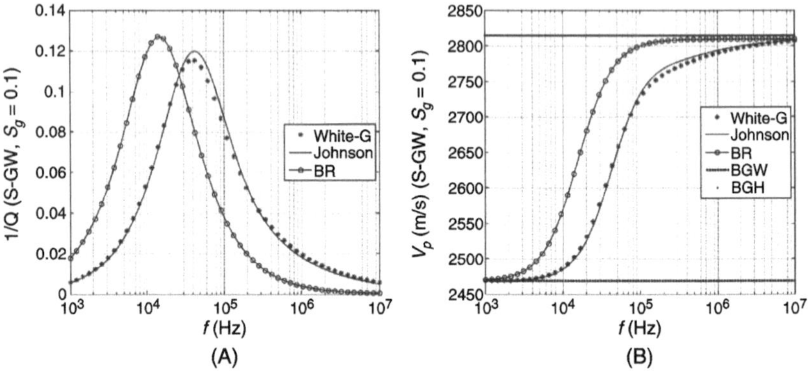

FIGURE 2.11 Inverse quality factor and phase velocity as a function of frequency for a sandstone saturated with gas and water. The porosity is 0.3, gas saturation is 0.1, and the outer radius is $R_b = 0.01$ m. S-GW indicates sandstone–gas (inner fluid)–water (outer fluid), where gas and water are inclusion and host, respectively.

The White and Johnson results are very similar, whereas those of the BR model are different. The BR model has two relaxation peaks in Figure 2.13, corresponding to the mesoscopic-loss and Biot mechanism, respectively.

Comparing Figures 2.11 and 2.14, it is clear that the anelastic effects are weaker in the case sandstone-oil (inner fluid)-water (outer fluid), whereas comparing Figures 2.11 and 2.15 shows that the location of the relaxation peaks and the velocity bounds change with the properties of the rock mineral.

2.4.2 Influence of Fluid Mobility

Comparing Figure 2.15 with Figures 2.11–2.14 shows that the relaxation peaks are shifted to the left on the frequency axis in the first plot. The main factor influencing this behavior is the lower permeability of the shale compared to the

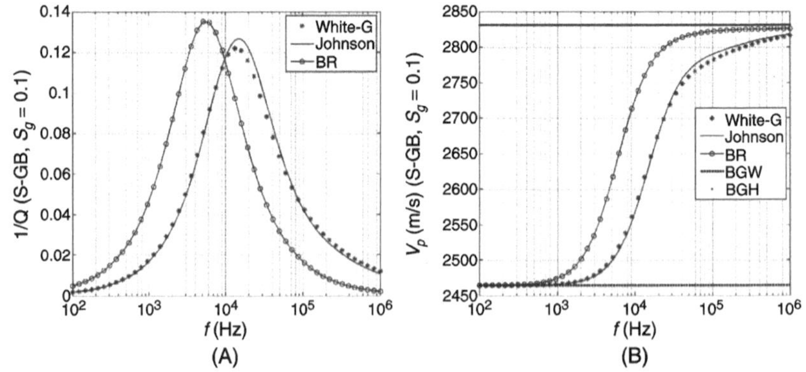

FIGURE 2.12 Inverse quality factor and phase velocity as a function of frequency for a sandstone saturated with gas and brine. The porosity is 0.3, gas saturation is 0.1, and the outer radius is $R_b = 0.01$ m. S-GB indicates sandstone–gas (inner fluid)–brine (outer fluid).

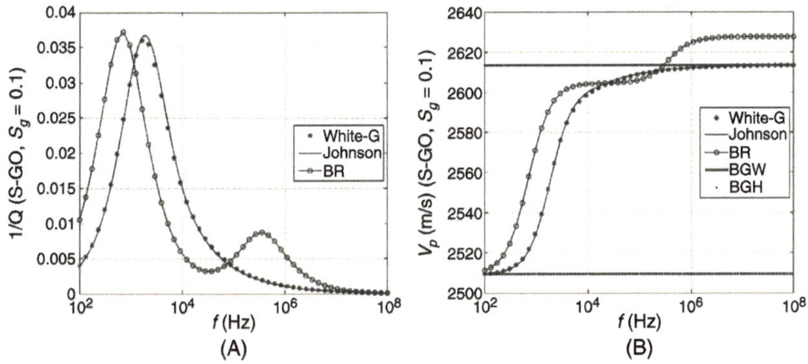

FIGURE 2.13 Inverse quality factor and phase velocity as a function of frequency for a sandstone saturated with gas and oil. The porosity is 0.3, gas saturation is 0.1, and the outer radius is $R_b = 0.01$ m. The relaxation peak of the BR attenuation curve at 10^6 Hz is induced by the macroscopic Biot loss. S-GO indicates sandstone–gas (inner fluid)–oil (outer fluid).

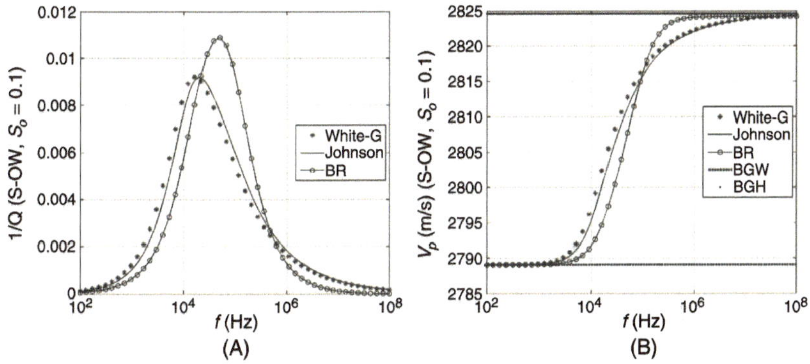

FIGURE 2.14 Inverse quality factor and phase velocity as a function of frequency for a sandstone saturated with oil and water. The porosity is 0.3, gas saturation is 0.1, and the outer radius is $R_b = 0.01$ m. S-OW indicates sandstone–oil (inner fluid)–water (outer fluid).

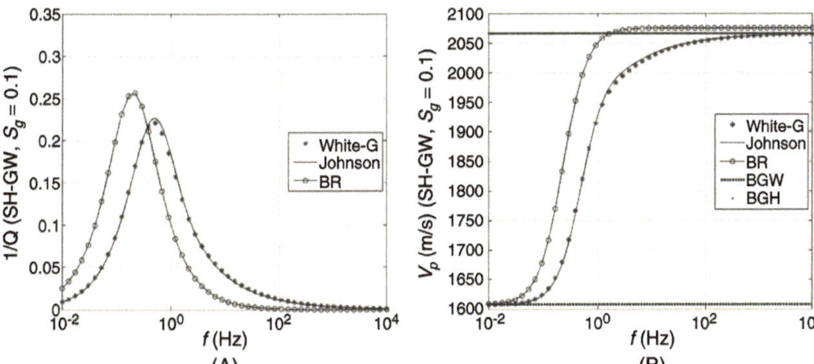

FIGURE 2.15 Inverse quality factor and phase velocity as a function of frequency for a shale saturated with gas and water. The porosity is 0.3, gas saturation is 0.1, and the outer radius is $R_b = 0.01$ m. SH-GW indicates shale–gas (inner fluid)–water (outer fluid).

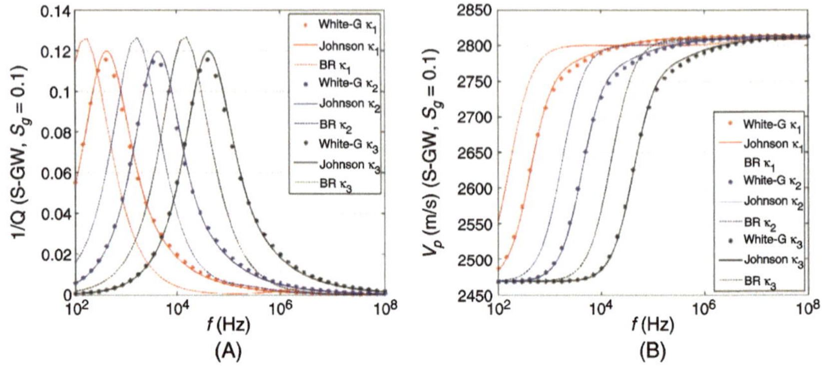

FIGURE 2.16 Inverse quality factor and phase velocity as a function of frequency for a sandstone saturated with gas (inner fluid), and water (outer fluid). The porosity is 0.3, gas saturation is 0.1 and the outer radius is $R_b = 0.01$ m. The permeability of the rock takes the following values: $\kappa_1 = 0.01$ D, $\kappa_2 = 0.1$ D, and $\kappa_3 = 1$ D.

sandstone. Viscosity has also an effect on the locations of the peaks as can be seen in Figures 2.11 and 2.13. If η is the fluid viscosity and κ the permeability, η/κ is the fluid mobility.

Figure 2.16 shows the inverse-quality factors and velocities as a function of frequency for $\kappa = 0.001$, 0.01, 0.1, and 1 D, and the other properties are those of Figure 2.11. The effect of viscosity is displayed in Figure 2.17, where we have considered different oil types, such as mazout and naphtha. Local fluid-flow peaks move to the left with decreasing permeability or increasing fluid viscosity of the fluid in the outer shell, whereas the amplitude of the peaks remains

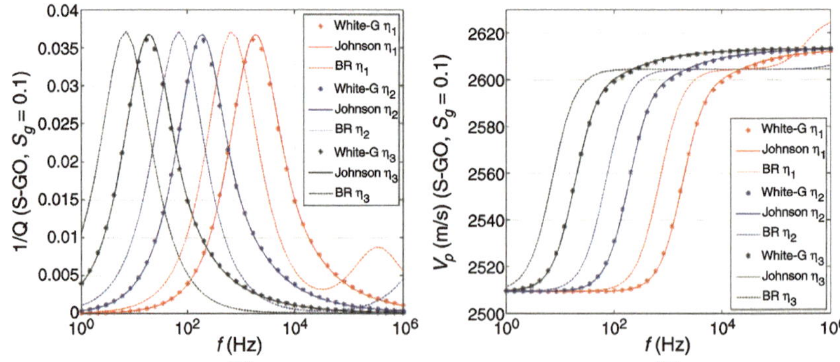

FIGURE 2.17 Inverse quality factor and phase velocity as a function of frequency for a sandstone saturated with gas (inner fluid) and oil (outer fluid). The porosity is 0.3, gas saturation is 0.1, and the outer radius is $R_b = 0.01$ m. The viscosity of the oil takes the following values: $\eta_1 = 0.01$ Pa·s, $\eta_2 = 0.1$ Pa·s, and $\eta_3 = 1$ Pa·s. The peak of the BR model in the ultrasonic band is the macroscopic Biot loss mechanism.

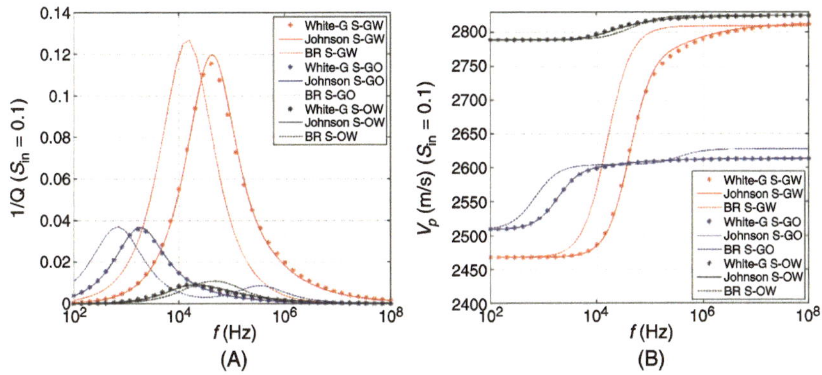

FIGURE 2.18 Inverse quality factor and phase velocity as a function of frequency for a sandstone saturated with two fluids. The porosity is 0.3, inclusion fluid saturation is 0.1, and the outer radius is $R_b = 0.01$ m. S-GW, gas–water; S-GO, gas–oil; S-OW, oil–water.

almost constant with fluid mobility. The peak frequencies are closely related to fluid mobility.

2.4.3 Influence of the Fluid Compressibility Ratio

Figure 2.18 displays the seismic properties for three fluid compressibility ratios. It shows that the ratio affects the peak amplitudes but not the peak locations. Basically a larger ratio implies stronger attenuation and velocity dispersion.

2.4.4 Influence of Rock Porosity

Rock porosity affects the bulk modulus of the frame, the rock density and the permeability. Figure 2.19 shows results for three porosity values. In this case, peak amplitudes and locations are affected. Higher porosity implies stronger attenuation and velocity dispersion.

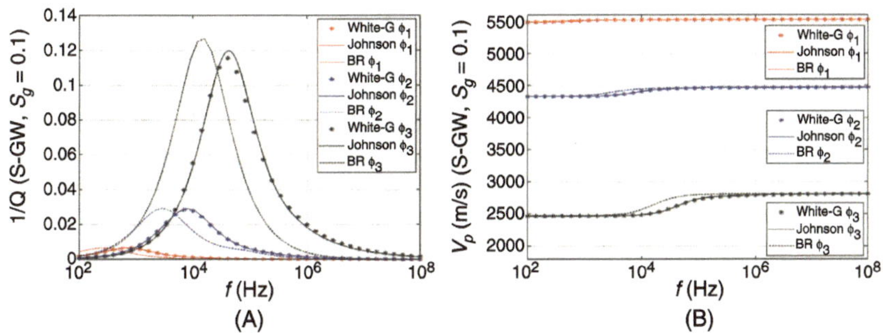

FIGURE 2.19 Inverse quality factor and phase velocity as a function of frequency for a sandstone saturated with gas and water. The gas saturation is 0.1 and outer radius is $R_b = 0.01$ m. Rock porosities are: $\phi_1 = 0.05$, $\phi_2 = 0.15$, and $\phi_{13} = 0.3$. S-GW denotes sandstone–gas (inner fluid)–water (outer fluid).

2.4.5 Influence of Saturation Degree

Figure 2.20 shows the seismic properties for different gas (water) saturations.

As shown in Figure 2.20A and B, attenuation and velocity dispersion decrease when gas saturation increases. Figures 2.20C and D, indicate that attenuation and dispersion at low gas saturation (0.01) is weaker than those at higher gas saturation (0.1). Figures 2.20E and F show the seismic properties as a function of saturation for three different frequencies. The behaviors are dissimilar.

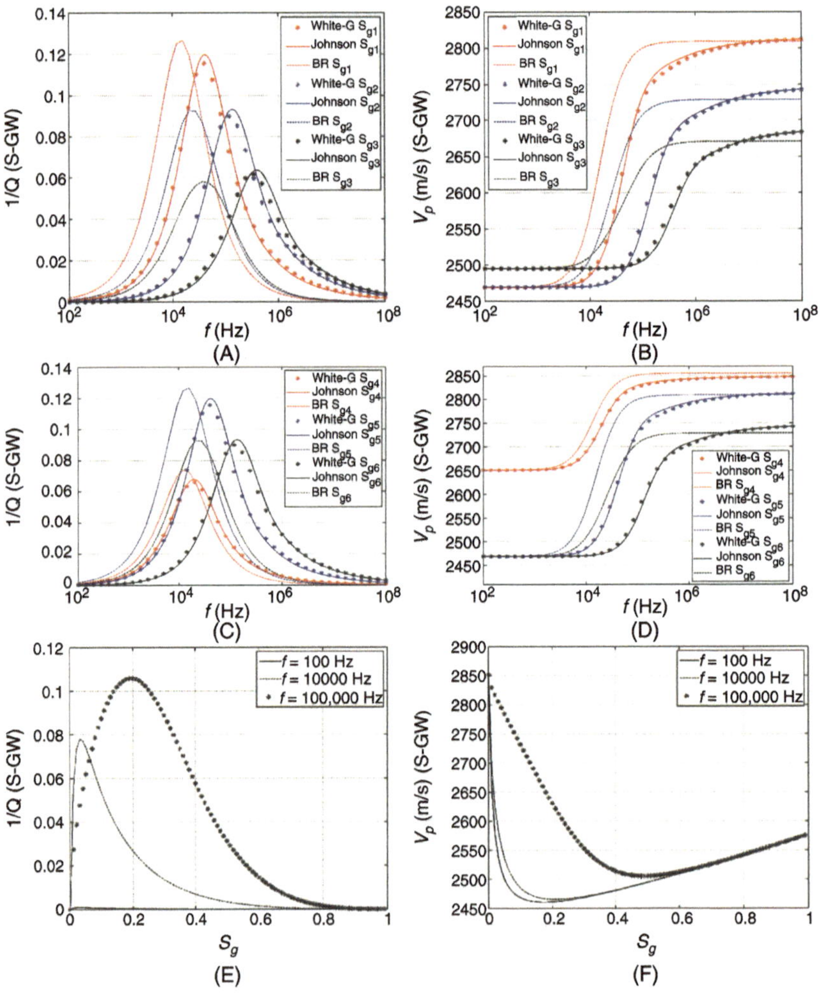

FIGURE 2.20 Inverse quality factor and phase velocity as a function of frequency for a sandstone saturated with gas (inner fluid) and water (outer fluid). Gas saturation varies as follows. (A, B) g1, gas saturation: 0.1; g2, gas saturation: 0.3; g3, gas saturation: 0.5. (C, D) g4, gas saturation: 0.01; g5, gas saturation: 0.1; g6, gas saturation: 0.3. The outer radius is R_b = 0.01 m. (E, F) S_g is the saturation of the inclusion fluid varying from 0 to1. Three frequencies are considered in this case.

In summary, fluid mobility affects the peak frequencies, whereas porosity has an influence on the peak attenuation amplitudes also. Fluid compressibility ratio and saturation degree in partially saturated rocks affect the peak-attenuation level. The BR model includes the Biot loss mechanism, whereas the White and Johnson models do not.

2.5 EFFECT OF INCLUSION PORE-FLUID: REFORMULATED BIOT–RAYLEIGH THEORY

To effectively describe wave propagation in a heterogeneous fluid-saturated porous medium, we further consider the influence of local-flow velocity field inside the inclusions and the case of liquid-pocket saturation (where the properties of pore fluid inside the inclusions cannot be neglected), theoretical equations are reformulated and extended.

Under the excitation of elastic waves, local fluid flow in a complex porous medium is a major cause of wave dispersion and attenuation. When wave propagation including this flow is described in a double-porosity medium, two porous skeletons are usually assumed, namely, that of the host rock and that of the inclusions. Previous models ignored the effect of the local-flow velocity field in the inclusions, and therefore they cannot properly describe the anelastic effects in case when inclusion fluid properties could not be neglected. Thus, the fluid kinetic energy of the inclusions needs to be considered. The local fluid flow governing equation is presented in Ba et al. (2011) by using Rayleigh's equation (Rayleigh, 1917) describing the spherical fluid oscillation.

In this section, we consider the local fluid-flow velocity fields inside and outside the inclusions, rederive the kinetic energy function and dissipation function based on the double-porosity model containing spherical inclusions, and obtain reformulated BR equations of wave propagation with the aid of Hamilton's principle.

The fluid kinetic energy in the inclusion, considering the same double-porosity medium model (Figure 2.2), is (Ba et al., 2014)

$$T_{\text{in}} = \frac{1}{2}\phi_{20}\rho_f \int_0^R 4\pi r^2 \left(\dot{r}_{\text{in}}\right)^2 dr = \frac{2}{5}\pi\phi_{20}\rho_f \left(\dot{R}\right)^2 R^3, \tag{2.46}$$

where the subscript "in" denotes the fluid saturating the inclusions.

On the other hand, the fluid kinetic energy outside the inclusion is (Ba et al., 2014)

$$T_{\text{out}} = \frac{2\pi\rho_f\phi_{20}^2}{\phi_{10}}\left(\dot{R}\right)^2 R^3 \tag{2.47}$$

where the subscript "out" denotes the fluid outside the inclusions.

On the basis of the kinetic energy of a single-porosity medium proposed by Biot (1962), the kinetic energy of a double-porosity medium is

$$
\begin{aligned}
T = {} & \frac{1}{2}\rho_{00}\left(\dot{u}_1^2+\dot{u}_2^2+\dot{u}_3^2\right)+\rho_{01}\left(\dot{u}_1\dot{U}_1^{(1)}+\dot{u}_2\dot{U}_2^{(1)}+\dot{u}_3\dot{U}_3^{(1)}\right) \\
& +\rho_{02}\left(\dot{u}_1\dot{U}_1^{(2)}+\dot{u}_2\dot{U}_2^{(2)}+\dot{u}_3\dot{U}_3^{(2)}\right)+\frac{1}{2}\rho_{11}\left(\dot{U}_1^{(1),2}+\dot{U}_2^{(1),2}+\dot{U}_3^{(1),2}\right) \\
& +\frac{1}{2}\rho_{22}\left(\dot{U}_1^{(2),2}+\dot{U}_2^{(2),2}+\dot{U}_3^{(2),2}\right)+T_{\mathrm{LFF}}
\end{aligned}
\tag{2.48}
$$

The LFF oscillation kinetic energy of a single spherical inclusion includes two parts, that is, the fluid kinetic energy inside and outside the inclusion, and can be expressed as (Ba et al., 2014)

$$
T_{\mathrm{LFF}}=\left(\frac{1}{30}+\frac{1}{6}\left(\frac{\phi_{20}}{\phi_{10}}\right)\right)\phi_2\phi_1^2\rho_f R_0^2\left(\dot{\zeta}\right)^2 \tag{2.49}
$$

The expression of the dissipation function of a double-porosity medium is

$$
D=\frac{1}{2}b_1\left(\dot{\mathbf{u}}-\dot{\mathbf{U}}^{(1)}\right)\left(\dot{\mathbf{u}}-\dot{\mathbf{U}}^{(1)}\right)+\frac{1}{2}b_2\left(\dot{\mathbf{u}}-\dot{\mathbf{U}}^{(1)}\right)\left(\dot{\mathbf{u}}-\dot{\mathbf{U}}^{(1)}\right)+D_{\mathrm{LFF}} \tag{2.50}
$$

where b_1 and b_2 are the dissipation coefficients outside and inside the inclusion, respectively, and D_{LFF} denotes the dissipation function caused by the local fluid-flow oscillation. Similarly, the dissipation function D_{LFF} of the entire system caused by the local fluid flow can be expressed as

$$
\begin{aligned}
D_{\mathrm{LFF}} & =\frac{1}{2}\phi_{20}^2\left(\frac{\eta}{\kappa_2}\right)\int_0^R 4\pi r^2\left(\dot{r}_{\mathrm{in}}\right)^2 dr+\frac{1}{2}\phi_{10}^2\left(\frac{\eta}{\kappa_1}\right)\int_R^{\infty} 4\pi r^2\left(\dot{r}_{\mathrm{out}}\right)^2 dr \\
& =\left(\frac{1}{30}\left(\frac{\eta}{\kappa_2}\right)+\frac{1}{6}\left(\frac{\eta}{\kappa_1}\right)\right)\phi_{20}\phi_2\phi_1^2 R_0^2\left(\dot{\zeta}\right)^2
\end{aligned}
\tag{2.51}
$$

The potential energy function is the same as that of the original BR theory (Ba et al., 2011). Then, taking into account the kinetic energy and the dissipation due to the local fluid flow inside the inclusion and introducing the corresponding Lagrange equations, the reformulated BR equations of motion for a double-porosity solid saturated with a single fluid can be expressed as (Ba et al., 2014)

$$
\begin{aligned}
& N\nabla^2\mathbf{u}+(A+N)\nabla e+Q_1\nabla\left(\xi_1+\phi_2\zeta\right)+Q_2\nabla\left(\xi_2-\phi_1\zeta\right) \\
& \quad =\rho_{00}\ddot{\mathbf{u}}+\rho_{01}\ddot{\mathbf{U}}^{(1)}+\rho_{02}\ddot{\mathbf{U}}^{(2)}+b_1\left(\dot{\mathbf{u}}-\dot{\mathbf{U}}^{(1)}\right)+b_2\left(\dot{\mathbf{u}}-\dot{\mathbf{U}}^{(2)}\right)
\end{aligned}
\tag{2.52}
$$

$$
Q_1\nabla e+R_1\nabla\left(\xi_1+\phi_2\zeta\right)=\rho_{01}\ddot{\mathbf{u}}+\rho_{11}\ddot{\mathbf{U}}^{(1)}-b_1\left(\dot{\mathbf{u}}-\dot{\mathbf{U}}^{(1)}\right) \tag{2.53}
$$

$$
Q_2\nabla e+R_2\nabla\left(\xi_2-\phi_1\zeta\right)=\rho_{02}\ddot{\mathbf{u}}+\rho_{22}\ddot{\mathbf{U}}^{(2)}-b_2\left(\dot{\mathbf{u}}-\dot{\mathbf{U}}^{(2)}\right) \tag{2.54}
$$

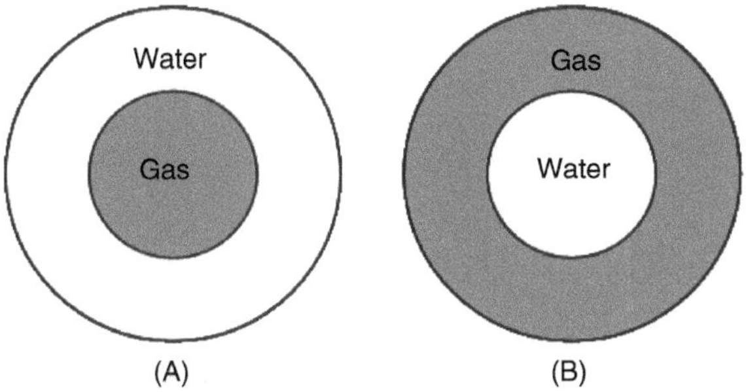

FIGURE 2.21 Double-porosity model with two fluids and one matrix (or frame). (A) "gas-in-water," where the gas-saturated porous frame is the inclusion, while the water-saturated frame is the host; (B) "water-in-gas," where the water-saturated frame is the inclusion.

$$\begin{aligned}&\frac{\phi_2}{15}\rho_f\phi_1^2R_0^2\left(\ddot{\zeta}\right)+\frac{\phi_2\phi_{20}}{3\phi_{10}}\rho_f\phi_1^2R_0^2\left(\ddot{\zeta}\right)\\&+\frac{1}{15}\phi_{20}\phi_2\phi_1^2R_0^2\frac{\eta}{\kappa_2}\left(\dot{\zeta}\right)+\frac{1}{3}\phi_{20}\phi_2\phi_1^2R_0^2\frac{\eta}{\kappa_1}\left(\dot{\zeta}\right)\\&=\phi_2\left(Q_1e+R_1\left(\xi_1+\phi_2\zeta\right)\right)-\phi_1\left(Q_2e+R_2\left(\xi_2-\phi_1\zeta\right)\right)\end{aligned}\quad(2.55)$$

where eqs. (2.52) ~ (2.54) are the dynamical wave-propagation equations, whereas eq. (2.55) is the LFF equation.

In a similar way, regions with unique matrix being saturated with different fluids can be treated as inclusions or host, according to the differences in density, elastic modulus, and viscosity of the pore fluid (see Figure 2.21). The wave-propagation equations for the two cases are basically the same, and the only difference is the LFF equation. After the reformulation, the LFF equation for a single-porosity solid saturated with two different fluids can be rewritten as:

$$\begin{aligned}&\frac{\phi_2}{15}\rho_{in}\phi_1^2R_0^2\left(\ddot{\zeta}\right)+\frac{\phi_2\phi_{20}}{3\phi_{10}}\rho_{out}\phi_1^2R_0^2\left(\ddot{\zeta}\right)\\&+\frac{1}{15}\phi_{20}\phi_2\phi_1^2R_0^2\frac{\eta_2}{\kappa_1}\left(\dot{\zeta}\right)+\frac{1}{3}\phi_{20}\phi_2\phi_1^2R_0^2\frac{\eta_1}{\kappa_1}\left(\dot{\zeta}\right)\\&=\left(Q_1e\phi_2+R_1\left(\xi_1+\phi_2\zeta\right)\phi_2-Q_2e\phi_1-R_2\phi_1\left(\xi_2-\phi_1\zeta\right)\right)\end{aligned}\quad(2.56)$$

Compared to the original BR theory, the reformulated LFF-governing equation is more reliable, since it takes into account the local-flow kinetic energy and the dissipation effects inside the inclusion.

Figure 2.22 displays an example by considering an "oil-in-water" double-porosity model. It shows the velocity dispersion and attenuation as a function of frequency corresponding to the original BR theory and the reformulated BR theory.

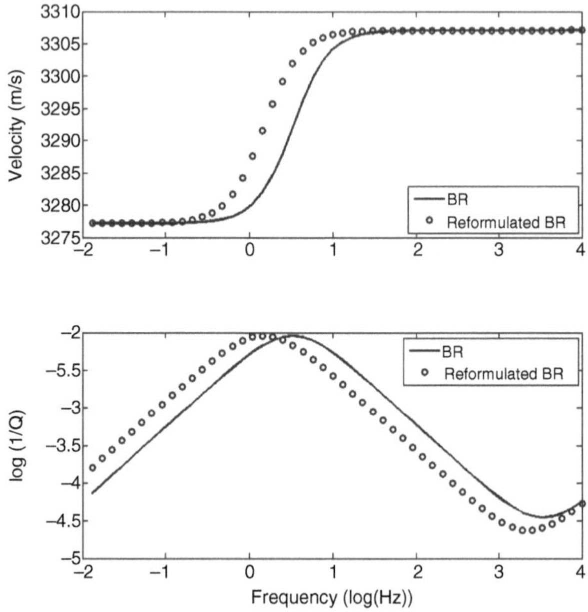

FIGURE 2.22 Numerical example of the "oil-in-water" double-porosity model (water saturation is 95%).

In the "oil-in-water" double-porosity model, the water-saturated porous solid is the host phase, whereas the oil-saturated porous solid is the inclusion. The difference resides in a shift of the relaxation peak to lower frequencies when using the reformulated BR theory. This difference is important and, as a consequence, it is not justifiable to neglect the LFF kinetic energy and dissipation of the inclusions.

In order to verify the applicability of the reformulated BR theory, the theoretical results are compared with ultrasonic experimental data performed on samples collected at the Sugaitebulake district near Aqsu. A cylindrical sample is measured at 800 KHz with a pore pressure of 10 MPa and a confining pressure of 80 MPa at room temperature. An imbibition technique is used to vary the water/oil saturation in the sample. The sample is mainly composed of tight dolomite. The permeability is 0.174 mD and the porosity is 0.0547. The average size of the inclusions is 0.035 mm, the bulk modulus of the rock skeleton K_d is 76.2 GPa and the shear modulus μ_d is 30.67 GPa. For dolomite, the grain bulk modulus K_s is 94.9 GPa, the grain shear modulus μ_s is 45 GPa and the grain density ρ_s is 2870 kg/m^3. The bulk modulus, viscosity, and density of water are 2.25 GPa, 0.001 Pa·s, and 990 kg/m^3 respectively, whereas for oil these properties are 0.6 GPa, 0.001 Pa·s, and 900 kg/m^3 respectively.

The theoretical results are based on the "oil-in-water" double-porosity model. Figure 2.23 shows the comparison, where the red circles represent the P-wave

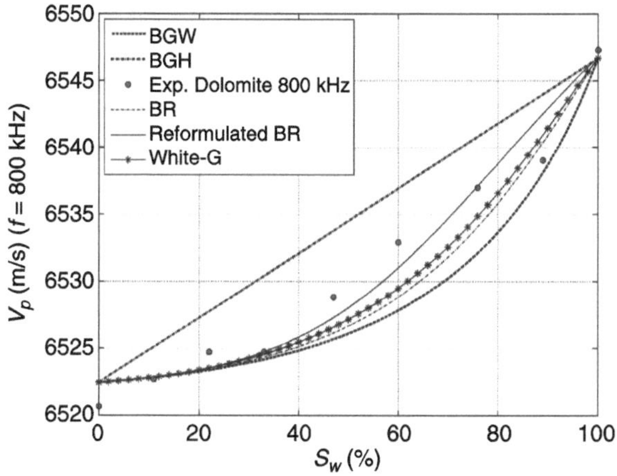

FIGURE 2.23 P-wave velocities as a function of water saturation (theoretical results and experimental measurements), where S_w denotes water saturation.

data measured in the laboratory. The five curves correspond to the BGW bound, the BGH bound, the White-G (generalized White's model), the BR model and the reformulated BR model. The comparison shows that the curve corresponding to the reformulated BR theory is in good agreement with the experimental data. In addition, this result validates the "oil-in-water" model as very close to the objective *in-situ* states of fluid distribution in reservoir rocks.

2.6 FLUID SUBSTITUTION IN PARTIALLY SATURATED SANDSTONES

In order to investigate theoretical fluid-substitution methods for sandstones, P- and S-wave velocity measurement are performed on sandstones, involving dry samples, full-brine saturation, full-oil saturation, and partially saturation. The Biot–Gassmann and constant-cement theories are compared with measured data.

A triaxial test system and a high-temperature and high-pressure system are employed. The P- and S-wave velocities are measured at hydrostatic pressure 25 MPa. Dry samples are measured at a pore pressure of 0.1 MPa, whereas saturated samples are measured at 10 MPa.

We perform experiments on cylindrical samples (25 mm in diameter and 50 mm in length) cut from Changyuan sandstones, collected from Daqing oilfield, northeast China. The samples are dried at 40°C under 40 MPa from 24 to 28 hours before the experiments. The fluids are injected at increasing pore pressures to reach full saturation. For samples saturated with an oil-brine mixture, brine is injected into oil-saturated samples to obtain the desired partial saturations.

Ultrasonic P- and S-wave velocities are measured on 65 sandstones at a frequency of 137 kHz, 50 measurements in dry sandstones, 36 in fine-grained

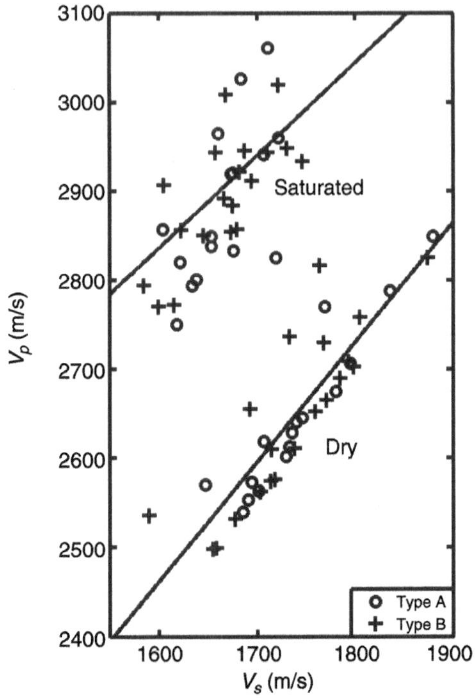

FIGURE 2.24 V_p–V_s relationships for 36 dry fine-grained sandstone samples. Type A and B are lithic sandstone and lithic arkose sandstone, respectively. The lines correspond to the best fit.

sandstones and 14 in medium-grained sandstones. The clay content ranges between 15% and 20%, quartz content from 40% to 50%, and porosity from 22% to 28%. The burial depth is 890–1000 m. For six samples, ultrasonic velocity measurements are performed from full-oil to full-brine saturation.

The 36 fine-grained samples are classified into two groups, type A and B. Type A is lithic sandstone and type B is lithic arkose sandstone. Figure 2.24 shows the ultrasonic measurements on the 36 fine-grained sandstones at dry and full-brine saturation conditions. The effective stresses are 25 MPa (dry) and 15 MPa (saturated).

The linear regression for the dry sandstones is $V_p = 1.344V_s + 311.3$ in m/s, with a correlation coefficient $r = 0.937$, whereas the relationship for the brine-saturated sandstones is $V_p = 1.039V_s + 1174$ in m/s, with $r = 0.81$. The measured V_p is smaller than that predicted by the empirical equations of Castagna and Backus (1993) or Han (1986). One reason can be the low effective pressure used in the experiments.

Figure 2.25 shows ultrasonic velocity measurements on 6 samples partially saturated with brine and oil at an effective stress of 15 MPa. These samples are mid- and fine-grained sandstones with porosities of 24.3%, 26.2%, 25.7%, 22.7%, 24%, and 28%, respectively. The velocities increase with brine saturation.

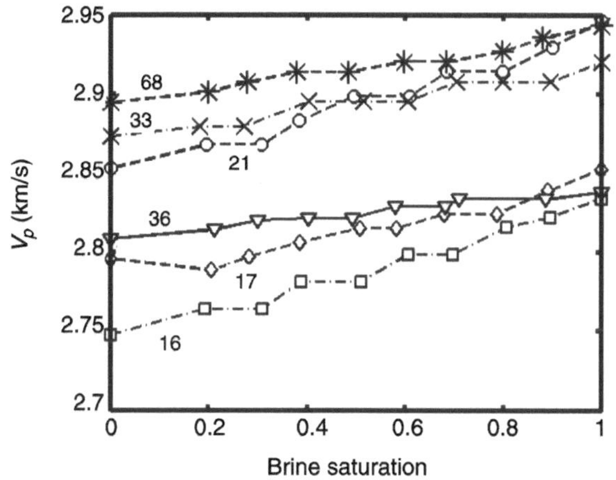

FIGURE 2.25 P-wave velocities as a function of brine saturation for six sandstone samples.

This increase in samples 16 and 21 from full-oil to full-brine saturation amounts to 85–95 m/s, respectively, whereas that of the other samples increase from 45 to 55 m/s. Samples 16 and 21 are medium-grained sandstones with larger pore size, whereas the other samples are fine-grained sandstones with more interstitial clay. Thus, we hypothesize that the grain size is an important factor affecting the velocity variations for partial-saturation conditions.

Next, we use the Biot–Gassmann theory (Gassmann, 1951, Biot, 1956) in fluid substitution and to predict the velocity dispersion. Figure 2.26 illustrates the measured and predicted P-wave velocities of 10 rock samples at full-brine saturation as a function of porosity. There are 3 medium-grained and

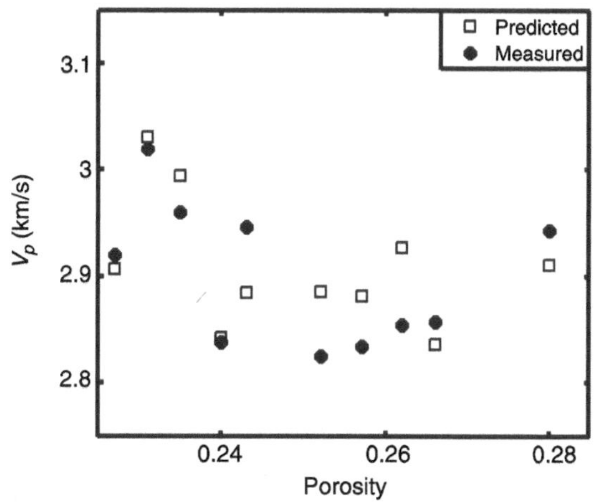

FIGURE 2.26 Predicted and measured P-wave velocities of brine-saturated samples versus porosity.

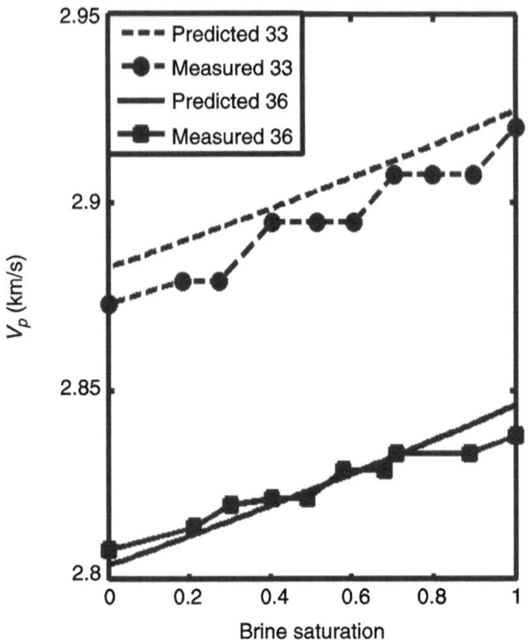

FIGURE 2.27 Predicted and measured P-wave velocities as a function of brine saturation.

7 fine-grained sandstones. The predictions are fairly close to the measurements for most of samples, with a maximum difference of 50 m/s between theory and measurements, and a relative error of 1.8%. Particularly, there are four samples whose differences are less than 10 m/s. However, the theoretical S-wave velocities are lower than the measurements (not shown here).

Figure 2.27 compares theoretical results and measurements for partially saturated samples 33 and 36. Sample 33 shows the higher difference (0.34%).

The constant-cement model (Avseth, 2000) is used to predict the dry-rock bulk modulus. This model assumes that sandstones of different porosity have the same amount of contact cement. It is a combination of the contact-cement and friable-sand models. We have

$$
\begin{aligned}
K_{\mathrm{dry}} &= \left[\frac{\phi/\phi_b}{K_b+4/3\mu_b}+\frac{1-\phi/\phi_b}{K+4/3\mu}\right]^{-1}-\frac{4}{3}\mu_b \\
\mu_{\mathrm{dry}} &= \left[\frac{\phi/\phi_b}{\mu_b+z}+\frac{1-\phi/\phi_b}{\mu+z}\right]^{-1}-z \qquad (2.57)\\
z &= \frac{\mu_b}{6}\left(\frac{9K_b+6\mu_b}{K_b+2\mu_b}\right)
\end{aligned}
$$

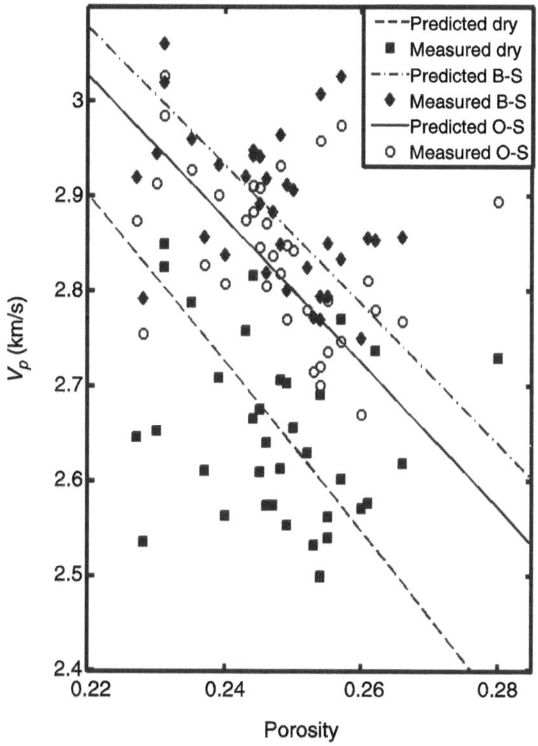

FIGURE 2.28 Theoretical and measured velocities of the fine-grained sandstones as a function of porosity. The prediction is obtained by combining the constant cement model and the Biot–Gassmann theory. B-S and O-S denote brine saturation and oil saturation, respectively.

where K_b and μ_b are the dry-rock bulk and shear moduli at the critical porosity ϕ_b, respectively, and K and μ are the bulk and shear moduli of the mineral phase, respectively.

Figure 2.28 shows the theoretical and measured velocities as a function of porosity for fine-grained samples, where the theory combines the constant-cement model and Biot–Gassmann equation. The effective moduli of the minerals used in the constant-cement model are calculated with the Hill average, considering fractions of quartz, feldspar, and clay. For most of the samples, the predictions are consistent with the data for both the dry and saturated samples. The correlation coefficient of the three saturation conditions between predictions and measurements are 0.6354, 0.6926, and 0.7637, respectively.

Grain size appears to be an important factor affecting the change of V_p from fully oil-saturated to fully brine-saturated status in Changyuan sandstones. The Biot–Gassmann theory can account for variations in P-wave velocities when the sandstones are saturated with brine. The theoretically predicted P-wave velocities were close to the measured data for most of the samples, with relative errors less than 1.8%. By estimating the dry-rock bulk modulus based on the constant-cement model, the relations of velocity to porosity are in good agreement with the measurements.

ACKNOWLEDGMENTS

This study is supported by the Natural National Science Foundation of China (41104066) and the Fluids/DHI Consortium (University of Houston). The authors are grateful to Haibo Zhao, Zhaobing Hao, Jiawei Liu, and Jack Dvorkin for their help in the relevant discussions, numerical examples, and rock tests.

REFERENCES

Avseth, P., 2000. Combining rock physics and sedimentology for seismic reservoir characterization of North Sea turbidite systems. PhD Dissertation. Stanford University, Stanford.

Ba, J., Cao, H., Yao, F., Nie, J., Yang, H., 2008a. Double-porosity rock model and squirt flow in the laboratory frequency band. Appl. Geophys. 5 (4), 261–276.

Ba, J., Carcione, J., Nie, J., 2011. Biot-Rayleigh theory of wave propagation in double-porosity media. J. Geophys. Res. 116 (B6), B06202.

Ba, J., Nie, J.X., Cao, H., Yang, H.Z., 2008b. Mesoscopic fluid flow simulation in double-porosity rocks. Geophys. Res. Lett. 35 (4), L04303.

Ba, J., Zhang, L., Sun, W., Hao, Z., 2014. Velocity field of wave-induced local fluid flow in double-porosity media. Science China Physics. Mech. Astron. 57 (6), 1020–1030.

Berryman, J.G., 1979. Theory of elastic properties of composite materials. Appl. Phys. Lett. 35 (11), 856–858.

Berryman, J.G., Wang, H.F., 2000. Elastic wave propagation and attenuation in a double-porosity dual-permeability medium. Int. J. Rock. Mech. Min. 37 (1), 63–78.

Biot, M.A., 1956. Theory of propagation of elastic waves in a fluid-saturated porous solid. I. Low-frequency range. J. Acoust. Soc. Am. 28, 168–178.

Biot, M.A., 1962. Mechanics of deformation and acoustic propagation in porous media. J. Appl. Phys. 33 (4), 1482–1498.

Biot, M.A., Willis, D.G., 1957. The elastic coefficients of the theory of consolidation. J. Appl. Mech. 24, 594–601.

Cadoret, T., Marion, D., Zinszner, B., 1995. Influence of frequency and fluid distribution on elastic wave velocities in partially saturated limestones. J. Geophys. Res. 100 (B6), 9789–9803.

Carcione, J.M., 2007. Wave Fields in Real Media: Wave Propagation in Anisotropic, Anelastic, Porous and Electromagnetic Media. Elsevier Science, Amsterdam.

Castagna, J.P., Backus, M.M., 1993. Offset-Dependent Reflectivity: Theory and Practice of AVO Analysis[M]. Investigations in Geophysics, No. 8. Society of Exploration Geophysicists, Tulsa.

Deresiewicz, H., Skalak, R., 1963. On uniqueness in dynamic poroelasticity. B. Seismol. Soc. Am 53 (4), 783–788.

Dutta, N., Odé, H., 1979a. Attenuation and dispersion of compressional waves in fluid-filled porous rocks with partial gas saturation (White model) – Part I: Biot theory. Geophysics 44 (11), 1777–1788.

Dutta, N., Odé, H., 1979b. Attenuation and dispersion of compressional waves in fluid-filled porous rocks with partial gas saturation (White model) – Part II: Results. Geophysics 44 (11), 1789–1805.

Dvorkin, J., Mavko, G., Nur, A., 1995. Squirt flow in fully saturated rocks. Geophysics 60 (1), 97–107.

Gassmann, F., 1951. Elastic waves through a packing of spheres. Geophysics 16 (4), 673–685.

Han, D., 1986. Effect of porosity and clay content on acoustic properties of sandstones and unconsolidated sediments. PhD Dissertation. Stanford University, Stanford.

Johnson, D., 1986. Recent developments in the acoustic properties of porous media. Front. Phys. Acoust. 93, 255–290.

Johnson, D.L., 2001. Theory of frequency dependent acoustics in patchy-saturated porous media. J. Acoust. Soc. Am. 110 (2), 682–694.

Knight, R., Nolen-Hoeksema, R., 1990. A laboratory study of the dependence of elastic wave velocities on pore scale fluid distribution. Geophys. Res. Lett. 17 (10), 1529–1532.

Lebedev, M., Toms-Stewart, J., Clennell, B., Pervukhina, M., Shulakova, V., Paterson, L., Gurevich, B., Wenzlau, F., 2009. Direct laboratory observation of patchy saturation and its effects on ultrasonic velocities. Leading Edge 28 (1), 24–27.

Liu, X., Greenhalgh, S., Zhou, B., 2009. Transient solution for poro-viscoacoustic wave propagation in double porosity media and its limitations. Geophys. J. Int. 178 (1), 375–393.

Müller, T.M., Gurevich, B., 2004. One-dimensional random patchy saturation model for velocity and attenuation in porous rocks. Geophysics 69 (5), 1166–1172.

Mavko, G., Mukerji, T., Dvorkin, J., 2003. The Rock Physics Handbook: Tools for Seismic Analysis in Porous Media. Cambridge University Press.

Murphy, W.F., 1982. Effects of microstructure and pore fluids on the acoustic properties of granular sedimentary materials. PhD Dissertation. Stanford University, Department of Geophysics, Stanford.

Pride, S.R., Berryman, J.G., 2003. Linear dynamics of double-porosity dual-permeability materials. I. Governing equations and acoustic attenuation. Phys. Rev. E 68 (3), 036603.

Pride, S.R., Berryman, J.G., Harris, J.M., 2004. Seismic attenuation due to wave-induced flow. J. Geophys. Res. 109 (B1), B01201.

Rayleigh, F., 1917. On the pressure developed in a liquid during the collapse of a spherical cavity. Lond. Edinb. Dubl. Phil. Mag. and J. Sci. 34 (200), 94–98.

Shapiro, S.A., Müller, T.M., 1999. Seismic signatures of permeability in heterogeneous porous media. Geophysics 64 (1), 99–103.

White, J., 1975. Computed seismic speeds and attenuation in rocks with partial gas saturation. Geophysics 40 (2), 224–232.

White, J., Mihailova, N., Lyakhovitsky, F., 1975. Low-frequency seismic waves in fluid-saturated layered rocks. J. Acoust. Soc. Am. 57, S30.

Chapter 3

Acoustics of Partially Saturated Rocks: Theory and Experiments

Tobias M. Müller[1], Eva Caspari[2], Qiaomu Qi[3], J. Germán Rubino[2], Danilo Velis[4], Sofia Lopes[3], Maxim Lebedev[3], Boris Gurevich[1,3]

[1]*Energy Flagship, CSIRO, Perth, Australia;*

[2]*Applied and Environmental Geophysics Group, University of Lausanne, Lausanne, Switzerland;*

[3]*Department of Exploration Geophysics, Curtin University, Perth, Australia;*

[4]*Facultad de Ciencias Astronómicas y Geofísicas, Universidad Nacional de La Plata, and CONICET, Argentina*

Chapter Outline

3.1 INTRODUCTION

A rock whose pore space is occupied by immiscible fluids is referred to as a partially saturated rock. Partial saturation arises when one fluid migrates (or is injected) into a rock partially replacing the in-situ fluid. Since the in-situ fluid will not be completely swept out, patches of the fluid phases will be formed. These patches not only exist within one pore but also can extend over many pores, as evidenced by direct laboratory observations (Lei and Xue, 2009; Lebedev et al., 2009a, 2009b; Alemu et al., 2013, Lopes et al., 2014). Acoustic monitoring of core flooding experiments reveals complex velocity-saturation relations (VSR) and attenuation–saturation relations (ASR).

At the field scale, patchy-saturation scenarios are typically linked to fluid injection into or production from hydrocarbon reservoirs and are thought to affect time-lapse seismic signals (Ajo-Franklin et al., 2013; Villaudy et al., 2013; Vasco et al., 2014). Strongest changes are observed near the injection well. This indicates that patchy saturation can affect sonic responses (Xue et al., 2006).

Seismic Exploration of Hydrocarbons in Heterogeneous Reservoirs. http://dx.doi.org/10.1016/B978-0-12-420151-4.00003-9

Strong amplitude distortions in the presence of gas chimneys (Yu et al., 2012) are possibly caused by patchy saturation.

Direct observations of fluid patches at the field scale are difficult to obtain. However, indirect evidence comes from flow simulations. For example, Sengupta et al. (2003) showed that the subseismic spatial distribution of fluids, not captured by traditional flow simulators, can influence the seismic response. They found that downscaling smooth saturation outputs from the flow simulator to a more realistic patchy distribution is required to provide a good quantitative match with the near- and far-offset time-lapse data, even though the fine details in the saturation distribution are below seismic resolution. Based on flow simulations of an outcrop analogue reservoir, Kirstetter et al. (2006) estimated fluid patch sizes for different production scenarios (e.g. water displaces oil) to be on the order of 1 m.

Therefore, predicting saturation effects on seismic responses requires a firm understanding of the underlying velocity and attenuation dependencies on the pore fluid distribution. Core flooding laboratory experiments and reservoir-scale injection/production operations differ by orders of magnitude in length scale. The former is typically monitored by ultrasonic waves and the latter by seismic waves. Further, the flow regimes leading to patchy saturation may significantly differ in both cases with respect to flow rates and driving forces. This makes the modeling and interpretation of VSRs and ASRs across the scales and in different frequency bands a challenging task. Yet, despite these striking differences, some aspects of wave attenuation can be explained in terms of a single mechanism, wave-induced pressure diffusion or wave-induced fluid flow (WIFF) (for a review, we refer to Müller et al., 2010).

In this chapter, we interpret laboratory experiments, analyze well log data and perform numerical simulations to reveal different facets of wave-induced pressure diffusion at different scales and frequencies. We focus on mesoscopic fluid patches. Mesoscopic refers to a length scale much larger than typical pore size and yet much smaller than the seismic wavelength. Typically, we envisage fluid patch sizes on the order of millimeter to centimeter. Such mesoscopic fluid patches are embedded in a porous background medium (Figure 3.1). The restriction to mesoscopic fluid patches enables us to use VSRs and ASRs that

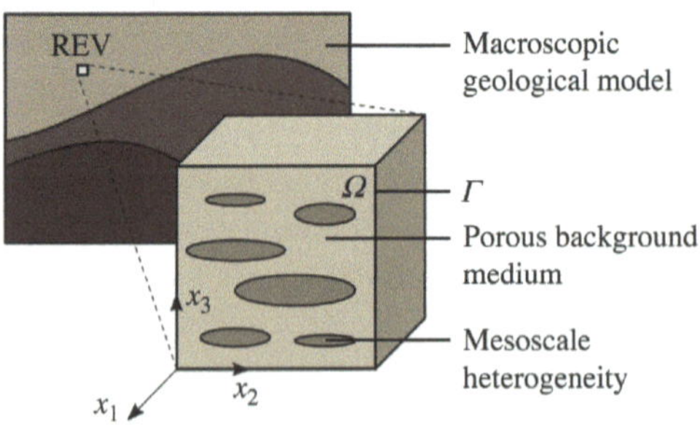

FIGURE 3.1 Elementary volume of a heterogeneous porous material containing mesoscale heterogeneities *(Wenzlau et al., 2010).*

can be consistently described within a single theoretical framework, that is, the theory of linear poroelasticity.

The chapter is outlined as follows. We first recall the basic notions of pressure diffusion and bounds for patchy saturation in the realm of linear poroelasticity and present theoretical models for patchy saturation based on random distributions of the fluid phases. In the section 'Laboratory Experiments', we review recent progress on the characterization of fluid patch evolution during fluid injection experiments. Further, we interpret measured acoustic signatures utilizing random medium models. WIFF in the sonic frequency band is then discussed utilizing time-lapse well logging data. In the section 'Signatures of Patchy Saturation in the Seismic Frequency Band', we employ a numerical upscaling technique to highlight WIFF implications on synthetic seismic zero-offset sections at reservoir scale. We present some concluding remarks with recommendations for future research in the section 'Perspectives for Future Research'.

3.2 FLUID PRESSURE DIFFUSION AND PATCHY SATURATION BOUNDS

If a patchy-saturated rock is probed by seismic waves, fluid pressure gradients develop across adjacent fluid patches. The induced pressure gradients scale with the fluid compressibility contrast. These induced pressure gradients vanish (relax) through fluid pressure diffusion, a process accompanied by friction and thus dissipation. Depending on the relation between the patch size, a, and the length scale λ_p, over which pressure diffusion takes place, different regimes occur. Moreover, we can distinguish different frequency regimes because this diffusion wavelength, similarly to a viscous skin depth, scales with the square root of frequency, $\lambda_p \sim \sqrt{1/\omega}$. In the low-frequency limit pressure gradients vanish, as there has been enough time in one wave cycle for the fluid pressure to equilibrate. Conversely, if the frequency is high enough, there will be no communication between different fluid pockets and the induced pressure gradients persist. This causes an overall stiffening of the rock and, hence, higher wave velocities. At intermediate frequencies, wave-induced pressure diffusion leads to dissipation and thus to attenuation and dispersion. This is known as WIFF. Note, however, that WIFF does not imply any net fluid flow through the rock.

In order to understand WIFF, let us first analyze the parameters that control fluid pressure diffusion. For a patchy-saturated rock, Norris (1993) shows that the fluid pressure equilibration process is governed by a diffusion equation with pressure diffusivity (sometimes referred to as hydraulic diffusivity):

$$D_p = \frac{\kappa}{\mu_f} N. \tag{3.1}$$

It is directly proportional to the flow permeability, κ, and inversely proportional to the shear viscosity of the fluid, μ_f. Since the rock matrix is deformable, the pressure diffusivity also depends on a combination of the fluid bulk modulus, K_f, solid-phase bulk modulus, K_s, drained rock bulk modulus, K_0, and shear modulus, μ. These poroelastic moduli together with the porosity of the rock, ϕ, are lumped into the modulus N. It is defined as

$$N = \frac{L}{H} M \tag{3.2}$$

and is inversely proportional to the fluid storage coefficient

$$\frac{1}{M} = \frac{\phi}{K_f} + \frac{\alpha - \phi}{K_s}. \tag{3.3}$$

The latter involves the Biot–Willis coefficient (Biot, 1962)

$$\alpha = 1 - \frac{K_0}{K_s} \tag{3.4}$$

relating the bulk modulus of the drained rock to the solid-phase bulk modulus. Therefore, in a deformable porous medium, the ability to store a fluid depends not only on the porosity and fluid bulk modulus but also on the drained rock and solid-phase bulk modulus. Furthermore, the modulus N depends on the ratio L/H that is the drained P-wave modulus

$$L = K_0 + \frac{4}{3}\mu \tag{3.5}$$

and the undrained P-wave modulus H. According to Gassmann's (1951) theory, the undrained P-wave modulus is

$$H = L + \alpha^2 M. \tag{3.6}$$

As both the Biot–Willis coefficient and the fluid storage coefficient are positive quantities, it follows that $H \geq L$. In the limiting case $K_s \gg K_f$, that is for example when the saturating fluid is a gas, we find $H \approx L$ and $N \approx \frac{K_f}{\phi}$. Then the pressure diffusivity becomes $D_p\big|_{K_s >> K_f} = \frac{\kappa}{\mu_f}\frac{K_f}{\phi}$ and is independent of the solid-phase and drained rock bulk moduli.

The fluid pressure diffusivity allows us to quantify the characteristic spatial domain and frequency associated with WIFF. The spatial domain over which the pressure diffusion process takes place is given by the diffusion wavelength

$$\lambda_p = \sqrt{\frac{D_p}{\omega}}. \tag{3.7}$$

The characteristic frequency associated with this pressure diffusion process is obtained from the resonance condition $a = \lambda_p$ and yields to

$$\omega_c \sim \frac{D_p}{a^2}. \tag{3.8}$$

The different frequency regimes of WIFF are characterized by distinct *P*-wave VSRs. If the porous rock frame itself is homogeneous and saturation is spatially variable, for low- and high-frequencies, that is, $\omega \ll \omega_c$ and $\omega \gg \omega_c$, exact bounds exist for the phase velocities. At intermediate wave frequencies, phase velocities are frequency dependent and lie between these limiting cases. If $\omega \approx \omega_c$, then the change in phase velocity for an incremental change in frequency is strongest and, according to the Kramers-Kronig relations, wave attenuation is maximum.

When the frequency of the incident wave is sufficiently low, $\omega \ll \omega_c$, so that the characteristic patch size of fluid heterogeneities is much smaller than the diffusion length

$$a \ll \lambda_p, \tag{3.9}$$

there is enough time for fluid pressure gradients to relax and equilibrate to a constant pressure. Wood's (1941) law for suspensions can be applied in this limit to determine an effective fluid bulk modulus

$$\frac{1}{K_{\mathrm{W}}} = \sum_{i=1}^{n} \frac{S_i}{K_{\mathrm{f}i}}. \tag{3.10}$$

This is the harmonic, saturation-weighted average over the fluid bulk moduli of n patches, wherein the individual patch bulk modulus is denoted as $K_{\mathrm{f}i}$ $(i = 1\ldots n)$ with saturation S_i and $\sum_{i=1}^{n} S_i = 1$. Once the effective bulk modulus K_{W} is obtained, Gassmann (1951) relations can be applied to estimate the low-frequency, undrained *P*-wave modulus for a partially saturated rock

$$H_{\mathrm{GW}} = L + \alpha^2 M(K_{\mathrm{W}}), \tag{3.11}$$

where $M(K_{\mathrm{W}})$ means that Wood's law is used in order to compute the fluid storage coefficient. This quasi-static limit is known as uniform saturation or Gassmann–Wood (GW) bound.

Conversely, when the wave frequency is sufficiently high, $\omega \gg \omega_c$, or equivalently, the characteristic patch size is larger than the diffusion length,

$$a \gg \lambda_p, \tag{3.12}$$

there is not enough time for pressure equilibration. Then the induced pressure gradients persist. That is to say, the patches will remain at different pressures. Application of Gassmann's theory to individual patches allows the undrained bulk modulus of each patch to be determined. According to Gassmann's equations, the saturated shear modulus of each patch is independent of the fluid bulk modulus. Thus, Hill's (1963) theorem can be applied to determine the overall undrained bulk modulus via harmonic averaging of the undrained P-wave moduli of the n patches

$$\frac{1}{H_{\mathrm{GH}}}=\sum_{i=1}^{n}\frac{S_i}{H(K_{\mathrm{fi}})}, \tag{3.13}$$

where $H(K_{\mathrm{fi}})$ denotes the undrained P-wave modulus computed for the ith patch having fluid bulk modulus K_{fi}. This high-frequency or no-flow limit is known as patchy or Gassmann–Hill (GH) bound. The elastic moduli in both, the low- and high-frequency limits, are given by real numbers. Johnson (2001) shows that

1. For any nonzero saturation, the GW moduli are always smaller than those for GH moduli,
2. For rock behavior consistent with the Biot–Gassmann theory, at intermediate frequencies, the VSR inevitably lies within these bounds.

Thus, for any intermediate saturation level, the phase velocity is a monotonically increasing function of frequency. The specific form of this VSR depends not only on the poroelastic parameters but also on the geometrical arrangement of the fluid patches. It is likely that in real rock formations and saturation conditions, these patches have irregular shapes and are distributed in a random fashion. Therefore, models describing the VSRs and ASRs for random patch distributions are of particular interest.

3.3 BIOT'S THEORY OF POROELASTICITY AND RANDOM PATCHY SATURATION MODELS

The GW equations are widely used in the petroleum industry for estimating seismic wave velocities in hydrocarbon reservoirs (Wang, 2001; Smith et al., 2003; Mavko et al., 2009). However, wave attenuation and dispersion associated with fluid pressure diffusion cannot be modeled with Gassmann's equations. To account for these dynamic effects, Biot's theory (Biot, 1956a, 1956b, 1962) of poroelasticity provides a general framework for modeling elastic wave propagation through fluid-saturated porous media. In the following we assume that the wave frequency is always much smaller than the Biot relaxation frequency ω_{i} so that the relative motion between fluid and solid is dominated by viscous friction. It can be shown that in this low-frequency (quasi-static) regime, the slow P-wave predicted by Biot degenerates to a diffusion wave with the wavenumber k_{p2}. Therefore, the analysis of attenuation because of WIFF can be thought of as the conversion (at interfaces or at spatial heterogeneities of the porous medium) between the propagating (fast) wave modes and the diffusive slow wave mode, which leads then to dissipation of wavefield energy. In the quasi-static frequency regime WIFF can be also analyzed in the framework of Biot's (1941) consolidation equations. Therein, an oscillating external load (implemented as oscillating boundary condition) can be used as a proxy for the fast *P*-wave. The subsequent coupled fluid pressure and solid stress diffusion is fully described by the consolidation equations.

Biot's (1941) consolidation equations, the so-called quasi-static poroelasticity equations, in the frequency domain are given by

$$\nabla \cdot \boldsymbol{\sigma} = 0, \tag{3.14}$$

$$i\omega \frac{\mu_f}{\kappa} \boldsymbol{w} = -\nabla p_f, \tag{3.15}$$

where $\boldsymbol{\sigma} \equiv (\sigma_{ij})$ is the total stress tensor, $\boldsymbol{w}$ the relative fluid–solid displacement, and p_f the fluid pressure. The first equation represents the stress equilibrium within the sample. The second equation is Darcy's law relating the fluid pressure gradient to the time derivative of the relative fluid–solid displacement (sometimes referred to as the filtration velocity). These two equations are coupled through the stress–strain relations

$$\sigma_{ij} = 2\mu\epsilon_{ij}\left(\boldsymbol{u}^s\right) + \delta_{ij}\left(\lambda_u \nabla \cdot \boldsymbol{u}^s - \alpha M \zeta\right), \tag{3.16}$$

$$p_f = -\alpha M \nabla \cdot \boldsymbol{u}^s + M\zeta, \tag{3.17}$$

where $\boldsymbol{u}^s$ denotes the average displacement vector of the solid phase, $\epsilon_{ij}(\boldsymbol{u}^s)$ is the strain tensor of the solid phase, and $\zeta = -\nabla \cdot \boldsymbol{w}$ represents the change in the fluid content. The remaining parameters can be written in terms of the poroelastic end-member properties. The undrained Lamé parameter is given by $\lambda_u = H - 2\mu$. These quasi-static poroelasticity equations can be solved by different means. In section 3.8, we will present an example of how WIFF attenuation and dispersion can be obtained for a complex fluid patch distribution using a numerical solution technique.

By introducing some simplifying assumptions, the poroelasticity equations can also be solved for complex fluid patch distributions by analytical methods. The continuous random media (CRM) models of Müller and Gurevich (2004) and Toms et al. (2007) provide approximate solutions of Biot's poroelasticity equations with randomly varying coefficients. Using the method of statistical smoothing, the so-called Bourret approximation (Karal and Keller, 1964), Müller and Gurevich (2005a, 2005b) analyzed the conversion scattering from fast *P*-waves into Biot's slow *P*-waves in randomly inhomogeneous porous media. This conversion scattering problem quantifies the dissipation of wave field energy due to energy transfer from the coherent component of the fast *P*-wave into the dissipative slow *P*-wave mode. For the case that inhomogeneities within the medium are only caused by partial saturation, the results describe the dependence of velocity and attenuation on saturation, frequency, and geometrical distribution of fluids, in particular the characteristic length scale of fluid patches.

The dynamic-equivalent *P*-wave modulus for a system of alternating layers with different fluids (say water and gas) and random thicknesses is a complex and frequency-dependent quantity that we denote as $\tilde{H}_{1D}(\omega)$. It can be expressed as a perturbation to the GW *P*-wave modulus, H_{GW} (Müller and Gurevich, 2004)

$$\tilde{H}_{1D}(\omega) = H_{GW}\left(1 - is\tilde{k}_{p2}\int_0^\infty B_{MM}(r)e^{i\tilde{k}_{p2}r}dr\right), \tag{3.18}$$

where $\tilde{k}_{p2}$ is the effective slow *P*-wave number

$$\tilde{k}_{p2} = \frac{\sqrt{\mu_{fw}N_w}\;S + \sqrt{\mu_{fg}N_g}\;(1-S)}{N_w\,S + N_g\,(1-S)}\sqrt{\frac{i\omega}{\kappa}} \tag{3.19}$$

and subscripts *w* and *g* refer to water (saturation *S*) and gas (saturation 1 – *S*), respectively. The function $B_{MM}(r)$ is the variance-normalized autocorrelation function of the randomly fluctuating fluid storage modulus *M*, that is, $B_{MM}(r=0)=1$, and vanishes for large correlation lags. The degree of inhomogeneity is quantified by the parameter

$$s = \frac{H_{GH}}{H_{GW}} - 1. \tag{3.20}$$

From equation 3.18 it becomes clear that for a uniformly saturated porous medium we have $s = 0$.

This 1D random patchy saturation model has been generalized to the 3D case by Toms et al. (2007). The dynamic-equivalent undrained *P*-wave modulus in 3D space obtained from the method of statistical smoothing is

$$\tilde{H}_{eff}(\omega) = H_0\left(1 - \Delta_2 - \Delta_1\tilde{k}_{p2}^2\int_0^\infty rB_{MM}(r)e^{i\tilde{k}_{p2}r}dr\right)^2. \tag{3.21}$$

Here, H_0 is the undrained *P*-wave modulus of the homogeneous poroelastic background medium supporting the slow *P*-wave $\tilde{k}_{p2} = \sqrt{i\omega / D_p}$. The perturbation coefficients Δ_1 and Δ_2 are given by

$$\Delta_1 = \frac{L}{H}\Delta_2, \qquad \Delta_2 = \frac{\alpha^2 M}{2H}\sigma_{MM}^2, \tag{3.22}$$

where σ_{MM}^2 is the normalized variance of the fluid storage coefficient. This undrained *P*-wave modulus is only valid for weak fluctuations in the fluid storage coefficient, which translates into weak contrast in the fluid bulk moduli. Since the exact low- and high-frequency limits for a partially saturated rock are known (i.e., H_{GW} and H_{GH}), this weak-contrast result for $\tilde{H}_{eff}(\omega)$ can be further improved. Toms et al. (2007) derive a re-scaled *P*-wave modulus for random patchy saturation

$$\tilde{H}_{3D}(\omega) = H_{GW}\left(1 + \frac{H_{GH} - H_{GW}}{H_{high} - H_{low}}\;\frac{\tilde{H}_{eff}(\omega) - H_{low}}{H_{GW}}\right). \tag{3.23}$$

This dynamic-equivalent *P*-wave modulus is consistent with the exact bounds for arbitrary large contrast in the fluid bulk moduli. It involves the low and high-frequency limits of $\tilde{H}_{eff}(\omega)$ which are given by

$$H_{low} = H_0(\Delta_2 - 1)^2, \quad H_{high} = H_0(1 - \Delta_2 + \Delta_1)^2. \tag{3.24}$$

For a formulation of this 3D random patchy saturation model to account for fractal fluid distribution, we refer to Müller et al. (2008).

Figure 3.2 shows *P*-wave attenuation and dispersion predicted by the 3D CRM model. In this example, we use poroelastic parameters resembling an

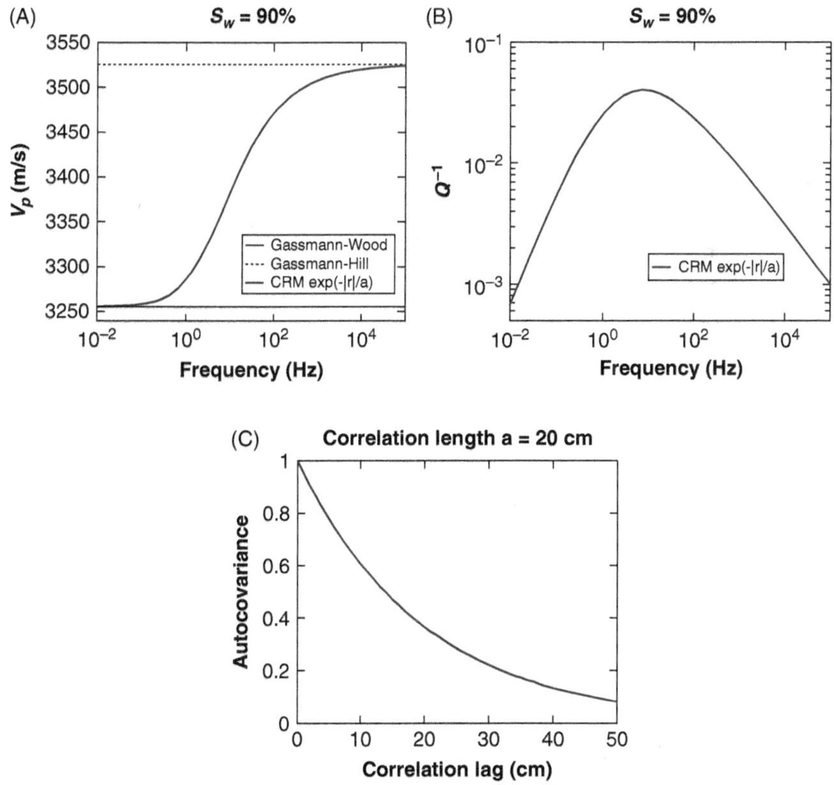

FIGURE 3.2 (A) *P*-wave dispersion and (B) attenuation expressed in terms of the reciprocal quality factor Q^{-1} predicted by the 3D CRM model for a water saturation of 90% in an air–water saturated sandstone. (C) In this example, the random fluid distribution is quantified by an exponential correlation function with correlation length of 20 cm.

air–water-saturated sandstone. The water patches are assumed to be characterized by an exponential correlation function with correlation length of 20 cm. We note that in these mesoscopic CRM models, any pressure equilibration processes linked to pore-scale heterogeneities are left out and their description would require an extension of the theoretical framework. For a discussion on these issues, we refer to Müller et al. (2010) and Müller and Sahay (2011).

Application of these 1D and 3D CRM models requires information on the statistical properties of fluid patch distributions. Such information can be obtained from CT images of partially saturated rock samples during core flooding. In the remainder of the chapter, we utilize either numerical simulation or the CRM models to analyze and interpret data examples.

3.4 LABORATORY EXPERIMENTS

While theoretical poroelastic models predict the seismic response for a given spatial distribution of fluid patches, the factors controlling the formation and evolution of these patches during reservoir production are less understood. In the laboratory setting, these factors can be studied in some detail through core flooding experiments. Previously reported laboratory observations demonstrate a qualitative link between fluid patch distribution and acoustic velocities (Cadoret et al., 1995, 1998; Monsen and Johnstad, 2005). X-ray tomographic images of Cadoret et al. (1998) show that imbibition experiments, where water displaces gas, produce regular patches of fluids distributed uniformly throughout the porous rock at high water saturations. Conversely, drainage or evaporative experiments, where the reverse fluid-substitution process occurs, produce gas clusters distributed nonuniformly throughout the porous rock at high water saturations. As drainage and imbibition produce different saturation patterns at the same level of saturation, differences in measured attenuation and phase velocity can be attributed to differences in fluid distribution. Indeed, phase velocities measured from drainage experiments are appreciably higher than those from imbibition experiments (Cadoret et al., 1995; Knight and Nolen-Hoeksema, 1990) and similarly for attenuation measurements (Cadoret et al., 1998). These laboratory evidences show that, estimates of attenuation and phase velocity are affected by the distribution of immiscible fluids.

Perhaps, the most detailed insights into the distribution of fluid patches and accompanying evolutions of ultrasonic P-wave velocity and attenuation with saturation can be gained from a recent series of laboratory experiments (Lei and Xue, 2009; Lopes and Lebedev, 2012; Alemu et al., 2013; Lopes et al., 2014). In these experiments, measurements of *P*-wave velocities and X-ray CT imaging is utilized to get a deeper insight into the factors influencing the patch distribution and the associated wave response. Lei and Xue (2009) injected (gaseous and supercritical) CO_2 into a water-saturated sandstone under controlled pressure and temperature conditions and monitored the saturation dependence of *P*-wave velocity and attenuation using a sensor array of ultrasonic transducers. They found that the VSRs and ASRs can be explained by WIFF assuming patch sizes on the order of a few millimeters. Lebedev et al. (2009a, 2009b) performed a combined fluid injection and ultrasonic experiment while monitoring fluid distribution using X-ray CT. For a low-permeability sandstone, they showed that, at low saturation, the velocity-saturation dependence can be described by the GW relationship (see below). At intermediate saturations, there is a transition behavior controlled by the fluid patch arrangement and fluid patch size that can be consistently modeled with the fluid pressure diffusion mechanism. Alemu et al. (2013) found for a similar experimental setup with a moderately layered sandstone that the *P*-wave amplitude is more sensitive to small changes in the fluid distribution than the *P*-wave velocity. Their measured VSRs fall between the GW and GH bounds indicating that attenuation and dispersion might be caused by WIFF.

To better understand and quantify the dependence of VSRs on the injection rate, Lopes et al. (2014) monitored forced imbibition experiments with variable injection rates using X-ray CT and active ultrasonic measurements. Here we focus on the experimental setups discussed by Lebedev et al. (2009a, 2009b) and Lopes et al. (2014). For both forced imbibition experiments at room temperature and atmospheric pressure, simultaneous measurements of *P*-wave velocities and rock sample X-ray CT imaging are carried out (Figure 3.3).

The CT imaging allows us to infer the fluid distribution inside the rock sample during saturation (water imbibition). The resolution of the medical CT scanner is $0.2 \times 0.2 \times 1$ mm^3 (voxel size). Each pixel of the CT scan has a "CT value," which is a proxy for the X-ray absorption, and is thus directly related to the density of the scanned sample given in Hounsfield units. The resolution is not sufficient to image the exact fluid patch geometry; however, it shows the overall character of the fluid distribution. Consecutive scans of the rock sample during injection reflect the presence of water. The fluid saturation for each pixel is estimated as a difference in average CT number (a value related to material density) between the saturated ($\mathrm{CT_w}$) and dried sample ($\mathrm{CT_{dry}}$) divided by the volume of porous fraction (ϕ)

$$S_w = \frac{\mathrm{CT_{dry}} - \mathrm{CT_w}}{1000\ \phi}\ . \qquad (3.25)$$

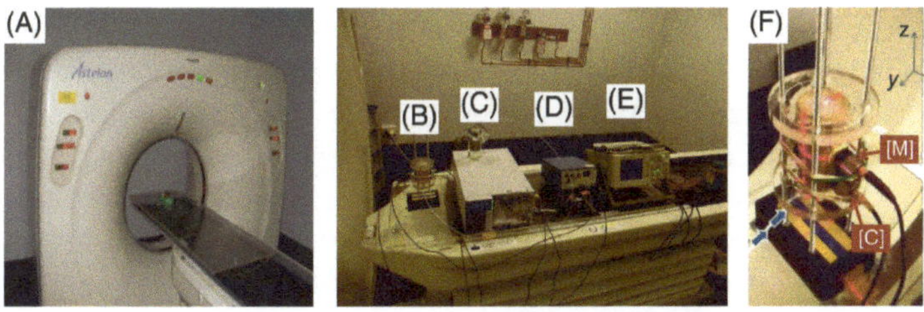

FIGURE 3.3 Experimental setup: (A) X-ray CT scanner; (B) sample set for vertical injection; (C) injection pump; (D) pulser/receiver, (E) oscilloscope, and (F) Savonnières limestone cylinder connected to two pairs of transducers – position C, close to the area of injection, and position M, in the middle of the sample. *(Adapted from Lopes et al., 2014.)*

TABLE 3.1 Petrophysical Properties of the Dry Rock Sample

	Casino Sandstone	Savonnieres Limestone
Bulk density, g/cm^3	2.2	1.947
Grain density, g/cm^3	2.65	2.65
Porosity, %	16.7	26.5
Permeability, mD	7.26	91.5

These experiments are performed on a cylindrical sample. The sample is sealed with a thin epoxy layer in order to prevent fluid leakage through the surface. Longitudinal (VP) and shear-wave (VS) velocities at 1 MHz are measured using broadband ultrasonic transducers. Velocities are obtained by first-break picking on the output signals. Intermediate aluminum "guide pins" are placed between the sample and transducers to secure sufficient and constant coupling, as well as to provide transparency for X-ray radiation.

In the experiments of Lebedev et al. (2009a, 2009b), the core sample is cut from a Casino sandstone specimen (Otway Basin, Australia). The petrophysical properties are shown in Table 3.1. The velocities are measured in the direction perpendicular to the core axis (perpendicular to the fluid flow). Two different saturation methods – referred to as dynamic and quasi-static saturation – are used in this study. In the dynamic saturation experiments, water is injected into the sample with an injection rate of 10 mL/day. In the quasi-static saturation experiments, the samples are saturated during a long period of time (up to 2 weeks) under reduced pressure in order to achieve near-isotropic fluid distribution for a given saturation level. A less than 1% difference in velocities measured in different directions indicates that fluid is uniformly distributed within the sample.

The fluid distribution in the dynamic saturation experiments (both spatial and time dependence) is characterized using X-ray CT (Figure 3.4). Scans 1–4 show the fluid distribution with increasing water saturation up to 50%. In scan 4, a clustering of the fluid patches spanning a few millimeters in diameter can be observed. Figure 3.4 shows the *P*-wave velocity as a function of saturation for the dynamic (empty symbols) and quasi-static saturation experiment (solid symbols). A transition from the GW to GH bound is clearly observed for water saturations of 40–50% for dynamic saturation and 60–70% for quasi-static saturation. This indicates that the process of fluid patch formation depends on the injection rate.

Lopes et al. (2014) performed a forced water imbibition experiment on Savonnieres limestone (Figure 3.5). The petrophysical properties are shown in Table 3.1. Water is injected at the bottom of the cylindrical sample and the sample is positioned vertically into the CT scanner. *P*-waves propagating in the direction perpendicular to the core axis are recorded close to the point of injection (Figure 3.6A). The imbibition experiment involves a 22-hour nonstop injection of distilled water subdivided into three phases (at a rate of 2 mL/h for

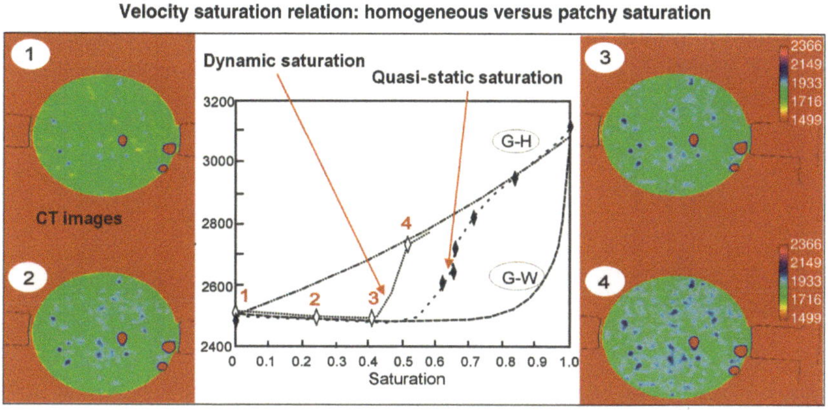

FIGURE 3.4 *P*-wave velocity versus water saturation for a forced imbibition experiment into a sandstone sample. Depending on the injection rate ("Quasi-static" = low or "Dynamic" = high), the VSR is different. However, both VSRs are bounded by the GW and GH relations. For the dynamic-saturation experiment, CT images indicate the accumulation and clustering of water patches with increasing saturation. *(Adapted from Lebedev et al., 2009a, 2009b).*

approximately 4 h, 0.2 mL/h for approximately 15 h, and 2 mL/h for approximately 3 h). The resulting waveforms reveal a reduction in amplitude with increasing water saturation and show sensitivity to injection rate changes (Figure 3.6B). Saturation maps generated from the CT images allow us to infer the statistical information about the fluid distribution, such as the characteristic size of fluid patches (Figure 3.6C). These saturation images suggest that fluid patches are formed at the millimeter scale. The patches extend over several

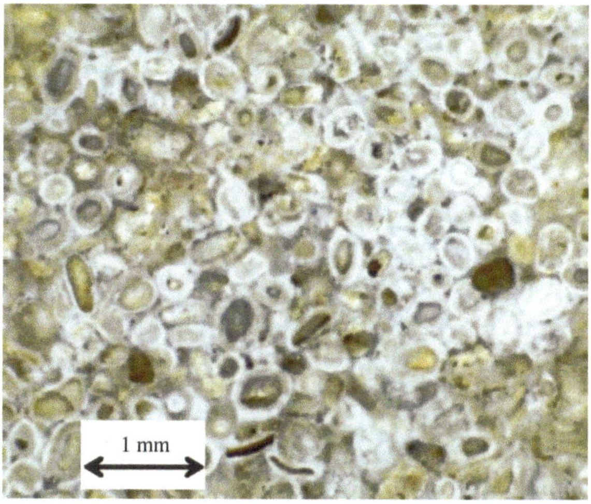

FIGURE 3.5 Microscopic view of the vuggy porosity of the Savonnieres limestone.

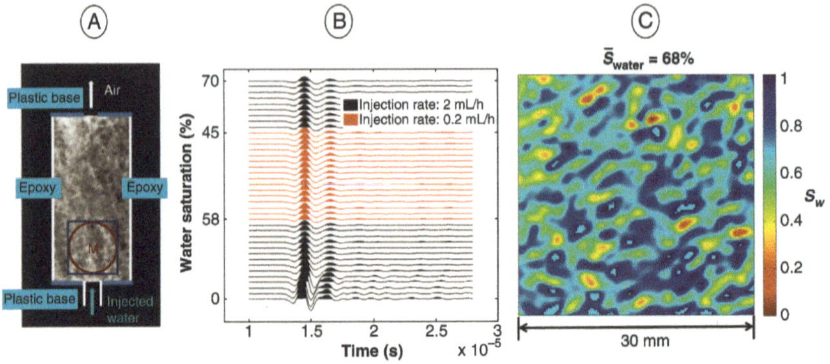

FIGURE 3.6 (A) Schematic setup of the dynamic imbibition experiment on Savonnières limestone (porosity 26.5% and permeability 91.5 mD). A CT scan in the axial direction of the sample is embedded in this sketch. Gray-color scale represents CT values, a proxy for X-ray absorption. Position of the ultrasonic transducers is marked by M. (B) Ultrasonic waveforms recorded at various levels of water saturation. The black waveforms are recorded when the injection rate is 2 mL/h. The red waveforms correspond to a injection rate of 0.2 mL/h. Note, the primary wave amplitude decays while saturation increases up to 58% and after switching the injection rate the amplitude increases while saturation decreases. (C) Water saturation map (average S_w = 68%) derived from the CT images. Millimeter-scale fluid patches appear. *(Modified from Qi et al., 2014)*

pores and grains, but are smaller than the dominant ultrasonic wavelength; thus, we can interpret them as mesoscopic-scale fluid patches.

A comparison of the observed VSR (for an injection rate of 2 mL) with the GW and GH bounds reveal that the data points fall in between these bounds. A possible reason for the deviation from these bounds is wave-induced pressure diffusion. In the next section, we will model the observed VSR by making use of the CT images to extract patch size and distribution.

3.5 LABORATORY DATA MODELING

The VSR of the Casino sandstone exhibits a transition from the GW to the GH bounds (Figure 3.4). Characteristics of this transition can be related to parameters controlling wave-induced pressure diffusion in poroelastic solid. There are two important length scales involved: the diffusion wavelength λ_p, which is on the order of 1 mm for this experiment, and the size of the fluid patches a. As mentioned above, the velocity value between the uniform and patchy saturation bound is controlled by the ratio of these two lengths. However, it is important to note that the patch size is likely to vary with the degree of saturation, and that this variation may itself be injection rate dependent. In CT scan 4 (more than 40% of water saturation), we observe some fluid clusters on the order of a few millimeters suggesting that $\lambda_p \ll a$ and hence the proximity to the GH bound is expected. Conversely, for small water saturations, the condition $\lambda_p \gg a$ seems to apply, and, therefore, the velocities follow the GW prediction. The transition

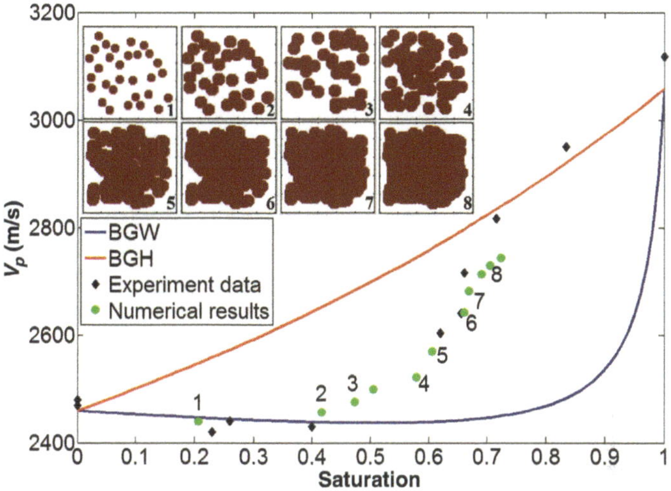

FIGURE 3.7 VSR determined from numerical simulations of wave propagation in a poroelastic solid with randomly distributed patches that cluster for larger saturation values (see inset). Experimentally determined velocities for the quasi-static injection experiment are also shown *(Lebedev et al., 2009a, 2009b)*.

between these two bounds occurs if $\lambda_p \approx a$. Then, the mechanism of wave-induced pressure diffusion takes place resulting in dispersion and attenuation of the ultrasonic signals.

To validate that the transition behavior in the VSR can be explained by the mechanism of wave-induced pressure diffusion, numerical simulations of wave propagation in a 2D poroelastic solid are performed using a finite difference solver of Biot's equations (Wenzlau and Müller, 2009). The measured material properties of the Casino sandstone and the wave frequency used in laboratory experiment are utilized. The varying degree of saturation is characterized by increasing the size of the randomly distributed fluid patches. It can be seen that for large saturations, the fluid patches in the numerical model begin to cluster (Figure 3.7) and are assumed to resemble the clustering observed in the CT scans (Figure 3.4). Despite the simplified numerical setup (2D modeling only and patch distribution not derived directly from the CT scans), the simulation results reproduce the overall behavior of the measured VSR (Figure 3.7). This confirms that the transition observed in the ultrasonic measurements can be attributed to the mechanism of wave-induced pressure diffusion.

In the experiment of Lopes et al. (2014), saturation maps generated from the CT images allow us to infer the statistical information about the fluid distribution, such as the characteristic size of fluid patches (Figure 3.8). These saturation images suggest that fluid patches are formed at the millimeter scale. Figure 3.9 shows the corresponding correlation functions extracted from the water saturation maps at different saturation levels. The autocovariance functions describe the

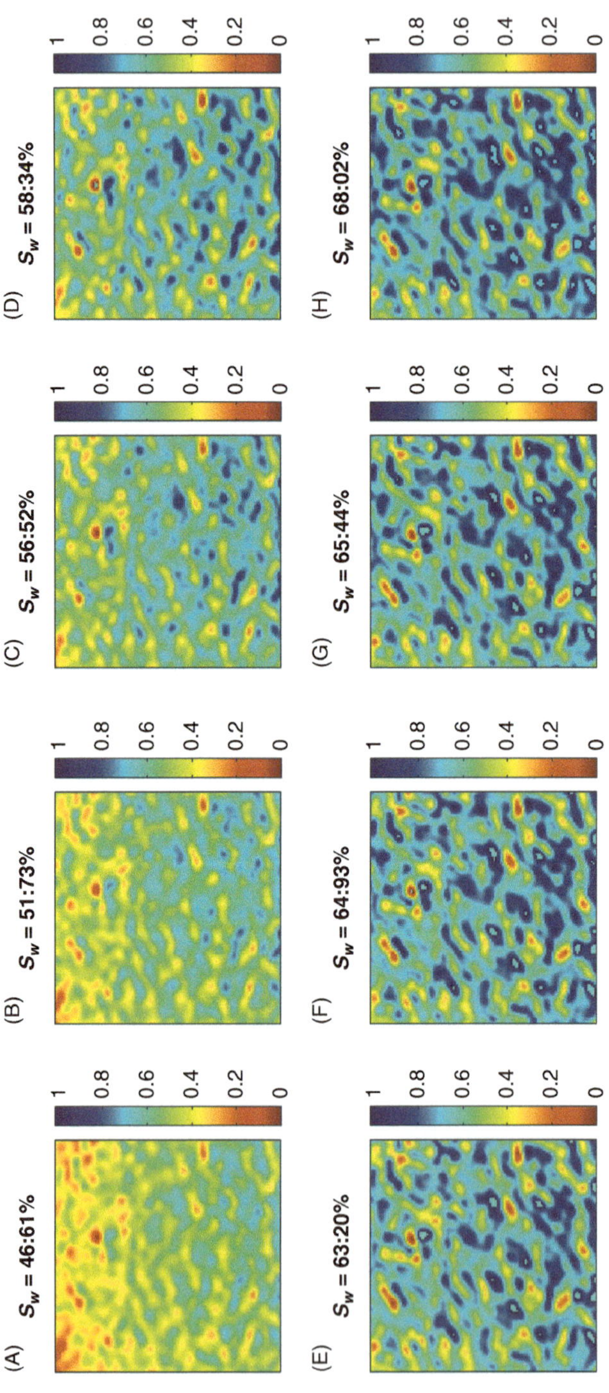

FIGURE 3.8 Water-saturation maps at different imbibition stages, where dark blue represents full water saturation and dark red represents full gas (air) saturation. The saturation maps are constructed from the raw CT scans for the monitoring area. The size of the maps is 3 × 3 cm. The color bar indicates water saturation.

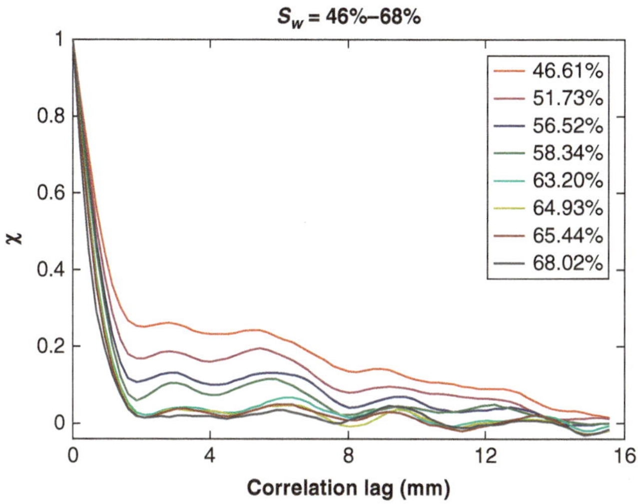

FIGURE 3.9 Fluid distribution characterization at different saturation levels using the normalized autocovariance function.

extent to which properties at two points are correlated in the system and thus are a measure of the characteristic length scale of fluid patches. For more information on how to characterize the fluid patch distribution in a quantitative way from the CT scans, we refer to Toms et al. (2009) and Qi et al. (2014). For the modeling, we choose four CT images consecutively scanned at injection rate $q = 2$ mL/h (before the rate change) and four at injection rate $q = 2$ mL/h (after rate change).

The recorded ultrasonic waveforms allow us to infer velocity and attenuation as a function of saturation (Figure 3.6B). A comparison of these observed VSR with the uniform and patchy saturation bounds reveals that the data points fall in between these bounds (Figure 3.10A). A possible explanation for the deviation from the bounds is wave-induced pressure diffusion. In order to understand if this mechanism can explain the observed VSR, we employ the continuous random medium model. The CRM model describes dispersion and attenuation caused by WIFF in a medium with fluid heterogeneities based on autocovariance functions (see Section 3.3). It is possible to fit the observed VSR trend using the CRM model with a single Debye correlation function and a correlation length of $a = 1.4$ mm (Figure 3.10A). However, the attenuation predicted by the CRM model is then twice as large as the measured one (Figure 3.10B). Furthermore, the correlation function extracted from the saturation maps yields velocity and attenuation estimates shown by the black circles in Figure 3.10, which predict lower velocities but even higher attenuation than in the observed data. It appears that there is a discrepancy that cannot be satisfactorily explained by the CRM model and hence by the mechanism of WIFF alone. A possible explanation for this mismatch is related to capillary effects that will be discussed in the next section.

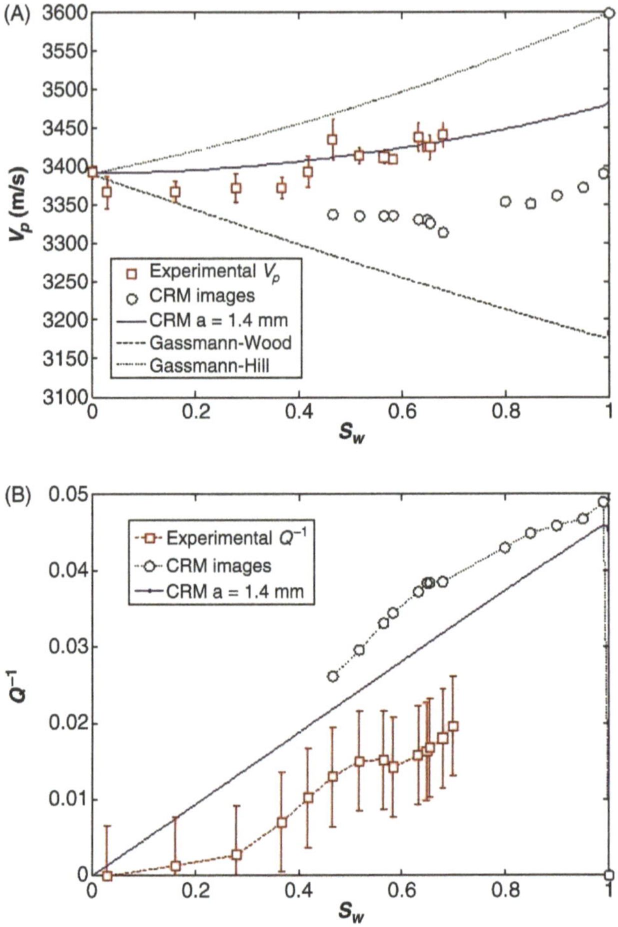

FIGURE 3.10 (A) VSRs and (B) ASRs. Red squares denote experimentally determined velocities from first-break picking and reciprocal quality factor Q^{-1} obtained by the spectral ratio method. The black circles are CRM predictions based on statistical attributes derived from the CT images. The solid blue lines are best-fit predictions using the CRM model with a constant correlation length, $a = 1.4$ mm.

3.6 PATCHY SATURATION AND TWO-PHASE FLOW CONCEPTS

Fluid patch formation and evolution are dictated by two-phase flow through porous rocks. This means that the acoustically monitored imbibition experiments possibly give us some insight into the relation between wave attenuation and two-phase flow concepts. As can be observed in Figure 3.6, injection rate changes during imbibition have a direct expression in ultrasonic waveforms. The corresponding P-wave velocity as a function of saturation and injection rate is summarized in Figure 3.11. It can be seen that the P-wave velocity and

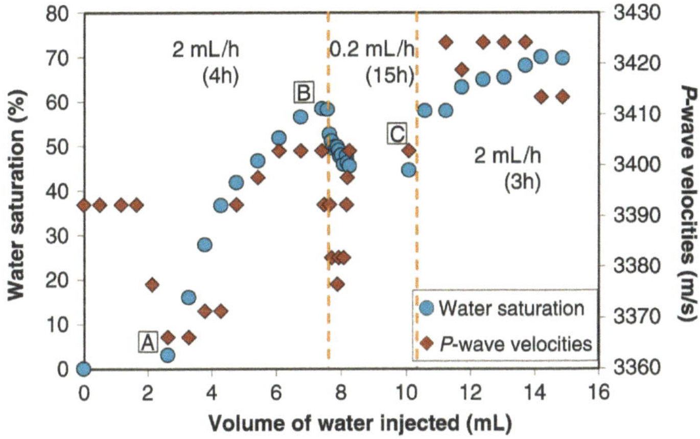

FIGURE 3.11 Water imbibition into dry limestone. *P*-wave velocity V_p water saturation S_w evolution with volume of injected water. Three distinctive points can be observed. (A) Once water enters the ultrasonically monitored position (Figure 3.6A), V_p and S_w increase. (B) After injection rate is decreased, V_p and S_w first decrease sharply and then gradually. (C) V_p and S_w increase again once the injection rate is switched back to its initial value. *(Modified after Lopes et al., 2014.)*

saturation respond in a relaxation-type manner to injection rate changes. This response can be explained by a redistribution of the forces driving the flow (Lopes et al., 2014). The interplay between capillary forces, viscous and gravity forces determine the flow regime in the rock. For the discussed experiments, capillary effects arising at fluid–fluid interfaces at the pore scale (Figure 3.11) seem to play an important role (Lopes et al., 2014 and Qi et al., 2014).

Then, for example, the question arises if capillary forces will produce a net effect on the macroscale so that WIFF signatures are affected. Here, we can make use of the conceptual model of Tserkovnyak and Johnson (2003) that assumes that capillary pressure gives rise to an interfacial tension across the fluid patch interfaces. The latter causes an overall stiffening and hence an elevated GW lower bound (Figure 3.12). Upon incorporating this capillarity effect into the CRM model, we obtain modified estimates for attenuation and dispersion (Qi et al., 2014a,b). Interestingly, this capillarity-extended CRM model is capable of modeling the VSR and ASR shown in Figure 3.10 in a consistent way. The capillarity-extended CRM model predicts higher velocities and lower attenuation than the CRM model with statistical information from the CT images, thereby offering a meaningful modeling choice.

3.7 FIELD-SCALE OBSERVATIONS

Laboratory experiments provide the most detailed information, but they are limited to the size of core samples. At the field scale, patchy-saturation scenarios are typically linked to fluid injection or production settings (Sengupta et al., 2003;

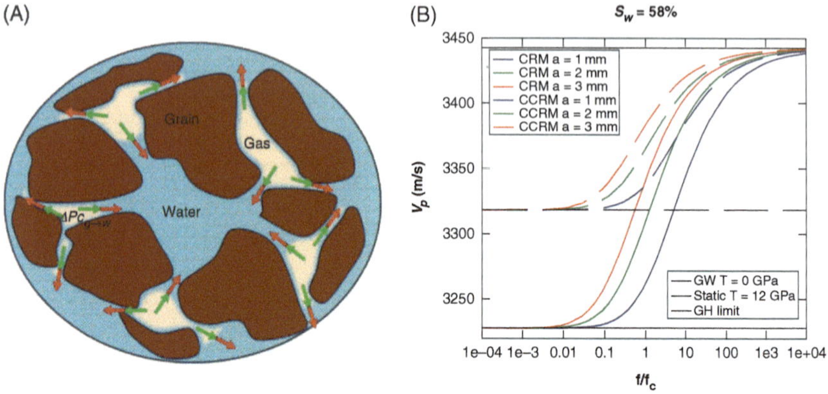

FIGURE 3.12 (A) Schematic of water patch formation. At pore scale, due to capillary forces, there is a pressure difference across the gas–water patch boundary. (B) Dispersion relations for the CRM model (solid line) and its capillarity extended version, CCRM (dashed line), for 3 different correlation lengths *a*. The static limit increases in presence of capillarity and renormalizes the dispersion relation depending on the capillarity stiffening parameter T (Qi et al., 2014b). GW and GH show the Gassmann–Wood and Gassmann–Hill limits, respectively. The critical frequency (f_c) is 180 Hz for this numerical example.

Kirstetter et al., 2006). Konishi et al. (2009) studied sonic and neutron logs obtained from a CO_2 sequestration test site in Japan. They found that *P*-wave velocity gradually decreases (almost linearly) with increasing CO_2 saturation and concluded that this behavior could indicate the presence of (mesoscopic) CO_2 patches. Caspari et al. (2011) obtained *P*-wave VSRs at the reservoir depth for two thin reservoir sections. Assuming a random distribution of CO_2 patches, they showed that WIFF can be evoked to explain these VSRs.

Direct observation of patchy saturation at the field scale remains elusive. Ajo-Franklin et al. (2013) presented results of a high-resolution time-lapse crosswell seismic survey for a pilot CO_2 injection experiment. Therein, the observed variations in *P*-wave velocity might be explained with a patchy saturation model. However, several sources of uncertainty remain. Lee and Collett (2009) reported *P*-wave velocity dispersion between surface seismic and logging frequencies on the order of 25% in gas-hydrate bearing sediments containing free gas. This dispersion is modeled using the gas pocket model assuming patches on the centimeter scale. Significant *P*-wave attenuation due to WIFF between sedimentary layers with and without gas hydrates has been reported by Gerner et al. (2007).

These observations indicate that the wave-induced pressure diffusion mechanism can affect wave propagation at the field scale. In the well log environment fluid patches can affect sonic waveforms (Xue et al., 2006). Time-lapse sonic and neutron logs from the Nagaoka CO_2 sequestration experiment (Konishi et al., 2009) provide an opportunity to study the VSR at in-situ conditions in the sonic log frequency band and estimate the corresponding attenuation (Xue et al., 2006; Konishi et al., 2009; Caspari et al., 2011). Here we focus on the results presented in Caspari et al. (2011).

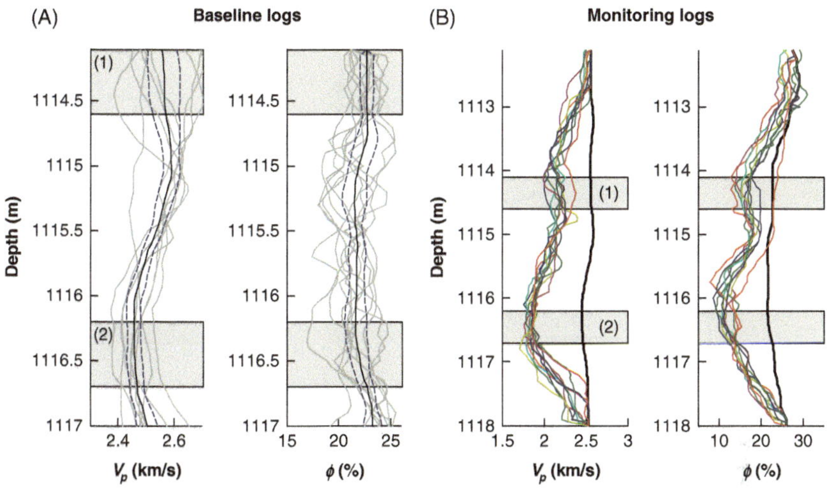

FIGURE 3.13 (A) Sonic and neutron porosity baseline logs 1–13 (gray lines), averaged baseline log (black line), and the corresponding standard variation (dashed blue line). (B) Sonic and neutron porosity monitoring logs 17–26 (colored lines) and averaged baseline log (black line). Gray boxes indicate the chosen depth intervals *(data from Konishi et al., 2009).*

At the Nagaoka CO_2 test site, 10,400 tons of supercritical CO_2 were injected in an onshore saline aquifer at a depth of about 1100 m. The target zone is a 12-m-thick porous sandstone interval. The time-lapse data of the observation well OB2 located 40-m-down dip from the injection well comprises 13 baseline, 23 monitoring (during injection), and 14 post-injection sonic and neutron porosity logs. The averages of the first 13 sonic and neutron porosity logs recorded before the CO_2 breakthrough are used as baseline data and the logs 17–26 as monitoring data (Figure 3.13). The CO_2 saturation is estimated from differences in the time-lapse neutron logs (Konishi et al., 2009), since the neutron log is sensitive to changes in the water content.

The log responses vary with depth in the reservoir zone indicating that this zone is heterogeneous. Therefore, a VSR, inferred from log responses of the whole reservoir zone, might be influenced by variations of rock properties with depth and will mask the true VSR. To eliminate the ambiguity caused by this effect, we analyze two 0.5-m-thick intervals with relatively small depth variations. Further, to minimize the problem of random fluctuations between subsequent runs of the log data, the time-lapse data points are smoothed in time for each individual depth (Caspari et al., 2011). A comparison of these data sets to known VSRs is shown in Figure 3.14. It can be observed that the data points for the two depth intervals fall between the bounds of uniform (GW) and patchy saturation (GH) indicating that WIFF might play a role.

To test this assumption, we model the VSR using the CRM model with varying patch sizes. The dry bulk modulus of the rock is calculated from the sonic baseline data for each depth using Gassmann's equation. The velocity-saturation

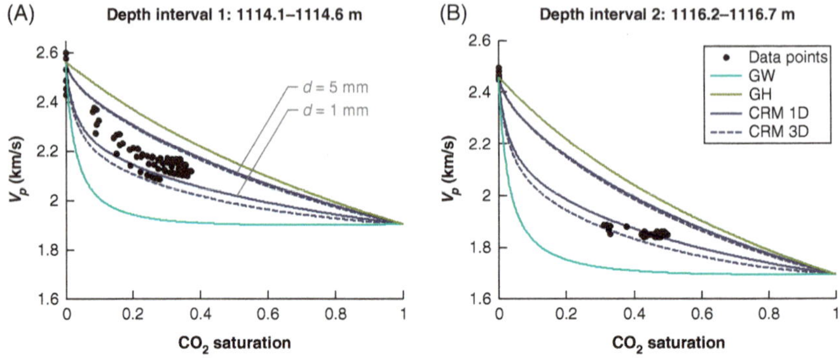

FIGURE 3.14 Comparison between VSR models and time-lapse log data for the (A) shallower and (B) deeper depth interval at sonic frequencies. The CRM models correspond to patch sizes *d* of 1 and 5 mm. GW, GH, and CRM are the Gassmann–Wood, Gassmann–Hill, and continuous random media models, respectively. *(Modified after Caspari et al., 2011.)*

trend of the smoothed data can be explained by the CRM models for a range of correlation lengths. This, in turn, allows us to infer characteristic patch sizes of CO_2 by fitting a CRM model to each data point (Figure 3.14). From this forward modeling, the patch size is estimated between 1 and 5 mm for the shallow interval, but mostly 1 mm in size for the deeper interval (Figure 3.15A). It demonstrates that the patch sizes may be different at different reservoir depths. The layer-dependent patch size variations can originate from variations of other poroelastic parameters (e.g., porosity) across the depth interval. Another possible explanation is that in a gas injection scenario lower CO_2 saturations (shallower depth interval) are more likely to show patchiness than higher CO_2 saturations. This emphasizes the importance of analyzing the VSR independently for different

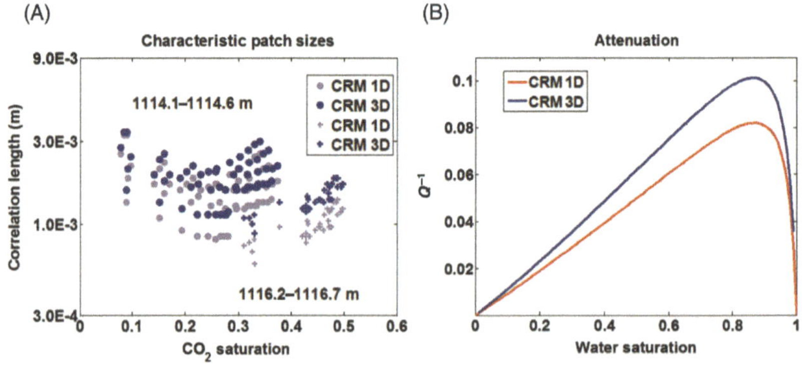

FIGURE 3.15 (A) Estimated correlation length (patch sizes) from the log data. Circles denote the shallow depth interval and cross the deeper one. (B) Attenuation versus water saturation for a patch size of 1 mm.

depth intervals. One difficulty in modeling the VSRs for partially saturated rocks lies in the fact that the VSR is controlled by the characteristic patch size, which itself can depend on saturation (Toms-Stewart et al., 2009 and Qi et al., 2014). However, the VSR trend in Figure 3.14 appears to be parallel to the CRM lines, indicating no distinct variation of the patch size with saturation.

As the estimated patch sizes are on the order of a few millimeters, they are larger than pore-scale features observed in thin sections (Xue et al., 2006). This indicates that wave-induced fluid pressure diffusion between mesoscopic inhomogeneities may occur at sonic frequencies and can be responsible for attenuation. Figure 3.15B shows attenuation as a function of water saturation for a patch size of 1 mm. It can be seen that the maximum attenuation occurs at low gas saturations (similar to the experimental data shown in the previous section).

The patch sizes estimated from sonic log data are on the same order as those observed in the laboratory ultrasound measurements of Lei and Xue (2009) on CO_2-saturated Tako sandstone. The main difference in petrophysical properties is that Tako sandstone is well cemented in contrast to the friable reservoir samples from the well OB2. However, the fluid pressure diffusion lengths $\lambda_p = 0.25$ mm (Tako sandstone – ultrasonic frequencies) and $\lambda_p = 0.65$ mm (Nagaoka OB2 well – sonic frequencies) are of the same order of magnitude. This indicates that similar patch sizes can play a role in ultrasonic and sonic-log measurements. It points to an intriguing possibility that laboratory fluid injection experiments may be used to simulate in-situ conditions in the sonic frequency band.

The sonic frequency band VSR results are limited by the penetration depth of the log measurements and do not permit any conclusion on fluid distribution on a larger scale, say meter and beyond. Fluid distributions on a larger scale in a reservoir setting and their effects on attenuation and dispersion at seismic frequencies are illustrated in a numerical modeling study in the next section.

3.8 SIGNATURES OF PATCHY SATURATION IN THE SEISMIC FREQUENCY BAND

In order to explore the potential implications of wave-induced pressure diffusion on surface seismic data, we consider the modeling scenario investigated by Rubino et al. (2011). It is inspired by the CO_2 storage project at the Sleipner field, Norway. Since 1996, millions of tons of CO_2 have been injected into the Utsira sand formation, a major saline aquifer (Chadwick et al., 2010).

Our aim is to simulate the propagation of seismic waves through a Utsira-like reservoir containing centimeter-scale CO_2 patches. Since the spatial scales at which WIFF occurs are much smaller than the predominant seismic wavelengths, corresponding numerical simulations of wave propagation based on the poroelasticity equations (Biot, 1962) turn out to be computationally too demanding. To alleviate this problem, a numerical upscaling procedure originally proposed by Rubino et al. (2009) and later improved by Rubino et al. (2013) is employed. This strategy allows us to determine an equivalent homogeneous

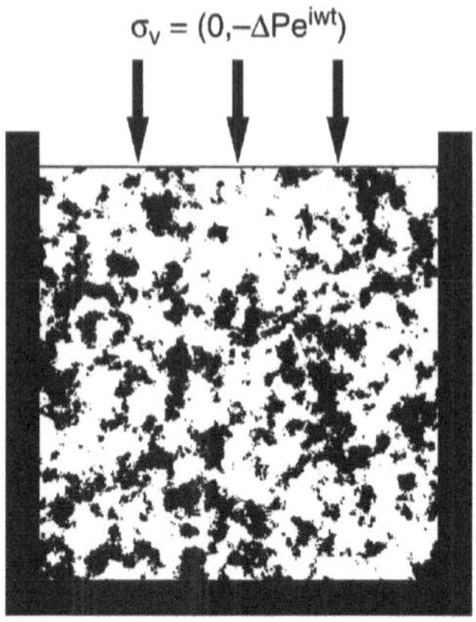

FIGURE 3.16 Schematic representation of an oscillatory compressibility test applied on a representative rock sample.

viscoelastic solid with seismic attenuation and velocity dispersion characteristics representative of the original heterogeneous fluid-saturated porous medium. Then, the seismic response accounting for all the WIFF effects is obtained by solving the viscoelastic equations of motion. This procedure is computationally feasible.

Figure 3.16 shows a schematic representation of the undrained oscillatory compressibility test involved in the numerical upscaling procedure. That is, we consider a rectangular representative rock sample containing mesoscopic-scale heterogeneities. This rock sample is subjected to a time-harmonic compression with constant amplitude of the form $\Delta Pe^{i\omega t}$ on its upper boundary, whereas no tangential forces act on the boundaries. In addition, the solid phase is neither allowed to move on the lower boundary nor to have horizontal displacements on the lateral boundaries. Finally, the fluid is not allowed to flow into or out of the sample, thereby yielding the undrained rock behavior. In order to estimate the behavior of the rock sample subjected to the oscillatory compressibility test, we solve Biot's (1941) consolidation equations in the space-frequency domain using a finite element procedure suggested by Rubino et al. (2009).

The resulting solid displacement field allows us to estimate an average vertical displacement $\Delta u(\omega)$ experienced by the top boundary of the sample. From this, we can obtain the amplitude of the resulting overall vertical strain experienced by the sample, $\varepsilon_v = \Delta u / L$ with L being its vertical side length. Under the assumption that the heterogeneous rock sample can be represented by an equivalent homogeneous viscoelastic solid, we can calculate the corresponding complex compressional velocity

$$\tilde{V}_p = \sqrt{\frac{\tilde{H}(\omega)}{\overline{\rho_b}}}, \tag{3.26}$$

where

$$\tilde{H}(\omega) = -\frac{\Delta P}{\varepsilon_v} \tag{3.27}$$

is the equivalent complex plane-wave modulus and $\overline{\rho_b}$ is the average bulk density. Finally, we can express the equivalent *P*-wave phase velocity $V_p(\omega)$ and inverse quality factor $Q_p^{-1}(\omega)$ as (Rubino et al., 2009)

$$V_p(\omega) = \left[\mathrm{Re}\left(\frac{1}{\tilde{V}_p}\right)\right]^{-1}, \tag{3.28}$$

$$\frac{1}{Q_p(\omega)} = \frac{\mathrm{Im}\left(\tilde{V}_p^2\right)}{\mathrm{Re}\left(\tilde{V}_p^2\right)}. \tag{3.29}$$

For the numerical simulations, we employ a poroelastic model of a homogeneous, poorly consolidated sandstone resembling the Utsira sand (Arts et al., 2004). We consider spatially variable CO_2-brine distributions wherein the fluids form randomly shaped patches. These binary fluid distributions are obtained by thresholding realizations of stochastic self-similar fractal fields derived from a von Karman statistical model with a correlation length of 10 cm.

Applying the described upscaling procedure to the poroelastic model in a Monte Carlo fashion allows us to obtain estimates for the statistical variability of attenuation and dispersion within the seismic frequency band for every tested overall CO_2 saturation. That is, for each set of overall CO_2 saturation and frequency, we generate a large number of realizations, *NE*, and compute the corresponding equivalent phase velocity and quality factor as dictated by the numerical upscaling procedure. Then, we compute their means and standard deviations that represent, for the considered frequency and CO_2 saturation, the statistical behavior of the samples under consideration. To ensure convergence of the Monte Carlo approach, we check that for each set of parameters the variances of the equivalent phase velocity and quality factor, as functions of the number of realization, stabilize at constant values after *NE* realizations.

Figure 3.17 shows the phase velocity and inverse quality factor resulting from the Monte Carlo analysis for various values of overall CO_2 saturations and frequencies in the range 0.1–100 Hz. We observe strong attenuation levels, with Q_P values even below 10, mainly for relatively low CO_2 saturations. Although not shown in Figure 3.17, for CO_2 saturations above 0.5, seismic attenuation tends to be less significant. Also, we see that significant velocity dispersion effects arise in the considered seismic band of frequencies, with velocity variations of about 10% for most considered CO_2 saturation values. As stipulated by the Kramers–Kronig relations (e.g., Mavko et al., 2009), such effects are more significant for relatively low CO_2 saturations, for which the attenuation

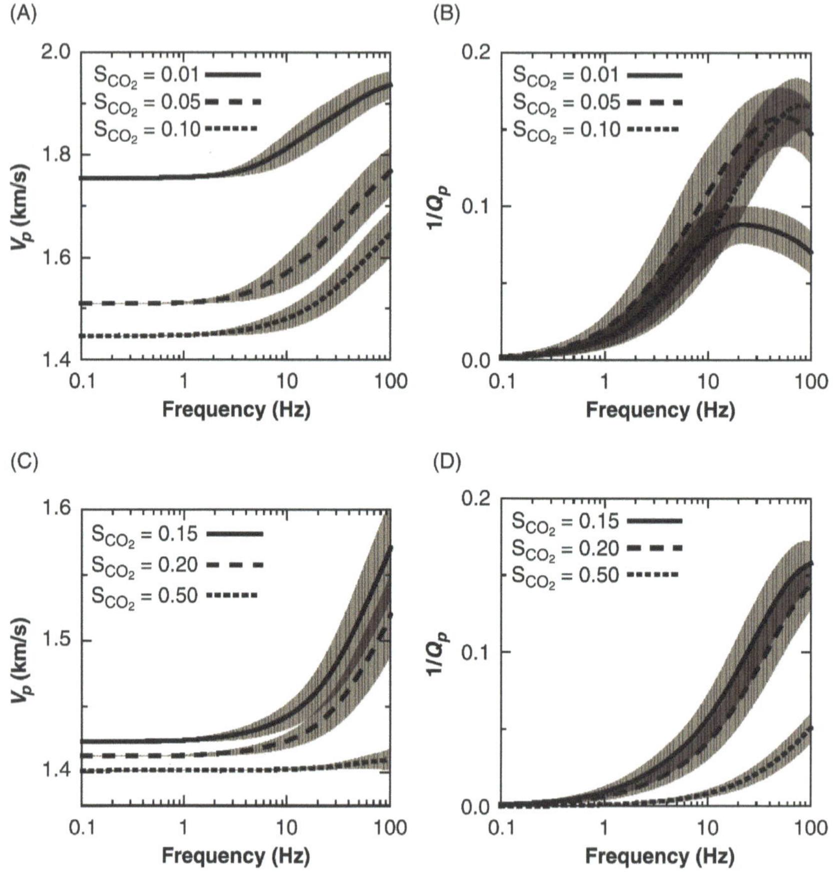

FIGURE 3.17 (A and C) Mean equivalent phase velocity and (B and D) inverse quality factor as functions of frequency for different overall CO_2 saturation levels. Gray-shaded areas indicate their corresponding standard deviation intervals. *(Modified after Rubino et al., 2011.)*

levels turned out to be very important. Conversely, for CO_2 saturations above 0.5, velocity dispersion effects are negligible. This analysis therefore suggests that WIFF effects may produce noticeable changes in the seismic responses of reservoir rocks containing centimeter-scale patches of CO_2, particularly at those parts of the plume where the CO_2 concentrations are relatively low.

As mentioned above, a poroelastic simulation at the reservoir scale is computationally expensive, especially in the presence of centimeter-scale fluid patches. For this reason, we take advantage of the dynamic equivalent viscoelastic solids resulting from the mean attenuation and phase velocity curves at each CO_2 saturation level to perform propagation of seismic waves accounting for the WIFF effects. That is, we use these viscoelastic solids to populate differently saturated regions of the geological model. Then, we solve the

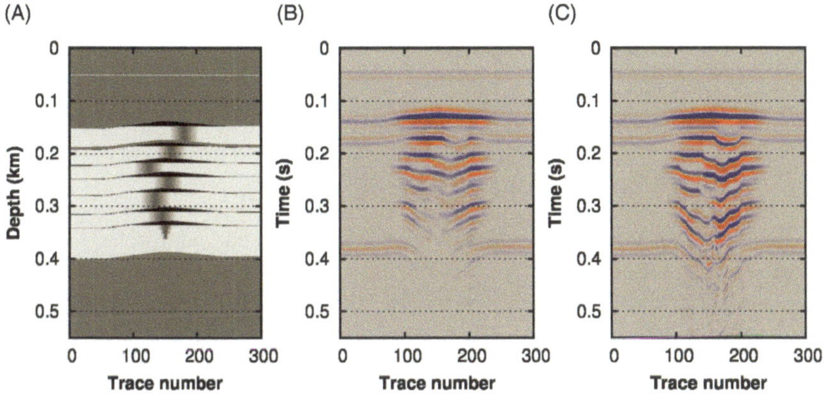

FIGURE 3.18 (A) Scheme of the geological model for the Sleipner field. The black thin areas indicate the main CO_2 component. The narrow gray-shaded areas indicate the presence of a diffuse CO_2 component. (B) Seismic response considering WIFF effects due to the presence of CO_2 patches. (C) Seismic response for an equivalent elastic model with *P*-wave velocity given by GW formula. *(Modified after Rubino et al., 2011.)*

viscoelastic wave equation in the resulting viscoelastic composite, which allows us to compute the corresponding synthetic seismograms in a computationally feasible manner. Figure 3.18A shows the conceptual geological model employed to represent a realistic CO_2 distribution. The CO_2 plume is located within a 250-m-thick Utsira-like sandstone embedded between two shale layers (top and bottom gray areas). Within the reservoir, the plume is partitioned into areas with high (about 0.9) and low (below 0.1) CO_2 saturations. The high-saturation zone (horizontal black thin regions) consists of seven layers with thicknesses of 8–10 m in their axial parts. The low-saturation (diffuse) zone has CO_2 concentrations gradually decreasing from 0.1 to 0 from the axial parts toward the lateral parts of the plume (gray-shaded areas). Figure 3.18B shows the corresponding zero-offset seismic sections (synthetics). For comparison, we include in Figure 3.18C the response obtained by replacing the viscoelastic solids with equivalent elastic solids. The latter have the same bulk density and shear velocity, but the *P*-wave velocity is given by GW formula. Examining the panels, we see strong amplitude distortions, mainly in the axial part of the plume, where most of the diffuse CO_2 accumulations concentrate. This result is in agreement with the curves shown in Figure 3.17, as WIFF effects turned out to be more important for low values of overall CO_2 saturation. By comparing Figure 3.18B and C, we can also observe kinematic changes. For example, the velocity pushdown produced by the presence of CO_2 is more significant when dispersion is neglected, that is, in the case of Figure 3.18C, where GW formula has been used.

The results of this analysis therefore suggest that WIFF due to the presence of centimeter-scale fluid patches may produce noticeable changes in the observed surface seismic data. Consequently, the nature of the fluid distributions as well as the proper modeling of the seismic response constitute key aspects that should be taken into account when characterizing reservoirs saturated with two immiscible fluids of differing compressibility.

3.9 PERSPECTIVES FOR FUTURE RESEARCH

ASRs and VSRs at different scales have been analyzed and interpreted in terms of the wave-induced pressure diffusion mechanism. It appears that some inferences about the fluid patch distribution can be made from these relations. Millimeter-size fluid patches affect attenuation in the ultrasonic and sonic frequency bands. Centimeter-size fluid patches can affect attenuation in the seismic frequency band provided that the fluid compressibility contrast is high, such as in brine–gas or brine–CO_2 systems.

There is always a certain ambiguity when relating wave attenuation to the presence fluid patches. This is because in real rock formations, it is likely that microscopic and macroscopic heterogeneities other than fluid patches will interfere with the wave-induced pressure diffusion across fluid patches. For example, capillary forces can prevent fluid mixing so that in a fractured rock, the equant pores and fractures are saturated with different fluids. Then attenuation and dispersion are controlled by a combination of the background-fracture and fluid–fluid compressibility contrasts. For certain combinations, this can even lead to a complete annihilation of WIFF (Kong et al., 2013).

Moreover, the formation and evolution of fluid patches are most likely linked to pore-scale heterogeneities. This aspect is related to spatial permeability variations. Modeling studies show that in presence of permeability fluctuations the diffusion length changes not simply with the square root of frequency (Caspari et al., 2013) and hence attenuation and dispersion also change (Rubino et al., 2013).

Laboratory experiments show that injection rate changes during core flooding affect ultrasonic waveforms. This can be an indication that capillarity alters the fluid patch distribution and thus affects ultrasonic wave attenuation. It is not clear if these findings have implications in the seismic frequency band. Further research is needed to understand the link between two-phase flow concepts and acoustic signatures. In the laboratory setting, measurements of capillary pressure and relative permeabilities might provide further insight.

ACKNOWLEDGMENTS

We wish to acknowledge the funding provided by the Australian Government through the CRC Program to support CO2CRC research, the Australian Research Council Discovery-Project–DP1096232, Australia-China Natural Gas Technology Fund (Contract Number CTR-DJ-04194-1), CSIRO, the Curtin Reservoir Geophysics Consortium (CRGC), and the CIPRS scholarship.

REFERENCES

Ajo-Franklin, J.B., Peterson, J., Doetsch, J., Daley, T.M., 2013. High-resolution characterization of a CO_2 plume using crosswell seismic tomography: Cranfield, MS, USA. Int. J. Greenhouse Gas Contr. 18, 497–509.

Alemu, B.L., Aker, E., Soldal, M., Johnsen, Ø., Aagaard, P., 2013. Effect of sub-core scale heterogeneities on acoustic and electrical properties of a reservoir rock: a CO_2 flooding experiment of brine saturated sandstone in a computed tomography scanner. Geophys. Prospect. 61, 235–250.

Arts, R., Eiken, O., Chadwick, A., Zweigel, P., van der Meer, L., Zinszner, B., 2004. Monitoring of CO_2 injected at Sleipner using time-lapse seismic data. Energy 29, 1383–1392.

Biot, M.A., 1941. General theory of three-dimensional consolidation. J. Appl. Phys. 12, 155–164.

Biot, M.A., 1956a. Theory of propagation of elastic waves in a fluid saturated porous solid. I-Low frequency range. J. Acou. Soc. Am. 28, 168–178.

Biot, M.A., 1956b. Theory of propagation of elastic waves in a fluid saturated porous solid. II. Higher frequency range. J. Acou. Soc. Am. 28, 179–191.

Biot, M.A., 1962. Mechanics of deformation and acoustic propagation in porous media. J. Appl. Phys. 33, 1482–1498.

Cadoret, T., Marion, D., B. Zinszner, B., 1995. Influence of frequency and fluid distribution on elastic wave velocities in partially saturated limestones. J. Geophys. Res. 100, 9789–9803.

Cadoret, T., Mavko, G., Zinszner, B., 1998. Fluid distribution effect on sonic attenuation in partially saturated limestones. Geophysics 63, 154–160.

Caspari, E., Müller, T.M., Gurevich, B., 2011. Time-lapse sonic logs reveal patchy CO_2 saturation in-situ. Geophys. Res. Lett. 38, L13301-1–L13301-4.

Caspari, E., Gurevich, B., Müller, T.M., 2013. Frequency-dependent effective hydraulic conductivity of strongly heterogeneous media. Phys. Rev. E 88, 042119.

Chadwick, A., Williams, G., Delepine, N., Clochard, V., Labat, K., Sturton, S., Buddensiek, M., Dillen, M., Nickel, M., Lima, A., Arts, R., Neele, F., Rossi, G., 2010. Quantitative analysis of time-lapse seismic monitoring data at the Sleipner CO_2 storage operation. Lead. Edge 29, 170–177.

Gassmann, F., 1951. Über die Elastizität poröser Medien. Vierteljahrsschrift der Naturforschenden Gesellschaft in Zürich 96, 1–23.

Gerner, A., Saenger, E.H., Shapiro, S.A., 2007. Attenuation of P-waves due to interlayer flow in hydrate-bearing sediments. J. Geophys. Eng. 4, 394–403.

Hill, R., 1963. Elastic properties of reinforced solids: some theoretical principles. J. Mech. Phys. Solids 11, 357–372.

Johnson, D. L., 2001, Theory of frequency dependent acoustics in patchy- saturated porous media. Journal of the Acoustical Society of America 110, 682–694.

Karal, F.C., Keller, J.B., 1964. Elastic, electromagnetic and other waves in random media. J. Math. Phys. 5, 537–547.

Kirstetter, O., Corbett, P., Somerville, J., 2006. Elasticity/saturation relationships using flow simulation from an outcrop analogue for 4D seismic modelling. Petroleum Geosci. 12, 205–219.

Knight, R., Nolen-Hoeksema, R., 1990. A laboratory study of the dependence of elastic wave propagation on porescale distribution. Geophys. Res. Lett. 17, 1529–1532.

Kong, L., Gurevich, B., Müller, T.M., Wang, Y., Yang, H., 2013. Effect of fracture fill on seismic attenuation and dispersion in fractured porous rocks. Geophys. J. Int. 195, 1679–1688.

Konishi, C., Azuma, H., Nobuoka, D., Xue, Z., Watanabe, J., 2009. CO2 saturation estimation from P-wave velocity changes by considering patchy saturation. Paper presented at Summer Research Workshop, Soc. of Explor. Geophys., Banff, AB, Canada, 23–27 August.

Lebedev, M., Toms-Stewart, J., Clennell, B., Shulakova, V., Pervukhina, M., Paterson, L., Müller, T.M., Gurevich, B., Wenzlau, F., 2009a. Direct laboratory observation of patchy saturation and its effects on ultrasonic velocities. Lead. Edge 28, 24–27.

Lebedev, M., Gurevich, B., Toms, J., Clennell, B., Pervukhina, M., Müller, T.M., 2009b. Laboratory Observation of Velocity-Saturation Relation Transition During Water Imbibition of Porous Rock: 4th Biot Conference on Poromechanics, June 08-10. Columbia University, New York, 941–946.

Lee, M.W., Collett, T.S., 2009. Unique problems associated with seismic analysis of partially gas-saturated unconsolidated sediments. Marine Petroleum Geol. 26, 775–781.

Lei, X., Xue, Z., 2009. Ultrasonic velocity and attenuation during CO2 injection into water-saturated porous sandstone: Measurements using difference seismic tomography. Phys. Earth Planets Interior 176, 224–234.

Lopes, S., Lebedev, M. 2012. Research note: Laboratory study of the influence of changing the injection rate on the geometry of the fluid front and on P-wave ultrasonic velocities in sandstone. Geophysical Prospecting 60, 572–580.

Lopes, S., Lebedev, M., Müller, T.M., Clennell, M.B., Gurevich, B., 2014. Forced imbibitions into a limestone: Measuring *P*-wave velocity and water saturation dependence on injection rate. Geophys. Prospect. 62, 1126–1142.

Mavko, G., Mukerji, T., Dvorkin, J., 2009. The Rock Physics Handbook, Tools for Seismic Analysis of Porous Media, second ed. Cambridge Univ. Press, Cambridge, UK.

Monsen, K., Johnstad, S.E., 2005. Improved understanding of velocity saturation relationships using 4D computer-tomography acoustic measurements. Geophys. Prospect. 53, 173–181.

Müller, T.M., Gurevich, B., 2004. One-dimensional random patchy saturation model for velocity and attenuation in porous rocks. Geophysics 69, 1166–1172.

Müller, T.M., Gurevich, B., 2005a. A first-order statistical smoothing approximation for the coherent wave field in random porous media. J. Acou. Soc. Am. 117, 1795–1805.

Müller, T.M., Gurevich, B., 2005b. Wave induced fluid flow in random porous media: Attenuation and dispersion of elastic waves. J. Acou. Soc. Am. 117, 2732–2741.

Müller, T.M., Toms-Stewart, J., Wenzlau, F., 2008. Velocity-saturation relation for partially saturated rocks with fractal pore fluid distribution. Geophys. Res. Lett. 26, L09306.

Müller, T.M., Sahay, P. N., Fast compressional wave attenuation and dispersion due to conversion scattering into slow shear waves in randomly heterogeneous porous media. J. Acou. Soc. Am. 129, 2785–2796.

Müller, T.M., Gurevich, B., Lebedev, M., 2010. Seismic wave attenuation and dispersion resulting from wave-induced flow in porous rocks - a review. Geophysics 75, 75A147-75A164.

Norris, A.N., 1993. Low-frequency dispersion and attenuation in partially saturated rocks. J. Acou. Soc. Am. 94, 359–370.

Qi, Q., Müller, T.M., Rubino, J.G., 2014. Seismic attenuation: effects of interfacial impedance on wave-induced pressure diffusion. Geophysical Journal International 199, 1677–1681.

Qi, Q., Müller, T.M., Gurevich, B., Lopes, S.C., Lebedev, M., Caspari, E., 2014. Quantifying effect of capillarity on dispersion and attenuation in patchy-saturated rocks. Geophysics 79, WB35–WB50.

Rubino, J.G., Ravazzoli, C.L., Santos, J.E., 2009. Equivalent viscoelastic solids for heterogeneous fluid-saturated porous rocks. Geophysics 74, N1–N13.

Rubino, J.G., Velis, D.R., Sacchi, M.D., 2011. Numerical analysis of wave-induced fluid flow effects on seismic data: Application to monitoring of CO2 storage at the Sleipner field. J. Geophys. Res. 116, B03306.

Rubino, J.G., Monachesi, L.B., Müller, T.M., Guarracino, L., Holliger, K., 2013. Seismic wave attenuation and dispersion due to wave-induced fluid flow in rocks with strong permeability fluctuations. J. Acou. Soc. Am. 134, 4742–4751.

Sengupta, M., Mavko, G., Mukerji, T., 2003. Quantifying subresolution saturation scales from time-lapse seismic data: A reservoir monitoring case study. Geophysics 68, 803–814.

Smith, T.M., Sondergeld, C.H., Rai, C.S., 2003. Gassmann fluid substitutions: a tutorial. Geophysics 68, 430–440.

Tserkovnyak, Y., Johnson, D.L., 2003. Capillary forces in the acoustics of patchy- saturated porous media. J. Acou. Soc. Am. 114, 2596–2606.

Toms, J., Müller, T.M., Gurevich, B., 2007. Seismic attenuation in porous rocks with random patchy saturation. Geophys. Prospect. 55, 671–678.

Toms-Stewart, J., Müller, T.M., Gurevich, B., Paterson, L., 2009. Statistical characterization of gas patch distributions in partially saturated rocks. Geophysics 74, WA51–WA64.

Vasco, D.W., Daley, T.M., Bakulin, A., 2014. Utilizing the onset of time-lapse changes: a robust basis for reservoir monitoring and characterization. Geophys J Int 197, 542–556.

Villaudy, F., Grochau, M.H., Benac, P.M., Lucet, N., Abreu, C.E.B.S., 2013. 4D simultaneous pre-stack inversion in an offshore carbonate reservoir. 13th International Congress of the Brazilian Geophysical Society, Rio de Janeiro, Brazil, August 26-29, Extended Abstract, 1-5.

Wang, Z., 2001. Fundamentals of seismic rock physics. Geophysics 66, 398–412.

Wenzlau, F., Müller, T.M., 2009. Finite-difference modeling of wave propagation and diffusion in poroelastic media. Geophysics 74, T55–T66.

Wenzlau, F., Altmann, J.B., Müller, T.M., 2010. Anisotropic dispersion and attenuation due to wave-induced fluid flow: quasi-static finite-element modeling in poroelastic solids. J. Geophys. Res. 115, B07204.

Wood, A.B., 1941. A textbook of sound. Bell, London.

Xue, Z., Tanase, D., Watanabe, J., 2006. Estimation of CO_2 saturation from time-lapse CO_2 well logging in an onshore aquifer, Nagaoka, Japan. Explorat. Geophys. 37, 19–29.

Yu, G., Shah, N., Robinson, M., Nghi, N.H., Nurhono, A.A., Thu, G.S., 2012, Resolution improvement and stratigraphic interpretation for the thin turbiditic sands in offshore Sabah, Malaysia – A pilot study: SEG, Expanded Abstract, 1–5.

Chapter 4

Fine Layering and Fractures: Effective Seismic Anisotropy

José M. Carcione
Istituto Nazionale di Oceanografia e di Geofisica Sperimentale, Trieste, Italy

Fracturing in surface rock is ubiquitous at all scales. One of the most important questions facing exploration geophysicists, as well as environmental geophysicists, is the extent of underground fracturing. This is largely because fractures can be the dominant mechanism for fluid mobility in some situations, while in others, fractures can be effective barriers to fluid flow.

Michael Schoenberg (1939–2008)

Chapter Outline

4.1 INTRODUCTION

Although many of the rocks in the Earth's crust are intrinsically anisotropic (for instance, many of the metamorphic and igneous rocks), their random orientation gives rise to a material that behaves isotropically when the dominant wavelength of the wavefield is long compared to the crystal dimensions. However, due to the compaction and the layered nature of sedimentary formations, transverse isotropy occurs when the wavelength is long enough compared with

Seismic Exploration of Hydrocarbons in Heterogeneous Reservoirs. http://dx.doi.org/10.1016/B978-0-12-420151-4.00004-0

the dimensions of the layers. Fine layering is one of the geological systems that frequently contribute to the formation of sedimentary basins. The constituents of a fine-layered medium are among the components of reservoir rocks: sandstones and limestones, for instance, which are the recipient rocks, and shales, which form the seal rocks and recently began to be sources of hydrocarbons.

Wave propagation effects in stratified media depend on the wavelength of the signal. For wavelengths short compared to the dimensions of the layers, scattering in the form of coda waves is present. Thus, these wavelengths show dispersive behavior, that is, the velocity is frequency dependent. On the other hand, for lossless (elastic) layers at long wavelengths or low frequencies, the medium behaves as a nondispersive, smoothed, transversely isotropic material. The first to obtain a solution to this problem is Bruggeman (1937). Other investigators, for example, Riznichenko (1949) and Postma (1955), studied the problem using different approaches and consider a two-constituent periodically layered medium. Later, Backus (1962) obtained the average elasticity constants in the case when the single layers are transversely isotropic with the symmetry axis perpendicular to the layering plane. Moreover, he assumed stationarity, that is, in a given length of composite medium much smaller than the wavelength, the proportion of each material, is constant (periodicity is not required). If each component layer is isotropic or transversely isotropic, the result is a transversely isotropic equivalent medium represented by five averaged elasticities and an average density.

Helbig (1984) analyzed the dispersion relation for SH waves in a periodic layered medium and in the long-wave equivalent transversely isotropic medium. He shows that the equivalence is valid for wavelengths larger than three times the spatial period of layering. Melia and Carlson (1984) made experimental laboratory measurements of compressional wave velocities parallel and perpendicular to a periodic stratified medium consisting of glass and epoxy layers. They concluded that the long-wave approximation holds when the ratio between the P-wave dominant pulse wavelength and the spatial period of the layering lies between 10 and 100, and that it is highest in the midrange of compositions, that is, half glass and half epoxy. Backus averaging for the lossless case has been numerically verified by Carcione et al. (1991). They found that the above-mentioned ratio depends on the contrast between the constituents. For instance, for a periodic sequence of epoxy-glass, it is around 8, and for the sandstone–limestone (which has a lower reflection coefficient), it is between 5 and 6. In any case, an optimal ratio can be found for which the equivalence between a finely layered medium and a homogeneous transversely isotropic medium is valid.

The averaging equations are further generalized by Schoenberg and Muir (1989) for anisotropic single constituents of a general symmetry other than transversely isotropic. They extend Backus approach to single layers of arbitrary anisotropy using group theory and a matrix formalism. This generalization has been questioned by Hudson and Crampin (1991), who argue that the theory cannot be applied to oblique sets of cracks (even for weak anisotropy), since the structure is no longer 1D and Backus's assumptions are invalidated (see response in Schoenberg and Muir, 1991). Carcione et al. (2012) have shown that the Schoenberg–Muir theory is valid from the kinematic (travel times) and dynamic (amplitudes) points of view, no matter how anisotropic are the single constituent layers. Even for constituent transversely isotropic layers with tilted axis, for example, obliquely aligned cracks (the monoclinic case), the theory performs equally well. The theory is not limited to thin layers but the anisotropy can be due to other causes, such as cracks and fractures. Helbig (1999) has solved the problem of Schoenberg and Muir (1989) without recourse to group theory, but he does not discuss its validity.

Recent advances in seismic exploration are aimed on property estimates of target zones using amplitude variations with offset (AVO) analysis. The broad field of applications reaches from ocean-bottom cable recordings over reservoir permeability estimates to carbon-dioxide storage and monitoring. These approaches are heavily affected by the presence of attenuation anisotropy. This has a significant impact on AVO analysis (e.g., Carcione et al., 1998). The physical reasons for wave attenuation and attenuation anisotropy are numerous and viscoelasticity is invoked in general, but recently a model based on wave-induced fluid flow (mesoscopic loss) seems to be able to describe the physics (see Müller et al., 2010). Interlayering could be a major agent besides viscosity, variable porosity, and saturation effects. In the presence of hydrocarbons, the media may show substantial attenuation properties and velocity dispersion (Carcione et al., 2003). This is the case with porous or cracked rocks such as sandstones and limestones, respectively, and even source shale formations with considerable fluid content.

The combination of effective anisotropy and attenuation implies anisotropic anelasticity of seismic waves, which means Q-anisotropy and anisotropic velocity dispersion. As a first simple model, anelasticity is well described by considering rocks as viscoelastic solids. It is important, particularly in exploration seismology, that the material exhibits causal behavior and an approximately constant Q factor as a function of frequency. Carcione (1992) generalized Backus averaging to the anelastic case, obtaining the first model for Q-anisotropy (see Carcione, 2007). The viscoelastic model is based on the Zener rheology. A series or a parallel connection of these elements can give any type of frequency-dependent Q factor. In particular, a continuous distribution of relaxation mechanisms is suitable for obtaining a nearly constant quality factor. For numerical simulations, a discrete set of elements is needed. The model has been further studied by Picotti et al. (2010, 2012), who have shown how to compute the stiffness components of the effective medium with quasistatic numerical experiments. Analyses on sequences of sandstone–limestone and shale–limestone with different degrees of anisotropy indicate that the quality factors of the shear modes are more anisotropic than the corresponding phase velocities, cusps of the shear wave are more pronounced for low frequencies and midrange proportions, and, in general, attenuation is higher in the direction perpendicular to layering or close to it, provided that the material with lower velocity is the more dissipative. Other alternative models of Q-anisotropy are proposed by Carcione and Cavallini (1994) and Carcione et al. (1998). Some of these models are analyzed in this book and a brief description can be found in Carcione (2007).

Wave-induced fluid flow occurs at mesoscopic scales (Pride et al., 2004). The P-wave induces a fluid–pressure difference at mesoscopic-scale inhomogeneities (larger than the pore size but smaller than the wavelength, typically tens of centimeters), generating fluid flow and slow (diffusion) Biot waves (continuity of pore pressure is achieved by energy conversion to slow P-waves that diffuse away from the interfaces). White et al. (1975) were the first to introduce the mesoscopic-loss mechanism based on approximations in the framework of Biot theory (White model). They considered porous and thin plane layers. This has been the first so-called "mesoscopic-loss" mechanism. The mesoscopic-loss theory has been further refined by other researchers. A review can be found in Carcione and Picotti (2006), Carcione (2007), Carcione et al. (2010), and Müller et al. (2010).

Attenuation in rocks may occur at other spatial scales: macroscopic and microscopic. The attenuation mechanism predicted by Biot theory (Biot, 1962; Carcione, 2007) has a macroscopic nature. It is the wavelength-scale equilibration between the peaks and troughs of the fast P-wave. The related relaxation peak is mainly located at the kilohertz's range and moves toward the high frequencies with increasing viscosity and decreasing permeability, which is the opposite behavior of the mesoscopic peak. On the other hand, microscopic models, such as the "squirt-flow" and "grain friction" models seem to be important at high (laboratory-scale) frequencies (Pride et al., 2004).

White et al. (1975) obtained the complex and frequency-dependent P-wave stiffness associated with the symmetry axis of thin poroelastic layers. The next step has been achieved by Gelinsky and Shapiro (1997) who obtained the relaxed and unrelaxed stiffnesses of the equivalent poro-viscoelastic medium. Finally, Krzikalla and Müller (2011) combined the two previous models to obtain the five complex and frequency-dependent stiffnesses of the equivalent medium. They have exploited the 1D character of the fluid pressure equilibration process between the poroelastic layers. Because the fluid-flow direction is perpendicular to the layering plane, there is only one relaxation function, corresponding to the symmetry-axis P-wave stiffness. Therefore, knowing this relaxation function and the high and low-frequency elastic limits of the stiffness tensor, Krzikalla and Müller (2011) obtained the five complex and frequency-dependent stiffnesses of the equivalent poro-viscoelastic medium.

In order to test the mesoscopic-loss theory, Carcione et al. (2011) performed numerical simulations using an upscaling procedure to obtain the complex stiffnesses of the effective transversely isotropic medium. It consists in the simulation of oscillatory compressibility and shear tests in the space-frequency domain, which enable one to obtain the complex stiffnesses. The method is illustrated in Picotti et al. (2010) and Santos et al. (2011) for single-phase media. It is based on a finite-element method in the frequency domain to compute the solutions of the associated boundary value problem. Carcione et al. (2011) obtained the quality factors and velocities as a function of frequency and propagation angle from the complex stiffnesses and composite density.

Seismic wave propagation through fractures and cracks is another important subject in exploration and production geophysics, earthquake seismology, and mining (Steen et al., 1998). Fractures constitute the sources of earthquakes (Daub and Carlson, 2010) and hydrocarbon and geothermal reservoirs are mainly composed of fractured rocks (Xu and Pruess, 2001). This makes characterizing the seismic properties of fractures essential in those fields of research. For instance, aligned vertical fractures can lead to azimuthal anisotropy (Schoenberg et al., 1999), particularly azimuthal dependence of velocity and attenuation, which is sensitive to key properties of the reservoir, such as fluid type, flow direction, porosity, permeability, and others. The acoustic characterization of fractures and cracks is important from the point of view of reservoir development (e.g., Gurevich et al., 2009). A dense set of planar fractures behaves as an effective long-wavelength transversely isotropic medium, leading to azimuthally varying velocity and attenuation of seismic waves. Modeling fractures requires a suitable interface model. Schoenberg (1980, 1983) proposed the so-called linear-slip model, based on the discontinuity of the displacement field across the interface (Schoenberg model). A generalization states that the stress components are proportional to the displacement and velocity discontinuities through the specific stiffnesses and specific viscosities, respectively (Carcione, 1996). Displacement discontinuities conserve energy and yield frequency-dependent reflection and transmission coefficients, while velocity discontinuities generate energy loss at the interface. The specific viscosity accounts for the presence of a liquid. Chichinina et al. (2009a) and Carcione et al. (2012) obtained analytical solutions for the transversely isotropic and viscoelastic medium and more general anisotropic cases, respectively, that is, the complex and frequency-dependent stiffness components.

Schoenberg model, generalized to the viscoelastic case, has been introduced into the wave equation and solved numerically by Carcione (1996, 2007). He used pseudospectral methods and domain decomposition (two grids) to model wave propagation and attenuation through a fracture. Carcione (1998) generalized the method to the anisotropic case. Zhang (2005) and Zhang and Gao (2009) implemented Schoenberg's model (purely elastic case) by using finite-volume and finite-element methods, respectively. All these schemes treat a single fracture as a nonwelded interface that satisfies the linear-slip displacement-discontinuity conditions instead of using equivalent medium theories or layers as, for instance, in Coates and Schoenberg (1995).

The case of planar fractures embedded in a poroelastic background medium is a particular case of the thin layer problem, where one of layers becomes extremely thin and compliant. Alternative models, based on a distributions of finite cracks, are given in Chapman (2003) and Galvin and Gurevich (2009).

A comprehensive review of the wave-induced fluid-flow loss mechanisms, with a discussion of the related relaxation frequencies, is given in the review paper by Müller et al. (2010). Readers can also refer to Mavko et al. (2009) for detailed mathematical expressions of the basic loss mechanisms. Regarding fractures in poroelastic media, Gurevich (2003) and Brajanovski, Gurevich and Schoenberg (2005) found the low and high-frequency limit elasticities of the equivalent transversely isotropic medium under the long-wavelength assumption. Moreover, they obtained the expression of the P-wave modulus for waves propagating normal to the fracture planes as a function of frequency. As in the case of thin layers, the approach assumes a 1D character of the fluid pressure equilibration process that generates diffusive modes from the fast P wave, that is, the fluid-flow direction is perpendicular to the fracture plane, and obtain the five complex and frequency-dependent stiffnesses of the equivalent fractured poro-viscoelastic medium.

Preface: In hydrocarbon exploration, earthquake seismology, and the field of mechanics (e.g., adhesive joints, frictional contacts, metallurgical defects, and composite materials), the theory of media in contact has important applications (Blitz and Simpson, 1996). For instance, knowledge of fracture orientation and density is essential to enhance oil and gas production (Liu and Martinez, 2013). In seismology, dry and partially saturated faults present in the Earth's crust may constitute earthquake sources (Scholz, 2002). A sharp interface is the brittle–ductile transition, which plays an important role in earthquake seismology and geothermal studies (Carcione and Poletto, 2013). It may be the cause that trigger large shallow earthquakes (Meissner and Strehlau, 1982). Another applications arises in hydraulic fracturing, where a fluid is injected through an oil well to open or generate a fracture in the direction of least tectonic stress. Active and passive seismic waves are used to monitor the location and shape of the fracture (Li et al., 1998). In contact mechanics, models describing imperfect interfaces are essential since it is necessary to describe the strength and fatigue resistance of interfaces between dissimilar and composite materials (Hertz, 1882; Johnson, 2003).

Fine layering is a very common geological system, which can be represented as an anisotropic medium when the signal wavelength is much longer than the thicknesses of the single layers. A set of thin plane layers behaves as an equivalent transversely isotropic medium with no scattering effects. The first to obtain a solution for this problem was Bruggeman (1937). In this case, there is continuity of displacement and stress at the interfaces. For wavelengths short compared to the dimensions of the layers, scattering in the form of coda waves is present. Thus, these wavelengths show dispersive behavior, that is, the velocity is frequency dependent. The problem is treated in detail, for instance, in Carcione (2007).

Fractures are media where imperfect bonding occurs. The first related theories were mainly based on the displacement discontinuity model (e.g., Newmark et al., 1951). Mindlin (1960) used it for the analysis of wave coupling in a plate with elastically restrained edges. Schoenberg (1980) introduced the linear-slip concept to model the displacement discontinuity at an imperfect interface in terms of two compliances, describing the normal and tangential behaviors. The wave reflection coefficients are frequency dependent even when the compliances are real and frequency independent. Pyrak-Nolte et al. (1990) proposed a nonwelded interface model based on the discontinuity of the displacement and the particle velocity across the interface. This model is already implicit in Schoenberg's theory if one considers complex and frequency-dependent compliances. The stress components are proportional to the displacement and velocity discontinuities through the specific stiffnesses and a specific viscosity, respectively. The specific viscosity may account for the presence of a liquid under saturated conditions. The liquid introduces a viscous coupling between the two surfaces of the fracture and enhances energy transmission, but at the same time this is reduced by viscous losses. Carcione (1996, 1998, 2007) considers several aspects of the problem of reflection and transmission from a nonwelded interface, including the numerical simulation of transient wave propagation. Moreover, as in the case of fine layering, a dense set of fracture behaves as an anisotropic medium when the wavelength is large compared to the fracture distance.

In this chapter, we consider the theoretical and numerical aspects of seismic wave propagation in fine layers and fractures, including the reflection and transmission coefficients of a single fracture. The analysis involves realistic rheologies, such as the presence of intrinsic attenuation and energy loss at the interfaces in the case of fractures, and mainly the effects of poroelasticity.

Notation: We denote the time variable by t, the frequency by f, the angular frequency by $\omega = 2\pi f$ and the position vector by $\mathbf{x} = (x, y, z) = (x_1, x_2, x_3)$. We indicate the spatial variables x, y, and z of a right-hand Cartesian system by the indices i,j,k = 1, 2, and 3, respectively, a partial derivative of order m with respect to a variable x_i with ∂_i^m, and a first and second time derivative with ∂_t and ∂_{tt}^2, respectively. The upper case indices $I, J, \ldots$ = 1, ..., 6 indicate the shortened matrix notation (Voigt's notation) where pairs of subscripts (i, j) are replaced by a single number (I or J) according to (11) → 1, (22) → 2, (33) → 3, (23) = (32) → 4, (13) = (31) → 5, (12) = (21) → 6. Matrix transposition is denoted by the superscript "T" (it is not indicated in two and three-components vectors), $\sqrt{-1}$ by i, time convolution by "$*$", complex conjugate by the superscript "*", the scalar and matrix products by the symbol "$\cdot$", the vector product by the symbol "$\times$", and unit vectors by $\hat{\mathbf{e}}_i$, i = 1,2,3 if referring to the Cartesian axes. The identity matrix in n-dimensional space is denoted by $\mathbf{I}_n$. The gradient, divergence, Laplacian and curl operators are denoted by grad [$\cdot$], div [$\cdot$], Δ [$\cdot$], and curl [$\cdot$], respectively. The operators Re($\cdot$) and Im($\cdot$) take the real and imaginary parts of a complex quantity (in some cases, the subindices "R" and "Im" are used). The Fourier-transform operator is denoted by $\mathcal{F}$ [$\cdot$] or a tilde above the function. The convention is

$$\tilde{g}(\omega) = \int_{-\infty}^{\infty} g(t) \exp(-\mathrm{i}\omega t) dt, \quad g(t) = \frac{1}{2\pi} \int_{-\infty}^{\infty} \tilde{g}(\omega) \exp(\mathrm{i}\omega t) dt.$$

The Einstein convention of repeated indices is assumed, but the notation $I(I)$ or $i(i)$ implies no summation. In general, we express vectors and column matrices (arrays) by bold and lower case letters and matrices and tensors by bold and upper case letters. In some cases, for simplicity, the same letter is used to denote different physical quantities.

4.2 THEORY OF WAVE PROPAGATION

The wave equation combines the stress–strain relation and the equations of momentum conservation. Then, measurable quantities such as the phase and energy velocities and the attenuation and quality factors can be obtained from the dispersion equation and energy considerations in terms of the medium properties and propagation directions.

In the following, we denote a transversely isotropic and viscoelastic medium with the acronym TIv. If the symmetry axis of rotational symmetry is vertical (i.e., along the z-direction), horizontal (along the x- or y-directions), or makes an angle γ with respect to the z-axis, the medium is termed VTIv, HTIv, and γTIv, respectively. With this convention VTI = 0TI and 90TI = HTI.

4.2.1 Stress–Strain Relation

The stress–strain relation for an isothermal, anisotropic viscoelastic (anelastic) medium can be written as

$$\sigma_{ij}(\mathbf{x},t) = \psi_{ijkl}(\mathbf{x},t) * \partial_t \varepsilon_{kl}(\mathbf{x},t) \tag{4.1}$$

(Christensen, 1982; Carcione, 2007, Section 2.1.1), where σ_{ij}, ψ_{ijkl}, and ε_{kl} are the stress, relaxation-tensor, and strain components, respectively. Let us consider time-harmonic fields of the form $\exp(\mathrm{i}\omega t)$. Using the shortened Voigt's notation, ψ_{ijkl} become the Voigt components ψ_{IJ} and equation (4.1) simplifies to

$$\begin{pmatrix} \sigma_{11} \\ \sigma_{22} \\ \sigma_{33} \\ \sigma_{23} \\ \sigma_{13} \\ \sigma_{12} \end{pmatrix} = \begin{pmatrix} p_{11} & p_{12} & p_{13} & p_{14} & p_{15} & p_{16} \\ p_{12} & p_{22} & p_{23} & p_{24} & p_{25} & p_{26} \\ p_{13} & p_{23} & p_{33} & p_{34} & p_{35} & p_{36} \\ p_{14} & p_{24} & p_{34} & p_{44} & p_{45} & p_{46} \\ p_{15} & p_{25} & p_{35} & p_{45} & p_{55} & p_{56} \\ p_{16} & p_{26} & p_{36} & p_{46} & p_{56} & p_{66} \end{pmatrix} \cdot \begin{pmatrix} \varepsilon_{11} \\ \varepsilon_{22} \\ \varepsilon_{33} \\ 2\varepsilon_{23} \\ 2\varepsilon_{13} \\ 2\varepsilon_{12} \end{pmatrix}, \tag{4.2}$$

where

$$p_{IJ}(\omega) = \int_{-\infty}^{\infty} \partial_t \psi_{IJ}(t) \exp(-\mathrm{i}\omega t) dt. \tag{4.3}$$

In compact notation, we have

$$\boldsymbol{\sigma} = \mathbf{P} \cdot \mathbf{e}, \quad (\sigma_I = p_{IJ} e_J), \tag{4.4}$$

where $\boldsymbol{\sigma}$ and $\mathbf{e}$ are the 1D arrays on the right-hand side and left-hand side of equation (4.2), respectively. In particular

$$\mathbf{e}^{\mathrm{T}} = [e_1, e_2, e_3, e_4, e_5, e_6]^{\mathrm{T}} = [\varepsilon_{11}, \varepsilon_{22}, \varepsilon_{33}, 2\varepsilon_{23}, 2\varepsilon_{13}, 2\varepsilon_{12}]^{\mathrm{T}}. \tag{4.5}$$

For anelastic media, the components of $\mathbf{P}$ are complex and frequency dependent. If we denote the elastic (or unrelaxed) stiffness components by c_{IJ}, then $p_{IJ}(\omega \to \infty) = c_{IJ}$.

4.2.2 Correspondence Principle

The correspondence principle allows us to obtain viscoelastic solutions from the corresponding elastic (lossless) solutions. The stress–strain relation (4.1) has the form (4.4) in the frequency domain. This form is similar to the stress–strain relation of linear elasticity theory, except that the stiffness moduli are complex and frequency dependent. Therefore, if the elastic solution is available, the viscoelastic solution is obtained by replacing the elasticity moduli with the corresponding viscoelasticity moduli (Golden and Graham, 1988, p. 68). The limitations of the principle in the case of heterogeneous media are analyzed by Mukherjee and Paulino (2003).

4.2.3 Snell's Law

A general plane-wave solution for the displacement vector of body waves is

$$\mathbf{u} = \mathbf{u}_0 \exp[\mathrm{i}(\omega t - \mathbf{k} \cdot \mathbf{x})], \tag{4.6}$$

where $\mathbf{u}_0$ represents a constant complex vector and

$$\mathbf{k} \equiv \boldsymbol{\kappa} - \mathrm{i}\boldsymbol{\alpha} \tag{4.7}$$

is the complex wavenumber vector or wavevector. The plane wave (4.6) is termed "inhomogeneous" since the propagation and attenuation directions defined by $\hat{\boldsymbol{\kappa}}$ and $\hat{\boldsymbol{\alpha}}$, respectively, do not coincide in general. When these vectors point in the same direction, the wave is called "homogeneous." We recall that when using complex notation for plane waves, the field variables are obtained as the real part of the corresponding wavefields.

A picture illustrating the reflection–transmission phenomenon is shown in Figure 4.1, where the vectors $\boldsymbol{\kappa}$, $\boldsymbol{\alpha}$, and $\hat{\mathbf{n}}$ need not be coplanar.

Depending on the nature of the incident waves, four or two waves are generated at the interface. If the two media are in welded contact, the boundary conditions are the continuity of the displacements (or particle velocities) and normal stresses across the interface, that is continuity of

$$u_i \text{ and } \sigma_{ij} n_j \tag{4.8}$$

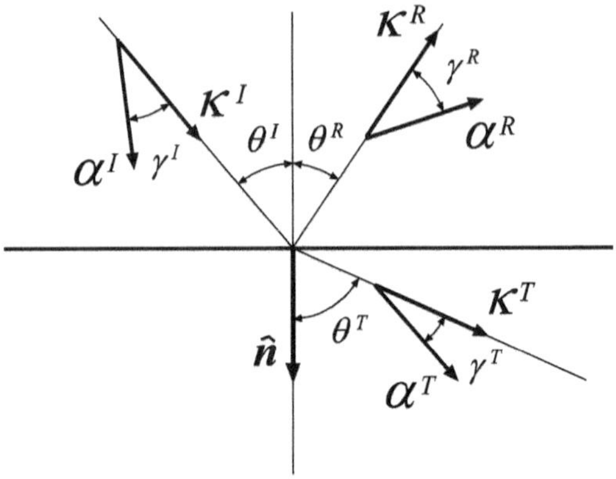

FIGURE 4.1 The reflection-transmission problem in viscoelastic media. The superscripts indicate incident (I), reflected (R), and transmitted (T).

(Auld, 1990, p. 124). This implies that the complex phase $\mathbf{k} \cdot \mathbf{x}$ at any point of the interface is the same for all the waves involved in the process, that is

$$\mathbf{k}^I \cdot \mathbf{x} = \mathbf{k}^R \cdot \mathbf{x} = \mathbf{k}^T \cdot \mathbf{x}, \quad \mathbf{x} \cdot \hat{\mathbf{n}} = 0, \tag{4.9}$$

or

$$\mathbf{k}^I \times \hat{\mathbf{n}} = \mathbf{k}^R \times \hat{\mathbf{n}} = \mathbf{k}^T \times \hat{\mathbf{n}}. \tag{4.10}$$

In terms of the complex wavevector $\mathbf{k}$ and slowness vector $\mathbf{s} = \mathbf{k}/\omega$, and identifying the interface with the plane $z = 0$, we have

$$k_1^I = k_1^R = k_1^T \quad \text{and} \quad k_2^I = k_2^R = k_2^T, \tag{4.11}$$

$$s_1^I = s_1^R = s_1^T \quad \text{and} \quad s_2^I = s_2^R = s_2^T. \tag{4.12}$$

This general form of Snell's law implies the continuity at the interface of the tangential component of the real and imaginary parts of the complex wavevector (or complex-slowness vector) and, therefore, the continuity of the tangential components of $\boldsymbol{\kappa}$ and $\boldsymbol{\alpha}$. This condition can be written as

$$\begin{aligned} &\kappa^I \sin\theta^I = \kappa^R \sin\theta^R = \kappa^T \sin\theta^T, \\ &\alpha^I \sin(\theta^I - \gamma^I) = \alpha^R \sin(\theta^R - \gamma^R) = \alpha^T \sin(\theta^T - \gamma^T). \end{aligned} \tag{4.13}$$

It is not evident from this equation that the reflection angle is equal to the incidence angle for waves of the same type. Because

$$k_3^2 = k^2 - (k_1^2 + k_2^2), \tag{4.14}$$

equation (4.11) implies $k_3^{I^2} = k_3^{R^2}$. Since the z-components of the incident and reflected waves should have opposite signs, we have

$$k_3^I = -k_3^R. \tag{4.15}$$

This relation and equation (4.11) imply $\kappa^I = \kappa^R$ and $\alpha^I = \alpha^R$, and from (4.13)

$$\theta^R = \theta^I \quad \text{and} \quad \gamma^R = \gamma^I. \tag{4.16}$$

Therefore, the reflected wave is homogeneous only if the incident wave is homogeneous.[1]

4.2.4 Wave Equation and Dispersion Equation

The strain–displacement relation can be written as

$$\mathbf{e} = \nabla^{\mathrm{T}} \cdot \mathbf{u}, \quad (e_I = \nabla_{Ij} u_j), \tag{4.17}$$

where

$$\mathbf{u} = (u_1, u_2, u_3) \tag{4.18}$$

is the displacement vector and

$$\nabla = \begin{pmatrix} \partial_1 & 0 & 0 & 0 & \partial_3 & \partial_2 \\ 0 & \partial_2 & 0 & \partial_3 & 0 & \partial_1 \\ 0 & 0 & \partial_3 & \partial_2 & \partial_1 & 0 \end{pmatrix} \tag{4.19}$$

is a gradient operator (Auld, 1990). On the basis of these definitions and in the absence of external body forces, the equation of momentum conservation is

$$\nabla \cdot \ \sigma = \rho \partial_{tt}^2 \mathbf{u}, \tag{4.20}$$

where ρ is the mass density (Carcione, 2007; eq. (1.28)).

Assuming time-harmonic fields, equations (4.4), (4.17), and (4.20) combine to give the wave equation

$$\nabla \cdot [\mathbf{P} \cdot (\nabla^{\mathrm{T}} \cdot \mathbf{u})] = -\rho\omega^2 \mathbf{u}, \tag{4.21}$$

or

$$\Gamma_\nabla \cdot \mathbf{u} + \rho\omega^2 \mathbf{u} = 0, \quad (\Gamma_{\nabla ij} u_j + \rho\omega^2 u_i) = 0, \tag{4.22}$$

where

$$\Gamma_\nabla = \nabla \cdot \mathbf{P} \cdot \nabla^{\mathrm{T}}, \quad (\Gamma_{\nabla ij} = \nabla_{iI} p_{IJ} \nabla_{Jj}) \tag{4.23}$$

is the 3×3 symmetric Kelvin–Christoffel differential-operator matrix.

1. Note that to obtain Snell's law, we have not used the assumption of isotropy. Thus, equation (4.13) is also valid for anisotropic anelastic media.

We consider plane waves of the form (4.6) propagating along the direction

$$\hat{\boldsymbol{\kappa}} = l_1\hat{\mathbf{e}}_1 + l_2\hat{\mathbf{e}}_2 + l_3\hat{\mathbf{e}}_3, \tag{4.24}$$

(or (l_1, l_2, l_3)), where l_1, l_2, and l_3 are the direction cosines. We have

$$\mathbf{k} = (k_1, k_2, k_3) = k(l_1, l_2, l_3) = k\,\hat{\boldsymbol{\kappa}} = (\kappa - \mathrm{i}\alpha)\hat{\boldsymbol{\kappa}}\ , \tag{4.25}$$

where $k = \kappa - \mathrm{i}\alpha$ is the magnitude of the complex wavevector and we have considered homogeneous viscoelastic waves for which the attenuation and propagation directions coincide. Then, the spatial differential operator (4.19) can be replaced by

$$\nabla \to -\mathrm{i}k \begin{pmatrix} l_1 & 0 & 0 & 0 & l_3 & l_2 \\ 0 & l_2 & 0 & l_3 & 0 & l_1 \\ 0 & 0 & l_3 & l_2 & l_1 & 0 \end{pmatrix} \equiv -\mathrm{i}k\mathbf{L}. \tag{4.26}$$

Substitution of this operator into the equation of motion (4.22) yields

$$k^2\boldsymbol{\Gamma}\cdot\mathbf{u} = \rho\omega^2\mathbf{u}, \qquad (k^2\Gamma_{ij}u_j = \rho\omega^2 u_i), \tag{4.27}$$

where

$$\boldsymbol{\Gamma} = \mathbf{L}\cdot\mathbf{P}\cdot\mathbf{L}^{\mathrm{T}}, \qquad (\Gamma_{ij} = l_{iI}p_{IJ}l_{Jj}) \tag{4.28}$$

is the symmetric Kelvin–Christoffel matrix.

4.2.5 Wave Velocities and Loss Factors

Defining the complex velocity as

$$v = \frac{\omega}{k}, \tag{4.29}$$

we find that equation (4.27) becomes an "eigenequation" (the Kelvin–Christoffel equation)

$$(\boldsymbol{\Gamma} - \rho v^2\mathbf{I}_3)\cdot\mathbf{u} = 0 \tag{4.30}$$

for the eigenvalues $(\rho v^2)_m$ and eigenvectors $(\mathbf{u})_m$, m = 1,2,3. The dispersion equation is given by

$$\det(\boldsymbol{\Gamma} - \rho v^2\mathbf{I}_3) = 0. \tag{4.31}$$

In explicit form, the components of the Kelvin–Christoffel matrix are

$$\Gamma_{11} = p_{11}l_1^2 + p_{66}l_2^2 + p_{55}l_3^2 + 2p_{56}l_2l_3 + 2p_{15}l_3l_1 + 2p_{16}l_1l_2,$$

$$\Gamma_{22} = p_{66}l_1^2 + p_{22}l_2^2 + p_{44}l_3^2 + 2p_{24}l_2l_3 + 2p_{46}l_3l_1 + 2p_{26}l_1l_2,$$

$$\Gamma_{33} = p_{55}l_1^2 + p_{44}l_2^2 + p_{33}l_3^2 + 2p_{34}l_2l_3 + 2p_{35}l_3l_1 + 2p_{45}l_1l_2,$$

$$\begin{aligned}\Gamma_{12} = p_{16}l_1^2 + p_{26}l_2^2 + p_{45}l_3^2 + (p_{46} + p_{25})l_2l_3 \\ + (p_{14} + p_{56})l_3l_1 + (p_{12} + p_{66})l_1l_2,\end{aligned} \quad (4.32)$$

$$\Gamma_{13} = p_{15}l_1^2 + p_{46}l_2^2 + p_{35}l_3^2 + (p_{45} + p_{36})l_2l_3 + (p_{13} + p_{55})l_3l_1 + (p_{14} + p_{56})l_1l_2,$$

$$\Gamma_{23} = p_{56}l_1^2 + p_{24}l_2^2 + p_{34}l_3^2 + (p_{44} + p_{23})l_2l_3 + (p_{36} + p_{45})l_3l_1 + (p_{25} + p_{46})l_1l_2.$$

The phase velocity is

$$\mathbf{v}_p = \left[\frac{\omega}{\mathrm{Re}(k)}\right]\hat{\kappa} = \left(\frac{\omega}{\kappa}\right)\hat{\kappa} = \left[\mathrm{Re}\left(\frac{1}{v}\right)\right]^{-1}\hat{\kappa}. \quad (4.33)$$

The slowness, defined as the inverse of the phase velocity, is

$$\mathbf{s}_R = \left(\frac{1}{v_p}\right)\hat{\kappa} = \mathrm{Re}\left(\frac{1}{v}\right)\hat{\kappa} \equiv \mathrm{Re}(s)\hat{\kappa}, \quad (4.34)$$

that is, its magnitude is the real part of the complex slowness $s = 1/v$.

According to the definition (4.7), the attenuation vector for homogeneous plane waves is

$$\alpha = -\mathrm{Im}(\mathbf{k}) = -\omega\,\mathrm{Im}\left(\frac{1}{v}\right)\hat{\kappa}. \quad (4.35)$$

The three solutions obtained from considering $m = 1,2,3$ in equation (4.31) correspond to the three body waves propagating in an unbounded homogeneous medium. At a given frequency ω, $v_p(l_1,l_2,l_3)$ defines a surface in the wavenumber space as a function of the direction cosines.

The wave or ray surface is defined as the energy velocity times one unit of time. The energy balance provides the necessary elements to compute this velocity. The whole demonstration is relatively lengthy and can be found in Carcione (2007). The wave surface depends on frequency, while the wavefront corresponds to the unrelaxed state, that is, to the high-frequency limit, since it involves the highest velocity. We recall that at infinite frequency, we have $p_{IJ} = c_{IJ}$.

4.2.5.1 Monoclinic Medium

Let us consider that the symmetry plane of a monoclinic medium is the (x,z)-plane. It is well known that in this plane there are one pure shear wave (SH) and two coupled waves (qP-qSV or qP-qS) (e.g., Carcione, 2007), where "H" and "V" stand for polarization in horizontal and vertical planes, respectively. The complex dispersion equations of the SH and qP-qSV waves are obtained from (4.31)

$$\begin{aligned}&\Gamma_{22} - \rho v^2 = 0,\\ &(\Gamma_{11} - \rho v^2)(\Gamma_{33} - \rho v^2) - \Gamma_{13}^2 = 0,\end{aligned} \quad (4.36)$$

respectively, where

$$
\begin{aligned}
\Gamma_{11} &= p_{11} l_1^2 + p_{55} l_3^2 + 2 p_{15} l_1 l_3, \\
\Gamma_{22} &= p_{66} l_1^2 + p_{44} l_3^2 + 2 p_{46} l_1 l_3, \\
\Gamma_{33} &= p_{33} l_3^2 + p_{55} l_1^2 + 2 p_{35} l_1 l_3, \\
\Gamma_{13} &= p_{15} l_1^2 + p_{35} l_3^2 + (p_{13} + p_{55}) l_1 l_3,
\end{aligned}
\tag{4.37}
$$

where $l_1 = \sin\theta$ and $l_3 = \cos\theta$, with θ the phase propagation angle (Carcione, 2007, eqs. 1.72 and 1.73 and corresponding principle).

If we label 1 the pure mode (the SH wave) and 2 and 3 the qS (or qSV) and qP waves, the corresponding complex velocities obtained from (4.36) are

$$
\begin{aligned}
\upsilon_1 &= \sqrt{\rho^{-1}(p_{66} l_1^2 + p_{44} l_3^2 + 2 p_{46} l_1 l_3)}, \\
\upsilon_2 &= (2\rho)^{-1/2} \sqrt{p_{11} l_1^2 + p_{33} l_3^2 + p_{55} + 2 l_1 l_3 (p_{15} + p_{35}) - A}, \\
\upsilon_3 &= (2\rho)^{-1/2} \sqrt{p_{11} l_1^2 + p_{33} l_3^2 + p_{55} + 2 l_1 l_3 (p_{15} + p_{35}) + A}, \\
A &= \sqrt{(\Gamma_{11} - \Gamma_{33})^2 + 4\Gamma_{13}^2}
\end{aligned}
\tag{4.38}
$$

(Carcione, 2007; eq. (4.31)).

The quality factor is defined as the ratio twice the strain energy to the dissipated energy over a cycle. It is given by

$$
Q = \frac{\mathrm{Re}(\upsilon^2)}{\mathrm{Im}(\upsilon^2)} \tag{4.39}
$$

(Carcione, 2007; eq. (4.92)). The values of the qP-wave quality factor along orthogonal directions are

$$
Q_{\mathrm{qP}}(\theta = \pi/2) = \frac{\mathrm{Re}(p_{11})}{\mathrm{Im}(p_{11})}, \quad Q_{\mathrm{qP}}(\theta = 0) = \frac{\mathrm{Re}(p_{33})}{\mathrm{Im}(p_{33})}, \tag{4.40}
$$

while those of the shear waves are

$$
\begin{aligned}
&Q_{\mathrm{qSV}}(\theta = \pi/2) = Q_{\mathrm{qSV}}(\theta = 0) = \frac{\mathrm{Re}(p_{55})}{\mathrm{Im}(p_{55})}, \\
&Q_{\mathrm{SH}}(\theta = 0) = \frac{\mathrm{Re}(p_{44})}{\mathrm{Im}(p_{44})}, \quad Q_{\mathrm{SH}}(\theta = \pi/2) = \frac{\mathrm{Re}(p_{66})}{\mathrm{Im}(p_{66})}.
\end{aligned}
\tag{4.41}
$$

Group and energy velocities are the same in lossless media and at the low and high-frequency limits in anelastic media (Carcione, 2007). As mentioned above, the wavefronts are determined by the unrelaxed (high-frequency limit) stiffness components c_{IJ}. The corresponding SH-wave unrelaxed energy velocity is

$$\mathbf{v}_e = \frac{1}{\rho v_p}[(c_{66}l_1 + c_{46}l_3)\hat{\mathbf{e}}_1 + (c_{44}l_3 + c_{46}l_1)\hat{\mathbf{e}}_3], \tag{4.42}$$

where v_p is obtained from equations (4.33) and $(4.38)_1$.

On the other hand, the qP- and qSV-wave unrelaxed energy-velocity components are

$$v_{e1} = \frac{(\Gamma_{33} - \rho v_p^2)(c_{11}l_1 + c_{15}l_3) + (\Gamma_{11} - \rho v_p^2)(c_{55}l_1 + c_{35}l_3) - \Gamma_{13}[2c_{15}l_1 + (c_{13} + c_{55})l_3]}{\rho v_p(\Gamma_{11} + \Gamma_{33} - 2\rho v_p^2)} \tag{4.43}$$

and

$$v_{e3} = \frac{(\Gamma_{33} - \rho v_p^2)(c_{55}l_3 + c_{15}l_1) + (\Gamma_{11} - \rho v_p^2)(c_{33}l_3 + c_{35}l_1) - \Gamma_{13}[2c_{35}l_3 + (c_{13} + c_{55})l_1]}{\rho v_p(\Gamma_{11} + \Gamma_{33} - 2\rho v_p^2)}. \tag{4.44}$$

where v_p is obtained from equations (4.33), $(4.38)_2$ (qSV), and $(4.38)_3$ (qP).

4.2.5.2 Orthorhombic Medium

Following Schoenberg and Helbig (1997) and Carcione (2007; eq. (1.190)), the dispersion equation (4.31) for orthorhombic media has the form

$$v^6 - a_2 v^4 + a_1 v^2 - a_0 = 0, \tag{4.45}$$

where

$$\begin{aligned} a_0 &= \Gamma_{11}\Gamma_{22}\Gamma_{33} - \Gamma_{12}^2\Gamma_{33} - \Gamma_{13}^2\Gamma_{22} - \Gamma_{23}^2\Gamma_{11} + \Gamma_{12}\Gamma_{23}\Gamma_{13} + \Gamma_{13}\Gamma_{12}\Gamma_{23}, \\ a_1 &= \Gamma_{22}\Gamma_{33} + \Gamma_{33}\Gamma_{11} + \Gamma_{11}\Gamma_{22} - \Gamma_{23}^2 - \Gamma_{13}^2 - \Gamma_{12}^2, \\ a_2 &= \Gamma_{11} + \Gamma_{22} + \Gamma_{33}, \end{aligned} \tag{4.46}$$

with

$$\begin{aligned} \Gamma_{11} &= p_{11}l_1^2 + p_{66}l_2^2 + p_{55}l_3^2, \\ \Gamma_{22} &= p_{66}l_1^2 + p_{22}l_2^2 + p_{44}l_3^2, \\ \Gamma_{33} &= p_{55}l_1^2 + p_{44}l_2^2 + p_{33}l_3^2, \\ \Gamma_{12} &= (p_{12} + p_{66})l_1l_2, \\ \Gamma_{13} &= (p_{13} + p_{55})l_3l_1, \\ \Gamma_{23} &= (p_{44} + p_{23})l_2l_3. \end{aligned} \tag{4.47}$$

As in monoclinic media, in the three symmetry planes of an orthorhombic medium, there is a pure shear wave (labeled 1 below) and two coupled waves. The corresponding complex velocities are given by a generalization of the lossless case to the anelastic case by using the correspondence principle (e.g., Carcione, 2007; Section 1.3.2):

(x, y)-plane:

$$\begin{aligned}
v_1 &= \sqrt{(\rho)^{-1}(p_{55}l_1^2 + p_{44}l_2^2)},\\
v_2 &= (2\rho)^{-1/2}\sqrt{p_{11}l_1^2 + p_{22}l_2^2 + p_{66} - A},\\
v_3 &= (2\rho)^{-1/2}\sqrt{p_{11}l_1^2 + p_{22}l_2^2 + p_{66} + A},\\
A &= \sqrt{[(p_{22} - p_{66})l_2^2 - (p_{11} - p_{66})l_1^2]^2 + 4[(p_{12} + p_{66})l_1 l_2]^2};
\end{aligned} \tag{4.48}$$

(x, z)-plane:

$$\begin{aligned}
v_1 &= \sqrt{(\rho)^{-1}(p_{66}l_1^2 + p_{44}l_3^2)},\\
v_2 &= (2\rho)^{-1/2}\sqrt{p_{11}l_1^2 + p_{33}l_3^2 + p_{55} - A},\\
v_3 &= (2\rho)^{-1/2}\sqrt{p_{11}l_1^2 + p_{33}l_3^2 + p_{55} + A},\\
A &= \sqrt{[(p_{33} - p_{55})l_3^2 - (p_{11} - p_{55})l_1^2]^2 + 4[(p_{13} + p_{55})l_1 l_3]^2};
\end{aligned} \tag{4.49}$$

(y, z)-plane:

$$\begin{aligned}
v_1 &= \sqrt{(\rho)^{-1}(p_{66}l_2^2 + p_{55}l_3^2)}\\
v_2 &= (2\rho)^{-1/2}\sqrt{p_{22}l_2^2 + p_{33}l_3^2 + p_{44} - A},\\
v_3 &= (2\rho)^{-1/2}\sqrt{p_{22}l_2^2 + p_{33}l_3^2 + p_{44} + A},\\
A &= \sqrt{[(p_{33} - p_{44})l_3^2 - (p_{22} - p_{44})l_2^2]^2 + 4[(p_{23} + p_{44})l_2 l_3]^2}.
\end{aligned} \tag{4.50}$$

In terms of angles, $l_1 = \sin\theta$ and $l_2 = \cos\theta$ in the (x, y)-plane, $l_1 = \sin\theta$ and $l_3 = \cos\theta$ in the (x, z)-plane, and $l_2 = \sin\theta$ and $l_3 = \cos\theta$ in the (y, z)-plane.

The complex velocities along the principal axes are:

(x, y)-plane:

$$\begin{aligned}
v_1(0^\circ) &= v_{SH}(0^\circ) = \sqrt{p_{44}/\rho}\\
v_1(90^\circ) &= v_{SH}(90^\circ) = \sqrt{p_{55}/\rho}\\
v_2(0^\circ) &= v_{qSV}(0^\circ) = \sqrt{p_{66}/\rho}\\
v_2(90^\circ) &= v_{qSV}(90^\circ) = \sqrt{p_{66}/\rho}\\
v_3(0^\circ) &= v_{qP}(0^\circ) = \sqrt{p_{22}/\rho}\\
v_3(90^\circ) &= v_{qP}(90^\circ) = \sqrt{p_{11}/\rho};
\end{aligned} \tag{4.51}$$

(x, z)-plane:

$$\begin{aligned}
v_1(0^\circ) &= v_{\mathrm{SH}}(0^\circ) = \sqrt{p_{44}/\rho} \\
v_1(90^\circ) &= v_{\mathrm{SH}}(90^\circ) = \sqrt{p_{66}/\rho} \\
v_2(0^\circ) &= v_{\mathrm{qSV}}(0^\circ) = \sqrt{p_{55}/\rho} \\
v_2(90^\circ) &= v_{\mathrm{qSV}}(90^\circ) = \sqrt{p_{55}/\rho} \\
v_3(0^\circ) &= v_{\mathrm{qP}}(0^\circ) = \sqrt{p_{33}/\rho} \\
v_3(90^\circ) &= v_{\mathrm{qP}}(90^\circ) = \sqrt{p_{11}/\rho}\ ;
\end{aligned} \tag{4.52}$$

(y, z)-plane:

$$\begin{aligned}
v_1(0^\circ) &= v_{\mathrm{SH}}(0^\circ) = \sqrt{p_{66}/\rho} \\
v_1(90^\circ) &= v_{\mathrm{SH}}(90^\circ) = \sqrt{p_{55}/\rho} \\
v_2(0^\circ) &= v_{\mathrm{qSV}}(0^\circ) = \sqrt{p_{44}/\rho} \\
v_2(90^\circ) &= v_{\mathrm{qSV}}(90^\circ) = \sqrt{p_{44}/\rho} \\
v_3(0^\circ) &= v_{\mathrm{qP}}(0^\circ) = \sqrt{p_{33}/\rho} \\
v_3(90^\circ) &= v_{\mathrm{qP}}(90^\circ) = \sqrt{p_{22}/\rho}\ .
\end{aligned} \tag{4.53}$$

The quality factor expressions for each symmetry plane are similar to the equations obtained for the symmetry plane of the monoclinic medium. Along pure mode directions, we have

$$Q_{I(I)} = \frac{\mathrm{Re}[p_{I(I)}]}{\mathrm{Im}[p_{I(I)}]}, \quad I = 1, \ldots, 6. \tag{4.54}$$

The energy velocity at the different symmetry planes can be obtained by using the equations reported in the next section, adapted to each propagation plane.

4.2.5.3 Transversely Isotropic Medium

We consider the (x, z)-plane. The complex velocities of a VTIv medium are obtained from equation (4.38) by setting $p_{46} = p_{35} = p_{15} = 0$ and considering that $p_{44} = p_{55}$,

$$\begin{aligned}
v_1 &= \sqrt{\rho^{-1}(p_{66}l_1^2 + p_{55}l_3^2)}, \\
v_2 &= (2\rho)^{-1/2}\sqrt{p_{11}l_1^2 + p_{33}l_3^2 + p_{55} - A}, \\
v_3 &= (2\rho)^{-1/2}\sqrt{p_{11}l_1^2 + p_{33}l_3^2 + p_{55} + A}, \\
A &= \sqrt{(\Gamma_{11} - \Gamma_{33})^2 + 4\Gamma_{13}^2},
\end{aligned} \tag{4.55}$$

where the three velocities correspond to the SH (1), qSV (2) and qP (3) waves.

The energy-velocity vector of the qP and qSV waves is given by

$$\frac{\mathbf{v}_e}{\upsilon_p} = (l_1 + l_3 \cot\psi)^{-1}\hat{\mathbf{e}}_1 + (l_1 \tan\psi + l_3)^{-1}\hat{\mathbf{e}}_3 \quad (4.56)$$

(Carcione 2007; eq. (6.158)), where

$$\tan\psi = \frac{\mathrm{Re}(\beta^* X + \xi^* W)}{\mathrm{Re}(\beta^* W + \xi^* Z)} \quad (4.57)$$

defines the angle between the energy-velocity vector and the z-axis,

$$\beta = \sqrt{A \pm B},$$

$$\xi = \pm \mathrm{pv}\sqrt{A \mp B}, \quad (4.58)$$

$$B = p_{11} l_1^2 - p_{33} l_3^2 + p_{55} \cos 2\theta,$$

where the upper and lower signs correspond to the qP and qSV waves, respectively. Moreover

$$W = p_{55}(\xi s_1 + \beta s_3),$$

$$X = \beta p_{11} s_1 + \xi p_{13} s_3, \quad (4.59)$$

$$Z = \beta p_{13} s_1 + \xi p_{33} s_3$$

(Carcione 2007; eqs. (6.121–6.123)), where $s = (s_1, s_3) = 1/\upsilon$ is the complex slowness and "pv" denotes the principal value, which has to be chosen according to established criteria (Sidler et al., 2008).

On the other hand, the energy velocity of the SH wave is

$$\mathbf{v}_e = \frac{\upsilon_p}{\rho \mathrm{Re}(\upsilon_1)} \left[l_1 \mathrm{Re}\left(\frac{p_{66}}{\upsilon_1}\right)\hat{\mathbf{e}}_1 + l_3 \mathrm{Re}\left(\frac{p_{55}}{\upsilon_1}\right)\hat{\mathbf{e}}_3 \right] \quad (4.60)$$

(Carcione, 2007; eq. (4.115)).

A useful equation states that the phase velocity is related to the energy velocity by

$$\upsilon_p = \upsilon_e \cos(\psi - \theta), \quad (4.61)$$

where $\upsilon_e = |\mathbf{v}_e|$ (Carcione, 2007; eq. (4.78)).

4.3 FINE LAYERING

Most geological formations can be described as plane thin layers, which refers to the case where the dominant wavelength of the pulse is much larger than the thicknesses of the single layers. When this occurs, the medium is transversely isotropic. The boundary conditions at the interface between two layers are those of welded contact, that is, continuity of displacement (or particle velocity) and stresses. The complexity of the problem for one interface in anelastic anisotropic media has been illustrated in Carcione (1997, 2007), including an energy balance. The case of thin multiple layers at low frequencies is referred to as Backus averaging. Backus (1962) generalizes preceding theories to random layering, where random refers to the medium properties or the layer thicknesses. In the following, we also consider the averaging for general material symmetries and the case of plane isotropic and poroelastic thin layers.

4.3.1 Backus Averaging

The periodic-layering approach (Bruggeman, 1937; Postma, 1955) has been generalized by Backus (1962), who obtains the equivalent (or effective) transversely isotropic (TI) medium when the single – planar – layers are transversely isotropic with their symmetry axis perpendicular to the layering plane (VTI). Backus assumes stationarity, that is, in a given length of composite medium much smaller than the wavelength, the proportion of each material is constant. Carcione et al. (1991) verified Backus averaging for the lossless case using numerical experiments. They found that the minimum P-wave wavelength to spatial period ratio depends on the contrast between the constituents. For a periodic sequence of epoxy-glass (high contrast), this ratio is approximately eight, while for a sandstone–limestone sequence, it is between five and six. Backus equations are further generalized by Schoenberg and Muir (1989) for anisotropic single constituents of a general nature (see Section 4.3.2).

The first seismic model including attenuation and describing Q-anisotropy has been proposed by Carcione (1992), who generalizes Backus averaging to the anelastic case. Each single medium of the fine-layering sequence is isotropic and anelastic with complex Lamé constants given by

$$\lambda(\omega)=\rho\left(c_{\mathrm{P}}^2-\frac{4}{3}c_{\mathrm{S}}^2\right)M_1(\omega)-\frac{2}{3}\rho c_{\mathrm{S}}^2 M_2(\omega) \quad \text{and} \quad \mu(\omega)=\rho c_{\mathrm{S}}^2 M_2(\omega) \tag{4.62}$$

(Carcione, 2007), where M_1 and M_2 are the dilatational and shear complex moduli, respectively, c_p and c_s are the elastic high-frequency (unrelaxed) compressional and shear-wave velocities. (In Carcione (1992), the relaxed (zero-frequency) moduli correspond to the elastic limit.)

Defining the P-wave modulus as

$$E=\lambda+2\mu, \tag{4.63}$$

the equivalent VTIv medium is defined by the following complex stiffness components or viscoelasticity constants:

$$\begin{aligned}
\overline{p}_{11} &= \langle E - \lambda^2 E^{-1} \rangle + \langle E^{-1} \rangle^{-1} \langle E^{-1} \lambda \rangle^2, \\
\overline{p}_{33} &= \langle E^{-1} \rangle^{-1}, \\
\overline{p}_{13} &= \langle E^{-1} \rangle^{-1} \langle E^{-1} \lambda \rangle, \\
\overline{p}_{55} &= \langle \mu^{-1} \rangle^{-1}, \\
\overline{p}_{66} &= \langle \mu \rangle,
\end{aligned} \tag{4.64}$$

where the weighted average of a quantity a is defined as

$$\langle a \rangle = \sum_{l=1}^{L} \mathrm{p}_l a_l \tag{4.65}$$

and p_l is the proportion of material l in a set of L different planar and parallel layers. The demonstration of equation (4.64) can be found in Postma (1955), Backus (1962) and Carcione (2007; Section 1.5). The other stiffness components are nil.

The moduli M_1 and M_2 can be described by mechanical models of viscoelasticity in terms of spring and dashpots. In particular, a continuous distribution of relaxation mechanisms, based on the standard linear solid, is useful to describe a constant Q medium. In this case

$$M_\nu(\omega) = \left(1 + \frac{2}{\pi Q_\nu} \ln \frac{\tau_2^{-1} + \mathrm{i}\omega}{\tau_1^{-1} + \mathrm{i}\omega} \right)^{-1}, \quad \nu = 1,2 \tag{4.66}$$

(Carcione, 2007; Chapter 6, eq. (2.210)), where τ_1 and τ_2 are time constants, with $\tau_2 < \tau_1$, Q_ν defines the value of the quality factor that remains nearly constant over the selected frequency range. At high frequencies, it is $M_\nu \to 1$. The P-wave and S-wave quality factors of each single isotropic layer are, respectively, given by

$$Q_\mathrm{P} = \frac{\mathrm{Re}(E)}{\mathrm{Im}(E)} \quad \text{and} \quad Q_\mathrm{S} = \frac{\mathrm{Re}(\mu)}{\mathrm{Im}(\mu)} \tag{4.67}$$

(Carcione, 2007).

If the single layers are viscoelastic and VTI, that is, VTIv, instead of isotropic, equation (4.64) generalizes to

$$\begin{aligned}
\overline{p}_{11} &= \langle p_{11} - p_{13}^2 p_{33}^{-1} \rangle + \langle p_{33}^{-1} \rangle^{-1} \langle p_{33}^{-1} p_{13} \rangle^2, \\
\overline{p}_{33} &= \langle p_{33}^{-1} \rangle^{-1},
\end{aligned}$$

$$\bar{p}_{13} = \langle p_{33}^{-1} \rangle^{-1} \langle p_{33}^{-1} p_{13} \rangle, \quad (4.68)$$

$$\bar{p}_{55} = \langle p_{55}^{-1} \rangle^{-1},$$

$$\bar{p}_{66} = \langle p_{66} \rangle,$$

where p_{IJ} are the complex and frequency-dependent stiffness components of the single layers. Equation (4.68) is the anelastic version of Backus equations (Carcione, 1992, 2007).

There are several models to describe a VTIv medium. Carcione and Cavallini (1994) and Carcione (2007; Section 4.1.2) considered the eigenstrain model, which can be used for an arbitrary medium symmetry, that is, from TI (five independent stiffnesses) to triclinic symmetry (21 independent stiffnesses). This attenuation model is based on the concept of eigenstiffness and eigenstrains, introduced by Lord Kelvin in the 19th century. In the lossless case, the strain energy is uniquely parameterized by six eigenstiffnesses or eigenvalues of the 6×6 elasticity tensor. From this fact and the correspondence principle, an anisotropic anelastic medium can essentially be described by six relaxation functions, which are the generalization of the eigenstiffnesses to the viscoelastic case.

In particular, the frequency-domain stiffnesses of a VTIv medium are

$$p_{11} = \lambda_1 (2 + a^2)^{-1} + \lambda_2 (2 + b^2)^{-1} + \lambda_4 / 2,$$

$$p_{12} = p_{11} - \lambda_4,$$

$$p_{33} = a^2 \lambda_1 (2 + a^2)^{-1} + b^2 \lambda_2 (2 + b^2)^{-1},$$

$$p_{13} = a \lambda_1 (2 + a^2)^{-1} + b \lambda_2 (2 + b^2)^{-1}, \quad (4.69)$$

$$p_{55} = \lambda_3 / 2,$$

$$p_{66} = \lambda_4 / 2,$$

where

$$a = \frac{4c_{13}}{c_{11} + c_{12} - c_{33} + \sqrt{c}}, \quad b = \frac{4c_{13}}{c_{11} + c_{12} - c_{33} - \sqrt{c}}, \quad (4.70)$$

and $\lambda_I(\omega)$, $I = 1, \ldots, 4$ are the complex and frequency-dependent eigenstiffnesses, given by

$$\lambda_1 = \frac{1}{2}(c_{11} + c_{12} + c_{33} + \sqrt{c}) M_1,$$

$$\lambda_2 = \frac{1}{2}(c_{11} + c_{12} + c_{33} - \sqrt{c}) M_2, \quad (4.71)$$

$$\lambda_3 = 2c_{55}M_2,$$

$$\lambda_4 = (c_{11} - c_{12})M_3,$$

with

$$c = 8c_{13}^2 + (c_{11} + c_{12} - c_{33})^2. \tag{4.72}$$

In isotropic elastic media, $\lambda_1 = 3c_{12} + 2c_{55}$ is three times the compression modulus and $\lambda_2 = \lambda_3 = 2c_{55}$ is twice the shear modulus. The kernels M_I, $I = 1,\ldots,3$ represent complex moduli dependent on frequency, of the form (4.66), for instance. The elastic case is obtained for $M_I \to 1$, when $\omega \to \infty$. For relatively weak anisotropy, the eigenvalues λ_1 and λ_2 are related to eigenstrains that consist of simultaneous changes in volume and shape, respectively, while λ_3 and λ_4 are related to pure shear eigenstrains; λ_3 and λ_4 determine the Q values of the shear waves along the principal axes.

Let us define the quality factor $q_I = \mathrm{Re}(M_I)/\mathrm{Im}(M_I)$. In weakly anisotropic viscoelastic media, q_1 is mainly associated with the dilatational component of the qP wave, q_2 with the qS wave and q_3 is the quality factor along the direction perpendicular to the symmetry axis. In an isotropic viscoelastic medium $q_1 = \mathrm{Re}(3p_{11} + 2p_{55})/\mathrm{Im}(3p_{11} + 2p_{55})$, that is, the quality corresponding to the compression modulus, and $q_2 = q_3 = \mathrm{Re}(p_{55})/\mathrm{Im}(p_{55})$, the quality factor corresponding to the shear modulus.

We consider alternating, periodic layers of a VTIv medium ($p_1 = 0.8$) and an isotropic elastic (lossless) medium ($p_2 = 0.2$). The VTIv medium has the following properties: $c_{11} = 38, c_{12} = 18, c_{13} = 5, c_{33} = 12, c_{55} = 3,\ c_{66} = 10,\ \rho = 2300$, $q_1 = 40, q_2 = 20, q_3 = 15, \tau_1 = 1/(0.2\pi)$ s, and $\tau_2 = 1/(4000\pi)$ s, where the stiffnesses are given in GPa and the density in kg/m^3. The isotropic and elastic medium is defined by $c_{11} = 20$ GPa, $c_{55} = 10$ GPa, and $\rho = 2100$ kg/m^3. Figure 4.2 shows the frequency dependency of the phase velocity and dissipation factor for $\theta = 0$.

The wavefronts and wave surfaces (curves) at 25 Hz are represented in Figure 4.3A and B, respectively. The wave front is associated with the higher energy velocity. In the elastic lossless case, all the wave surfaces have the same velocity – there is no velocity dispersion – and the wave front is the wave surface. In anelastic media, the wave front is the wave surface associated with the unrelaxed energy velocity.

The attenuation curves are shown in Figure 4.4, where the triplication of the qS wave implies the same phenomenon when the curves are represented as a function of the energy (ray) angle ψ. As can be seen, in this case, the attenuation of the qP wave is higher along the direction perpendicular to bedding planes, while the SH wave shows the opposite trend. However, this behavior is not general and depends on the values of the elasticity constants and specific attenuation model (Carcione and Cavallini, 1995; Carcione et al., 1998).

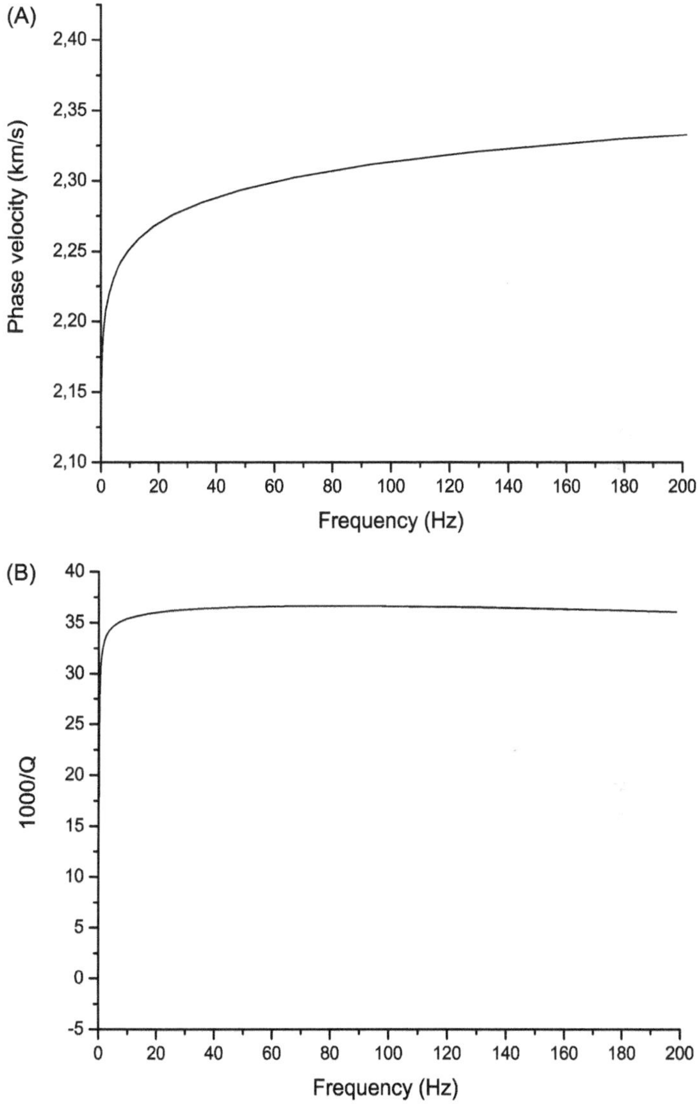

FIGURE 4.2 P-wave phase velocity (A) and dissipation factor (B) as a function of frequency, along the direction perpendicular to the layering plane (the corresponding stiffness is p_{33}). The Q factor is almost constant in the seismic exploration frequency band (approximately from 5 to 200 Hz).

4.3.1.1 Source Rocks

Most hydrocarbon source rocks are laminated structures composed of organic matter (kerogen and oil) and illite/smectite layers. Backus averaging and Gassmann equation (see Section 4.3.4) can be applied in this case to obtain the

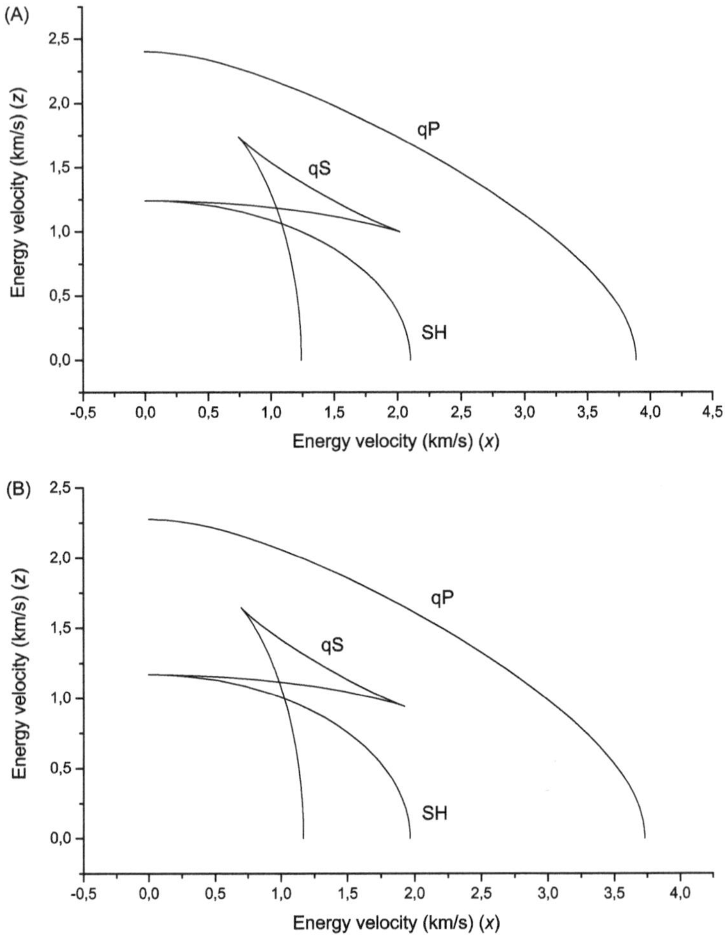

FIGURE 4.3 Energy velocities representing the wave fronts (A) and the wave surfaces (curves) at 25 Hz (B). The wave front is the high-frequency signal, having the maximum velocity, as can be seen in the plots, when compared to the 25 Hz curves.

properties of the equivalent medium (see Figure 4.5). Following Vernik and Nur (1992) and Carcione (2000), we assume that the immature rock is a two-layer composite made of illite and kerogen, with no additional porefluid. We consider illite transversely isotropic, kerogen isotropic, and both materials viscoelastic. Use of Backus averaging yields a transversely isotropic equivalent medium described by the five complex equivalent stiffnesses given in equation (4.68). Denoting the proportion of volumetric kerogen content by K, the weighted average of any physical quantity a in equation (4.68) is defined as

$$\langle a \rangle = (1 - \mathrm{K})a_{\mathrm{i}} + \mathrm{K}a_{\mathrm{k}}, \tag{4.73}$$

where the subscripts "i" and "k" here indicate illite and kerogen, respectively.

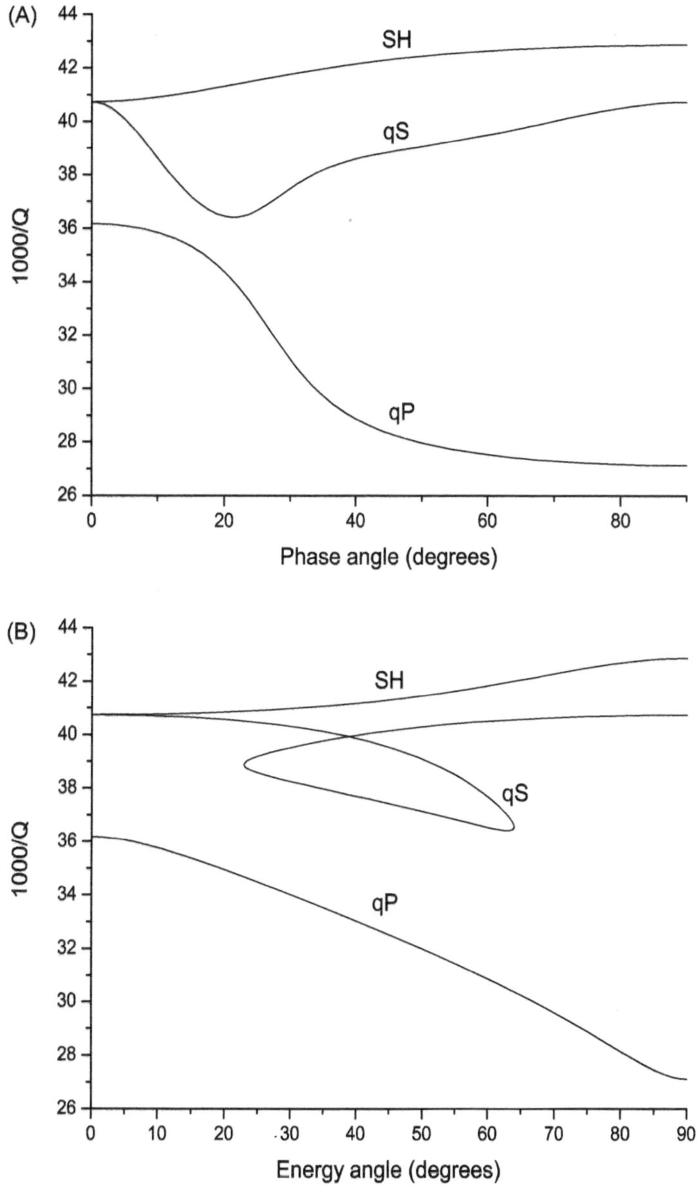

FIGURE 4.4 Dissipation factors as a function of the phase (A) and ray (B) angles at 25 Hz.

The properties of illite can be described by equation (4.69). As stated by Vernik and Nur (1992), the organic material can be viewed as an amorphous (isotropic) substance subject to liquefaction with increasing temperature and pressure (Carcione, 2000; Pinna et al., 2011; Carcione et al., 2011). After the

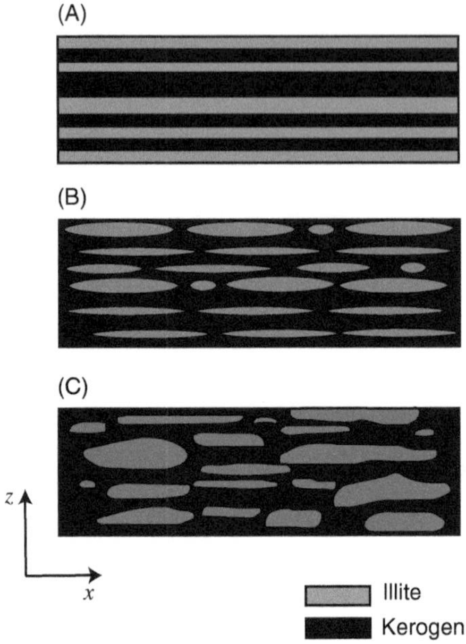

FIGURE 4.5 Schematic fabric topology of transversely isotropic kerogen-rich shales, according to Backus model (A), modified Backus model (B) (see eq. (4.80)) and Gassmann model (C) (see Section 4.2.4). The z-direction corresponds to the symmetry axis.

kerogen to oil partial conversion and assuming an oil saturation S, the stiffnesses of the mixture can be calculated by using an acoustic model developed by Kuster and Toksöz (1974), which assumes noninteracting spherical oil bubbles. The unrelaxed velocities are obtained from

$$\frac{\rho(c_P^2 - 2c_S^2)}{K_k} = \frac{1 + [4\mu_k(K_o - K_k)/(3K_o + 4\mu_k)K_k]S}{1 - [3(K_o - K_k)/(3K_o + 4\mu_k)]S} \tag{4.74}$$

and

$$\frac{\rho c_P^2}{\mu_k} = \frac{(1 - S)(9K_o + 8\mu_k)}{9K_k + 8\mu_k + S(6K_k + 12\mu_k)}, \tag{4.75}$$

where K_k and K_o denote the bulk moduli of kerogen and oil, respectively, μ_k is the shear modulus of kerogen, and $\rho = (1 - S)\rho_k + S\rho_o$. Then, each layer of the mixture is described by the complex Lamé parameters given by equation (4.62).

The volumetric kerogen content (in percent) can be transformed to total organic carbon (TOC, weight percent) and vice versa, using

$$\text{TOC} = \frac{0.75\rho_k \text{K}}{\rho_i - \text{K}(\rho_i - \rho_k)}. \tag{4.76}$$

where 0.75 is a constant relating TOC to kerogen weight percent (Vernik and Nur, 1992).

The anisotropy parameters, introduced by Thomsen (1986), are generalized by Carcione (2000) as

$$\varepsilon = \frac{\bar{p}_{11\mathrm{R}} - \bar{p}_{33\mathrm{R}}}{2\bar{p}_{33\mathrm{R}}},$$

$$\gamma = \frac{\bar{p}_{66\mathrm{R}} - \bar{p}_{55\mathrm{R}}}{2\bar{p}_{55\mathrm{R}}}, \quad (4.77)$$

$$\delta = \frac{(\bar{p}_{13\mathrm{R}} + \bar{p}_{55\mathrm{R}})^2 - (\bar{p}_{33\mathrm{R}} - \bar{p}_{55\mathrm{R}})^2}{2\bar{p}_{33\mathrm{R}}(\bar{p}_{33\mathrm{R}} - \bar{p}_{55\mathrm{R}})},$$

where the subindex "R" denotes the real part. Equivalent parameters to ε and γ, quantifying the anisotropic attenuation, can be defined as

$$\varepsilon_Q = \frac{Q_{11} - Q_{33}}{2Q_{33}} = \frac{1}{2}\left[\frac{\bar{p}_{33\mathrm{Im}}}{\bar{p}_{11\mathrm{Im}}}(2\varepsilon + 1) - 1\right],$$

$$\gamma_Q = \frac{Q_{66} - Q_{55}}{2Q_{55}} = \frac{1}{2}\left[\frac{\bar{p}_{55\mathrm{Im}}}{\bar{p}_{66\mathrm{Im}}}(2\gamma + 1) - 1\right], \quad (4.78)$$

where

$$Q_{I(I)} = \frac{\bar{p}_{I(I)\mathrm{R}}}{\bar{p}_{I(I)\mathrm{Im}}} \quad (4.79)$$

are the quality factors along the "crystal" axes and the subindex "Im" denotes the imaginary part. Note that, for instance, when $\bar{p}_{11\mathrm{Im}} = \bar{p}_{33\mathrm{Im}}$, then $\varepsilon_Q = \varepsilon$, which does not imply isotropic attenuation.

Backus averaging overestimates the velocities along the layering plane of source rocks (Vernik and Nur, 1992; Carcione, 2000), since illite has a lenticular textural pattern, as shown in Figure 4.5B (Vernik and Nur, 1992; Carcione, 2000). To obtain a better fit to the experimental data, we assume that the stiffnesses of illite "parallel to bedding" are affected as

$$p_{11\mathrm{i}} \to \langle p_{11} \rangle \text{ and } p_{66\mathrm{i}} \to \langle p_{66} \rangle, \quad (4.80)$$

which incorporate the respective local constants of both illite and kerogen.

We consider the marine Kimmeridge shale of the Draupne formation, located between 3.48 and 3.58 km depth in the central Viking Graben of the North Sea. According to the theory developed by Carcione (2000), if a fraction F of initial kerogen content K converted to oil generates a pore-pressure increase Δp_f, both quantities are related as

$$\Delta p_f = \frac{(D-1)F}{c_\mathrm{p} + c_\mathrm{k} - F[c_\mathrm{p} + c_\mathrm{k} - D(c_\mathrm{p} + c_\mathrm{o})]}, \quad (4.81)$$

TABLE 4.1 Material Properties

Medium	V_{11} (km/s)	V_{33} (km/s)	V_{55} (km/s)	V_{66} (km/s)	V_{13} (km/s)	ρ (g/cm^{-3})	q_1	q_2
Illite*	4.7	4.36	2.46	2.77	2.43	2.7	270	200
Kerogen*	2.6	2.6	1.2	1.2	1.97	1.4	30	20
Oil†	0.73	0.73	0	0	0.73	0.9	10	–

*Vernik (1994).
†Oil/bitumen (McCain, 1984).

where c_p is the pore compressibility, $c_\mathrm{k} = 1/K_\mathrm{k}$ is the kerogen compressibility, $c_\mathrm{o} = 1/K_\mathrm{o}$ is the oil compressibility, and $D = \rho_\mathrm{k}/\rho_\mathrm{o}$. We consider that the pore space stiffness c_p^{-1} decreases linearly with kerogen content, $c_\mathrm{p}[\mathrm{MPa}^{-1}] = (2381 - 5357\ \mathrm{K})^{-1}$.

The kerogen and oil proportions can be expressed as a function of pore pressure as

$$\phi_\mathrm{k} = \mathrm{K}(1-F)(1-c_\mathrm{k}\Delta p_f), \quad \phi_\mathrm{o} = \mathrm{K}FD(1-c_\mathrm{o}\Delta p_f), \tag{4.82}$$

respectively. The illite proportion is $\phi_\mathrm{i} = 1 - \phi_\mathrm{k} - \phi_\mathrm{o}$ and the oil saturation is $S = \phi_\mathrm{o} / (\phi_\mathrm{k} + \phi_\mathrm{o})$.

The material properties are given in Table 4.1, where $c_{IJ} = \rho V_{IJ}^2$ are the unrelaxed stiffnesses of each single medium, and q_1 and q_2 are the quality-factor parameters related to dilatational and shear deformations.

Figure 4.6 shows the P-wave velocity as a function of the kerogen content (solid lines), compared to experimental data obtained by Vernik (1992). The dashed lines correspond to a kerogen-oil conversion process for Δp_f = 30 MPa, where the kerogen content represents the initial value. As can be appreciated, the model provides a fairly good fit to the experimental values. It can be shown that maximum stiffness anisotropy is obtained for a kerogen content of approximately 30% (Carcione, 2000). Note that Thomsen's coefficients relate to the stiffness anisotropy and not the velocity anisotropy, which is generally lower than the stiffness anisotropy. As expected, the velocities in the presence of oil (dashed lines) are lower than those corresponding to 100% kerogen (solid lines).

More examples can be found in Carcione, 2000; Carcione et al., 2011; and Pinna et al., 2011.

4.3.2 Schoenberg–Muir Averaging

Schoenberg and Muir (1989) generalized Backus approach to single elastic layers of arbitrary anisotropy using group theory and a matrix formalism. We extend their theory to the anelastic case by using the correspondence principle, by

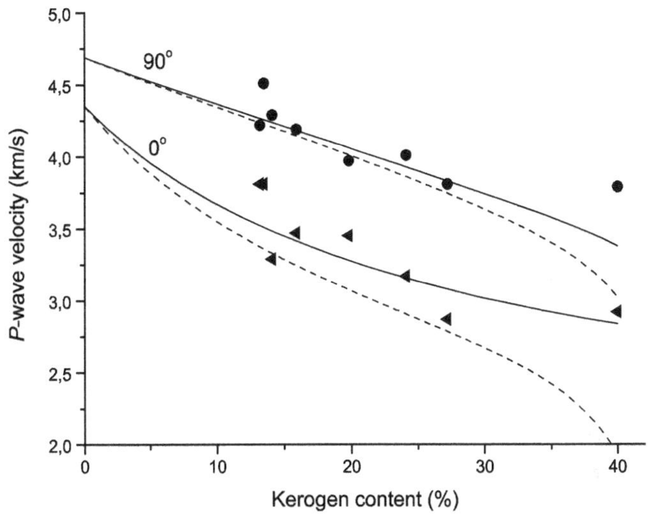

FIGURE 4.6 P-wave bedding-normal (0°) velocities (circles) and bedding-parallel (90°) velocities (triangles). The best fit of these experimental data (solid lines) defines the properties of illite and kerogen, shown in Table 4.1. The dashed lines correspond to a kerogen-oil conversion process for $\Delta p_f = 30$ MPa.

which the real and frequency-independent elasticity constants can be replaced by complex and frequency-dependent stiffness components.

Let us consider a finely layered medium composed of arbitrarily anisotropic layers, with the z-axis perpendicular to the layering plane (examples are shown in Figure 4.7). Each layer is defined by the density ρ, the proportion p_n, and the stiffness components p_{IJ}.

The stress–strain relation (4.2) of each layer can be rewritten in the frequency domain as

$$\begin{pmatrix} \sigma_1 \\ \sigma_2 \\ \sigma_6 \\ \\ \sigma_3 \\ \sigma_4 \\ \sigma_5 \end{pmatrix} = \begin{pmatrix} \sigma_{11} \\ \sigma_{22} \\ \sigma_{12} \\ \\ \sigma_{33} \\ \sigma_{23} \\ \sigma_{13} \end{pmatrix} = \begin{pmatrix} p_{11} & p_{12} & p_{16} & p_{13} & p_{14} & p_{15} \\ p_{12} & p_{22} & p_{26} & p_{23} & p_{24} & p_{25} \\ p_{16} & p_{26} & p_{66} & p_{36} & p_{46} & p_{56} \\ \\ p_{13} & p_{23} & p_{36} & p_{33} & p_{34} & p_{35} \\ p_{14} & p_{24} & p_{46} & p_{34} & p_{44} & p_{45} \\ p_{15} & p_{25} & p_{56} & p_{35} & p_{45} & p_{55} \end{pmatrix} \begin{pmatrix} e_1 \\ e_2 \\ e_6 \\ \\ e_3 \\ e_4 \\ e_5 \end{pmatrix}. \tag{4.83}$$

The stiffness matrix involved in equation (4.83) can be rewritten in terms of four submatrices as

$$\begin{pmatrix} P_{TT} & P_{TN} \\ \\ P_{TN}^{\mathrm{T}} & P_{NN} \end{pmatrix}. \tag{4.84}$$

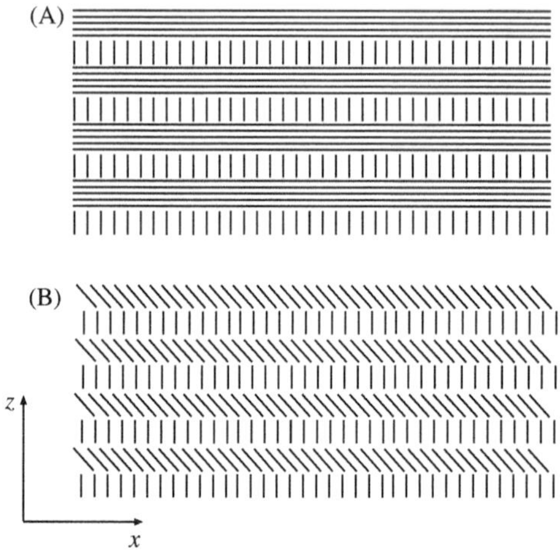

FIGURE 4.7 Stack of thin (compared to the wavelength) strata composed of HTI and VTI layers (A) and 45TI and VTI layers (B). The percentage of each constituent is assumed to be stationary with respect to the vertical coordinate.

Following Schoenberg and Muir (1989), the equivalent homogeneous medium is defined by

$$\begin{pmatrix} \overline{P}_{TT} & \overline{P}_{TN} \\ \overline{P}_{TN}^{\mathrm{T}} & \overline{P}_{NN} \end{pmatrix}, \tag{4.85}$$

where

$$\begin{aligned} \overline{P}_{NN} &= \langle P_{NN}^{-1} \rangle^{-1}, \\ \overline{P}_{TN} &= \langle P_{TN} P_{NN}^{-1} \rangle \overline{P}_{NN}, \\ \overline{P}_{TT} &= \langle P_{TT} \rangle - \langle P_{TN} P_{NN}^{-1} P_{NT} \rangle + \overline{P}_{TN} \langle P_{NN}^{-1} P_{NT} \rangle. \end{aligned} \tag{4.86}$$

Let us consider a specific example, namely, a periodic system composed of VTIv and HTIv layers, where the HTIv medium, labeled 2, is the same VTIv medium, labeled 1, but the symmetry axis is horizontally oriented.

A VTIv medium with symmetry axis along the z-axis has the following stiffness matrix:

$$\mathbf{P} = \begin{pmatrix} p_{11} & p_{12} & p_{13} & 0 & 0 & 0 \\ p_{12} & p_{11} & p_{13} & 0 & 0 & 0 \\ p_{13} & p_{13} & p_{33} & 0 & 0 & 0 \\ 0 & 0 & 0 & p_{55} & 0 & 0 \\ 0 & 0 & 0 & 0 & p_{55} & 0 \\ 0 & 0 & 0 & 0 & 0 & p_{66} \end{pmatrix}, \quad 2p_{66} = p_{11} - p_{12}. \tag{4.87}$$

A clockwise rotation of the vertical symmetry axis through an angle γ about the y-axis has the orthogonal transformation matrix

$$\begin{pmatrix} \cos\gamma & 0 & \sin\gamma \\ 0 & 1 & 0 \\ -\sin\gamma & 0 & \cos\gamma \end{pmatrix}. \tag{4.88}$$

The corresponding Bond transformation matrix is

$$\mathrm{M} = \begin{pmatrix} \cos^2\gamma & 0 & \sin^2\gamma & 0 & \sin(2\gamma) & 0 \\ 0 & 1 & 0 & 0 & 0 & 0 \\ \sin^2\gamma & 0 & \cos^2\gamma & 0 & -\sin(2\gamma) & 0 \\ 0 & 0 & 0 & \cos\gamma & 0 & -\sin\gamma \\ -\frac{1}{2}\sin(2\gamma) & 0 & \frac{1}{2}\sin(2\gamma) & 0 & \cos(2\gamma) & 0 \\ 0 & 0 & 0 & 0 & 0 & \cos\gamma \end{pmatrix} \tag{4.89}$$

(Auld, 1990; Carcione, 2007). Then, the stiffness matrix with the rotated symmetry axis is given by

$$\mathbf{P}' = \mathbf{M} \cdot \mathbf{P} \cdot \mathbf{M}^{\mathrm{T}} \tag{4.90}$$

(Auld, 1990; Carcione, 2007).

Using (4.90), an angle $\gamma = \pi/2$ transforms a VTIv medium into an HTIv medium

$$\begin{pmatrix} p_{11} & p_{12} & p_{13} & 0 & 0 & 0 \\ p_{12} & p_{11} & p_{13} & 0 & 0 & 0 \\ p_{13} & p_{13} & p_{33} & 0 & 0 & 0 \\ 0 & 0 & 0 & p_{55} & 0 & 0 \\ 0 & 0 & 0 & 0 & p_{55} & 0 \\ 0 & 0 & 0 & 0 & 0 & p_{66} \end{pmatrix} \rightarrow \begin{pmatrix} p_{33} & p_{13} & p_{13} & 0 & 0 & 0 \\ p_{13} & p_{11} & p_{12} & 0 & 0 & 0 \\ p_{13} & p_{12} & p_{11} & 0 & 0 & 0 \\ 0 & 0 & 0 & p_{66} & 0 & 0 \\ 0 & 0 & 0 & 0 & p_{55} & 0 \\ 0 & 0 & 0 & 0 & 0 & p_{55} \end{pmatrix}. \tag{4.91}$$

According to equations (4.83), (4.84), and (4.91),

$$P_{TT}^{(1)} = \begin{pmatrix} p_{11} & p_{12} & 0 \\ p_{12} & p_{11} & 0 \\ 0 & 0 & p_{66} \end{pmatrix}, \quad P_{TN}^{(1)} = \begin{pmatrix} p_{13} & 0 & 0 \\ p_{13} & 0 & 0 \\ 0 & 0 & 0 \end{pmatrix}, \quad P_{NN}^{(1)} = \begin{pmatrix} p_{33} & 0 & 0 \\ 0 & p_{55} & 0 \\ 0 & 0 & p_{55} \end{pmatrix}, \tag{4.92}$$

and

$$P_{TT}^{(2)} = \begin{pmatrix} p_{33} & p_{13} & 0 \\ p_{13} & p_{11} & 0 \\ 0 & 0 & p_{55} \end{pmatrix}, \quad P_{TN}^{(2)} = \begin{pmatrix} p_{13} & 0 & 0 \\ p_{12} & 0 & 0 \\ 0 & 0 & 0 \end{pmatrix}, \quad P_{NN}^{(2)} = \begin{pmatrix} p_{11} & 0 & 0 \\ 0 & p_{66} & 0 \\ 0 & 0 & p_{55} \end{pmatrix}, \tag{4.93}$$

with $p_{66} = \frac{1}{2}(p_{11} - p_{12})$.

From equations (4.85) and (4.86), the equivalent medium composed of VTIv–HTIv layers has orthorhombic symmetry and is defined by the following symmetric stiffness matrix:

$$\begin{pmatrix} \frac{p_{11}+p_{33}}{2} & \frac{p_{12}+p_{13}}{2} & p_{13} & 0 & 0 & 0 \\ * & p_{11}-\frac{(p_{12}-p_{13})^2}{2(p_{11}+p_{33})} & \frac{p_{11}p_{13}+p_{12}p_{33}}{p_{11}+p_{33}} & 0 & 0 & 0 \\ * & * & \frac{2p_{11}p_{33}}{p_{11}+p_{33}} & 0 & 0 & 0 \\ * & * & * & \frac{2(p_{11}-p_{12})p_{55}}{D} & 0 & 0 \\ * & * & * & * & p_{55} & 0 \\ * & * & * & * & * & \frac{D}{4} \end{pmatrix}, \tag{4.94}$$

where

$$D = p_{11} - p_{12} + 2p_{55}.$$

Figure 4.8 shows the energy velocities and quality factors of the orthorhombic medium at the (x,y)-plane, where p_{11} = (21.6,0.64), p_{12} = (9.1,0.08), p_{13} = (4.2,0.06), p_{33} = (17,0.6), p_{55} = (3.1,0.12), and ρ = 2260 kg/m^3, where the stiffness components are given in GPa. The p_{IJ} values corresponds to the equivalent VTIv medium obtained in Section 4.3.1. The same example, but for the lossless case, is illustrated in Carcione et al. (2012). They also considered a periodic system of VTI and 45TI layers, where the 45TI medium is the same VTI medium, but rotated by 45°.

Simulations performed by Carcione et al. (2012) revealed that the theory performs very well, no matter the degree of anisotropy. The only limitation is that the layers have to be very thin compared to the minimum wavelength of the signal; to be rigorous, the theory is exact at zero frequency. The theory performs equally well for amplitudes, going beyond the expectations of the authors, since they state (Schoenberg and Muir, 1989): "The real limitation is the long-wavelength one, which says, in effect, that it is concerned with kinematics arrival times-alone. It does not address the important dynamical question of how high-frequency energy is lost from the coherent to the scattered field."

The Schoenberg–Muir averaging has been questioned with the argument that it cannot be applied to oblique sets of cracks (even for weak anisotropy), since the structure is no longer 1D and Backus's assumptions are invalidated (see Carcione et al., 2012). As stated above, the simulations show that the average is valid from the kinematic (traveltimes) and dynamic (amplitudes) points

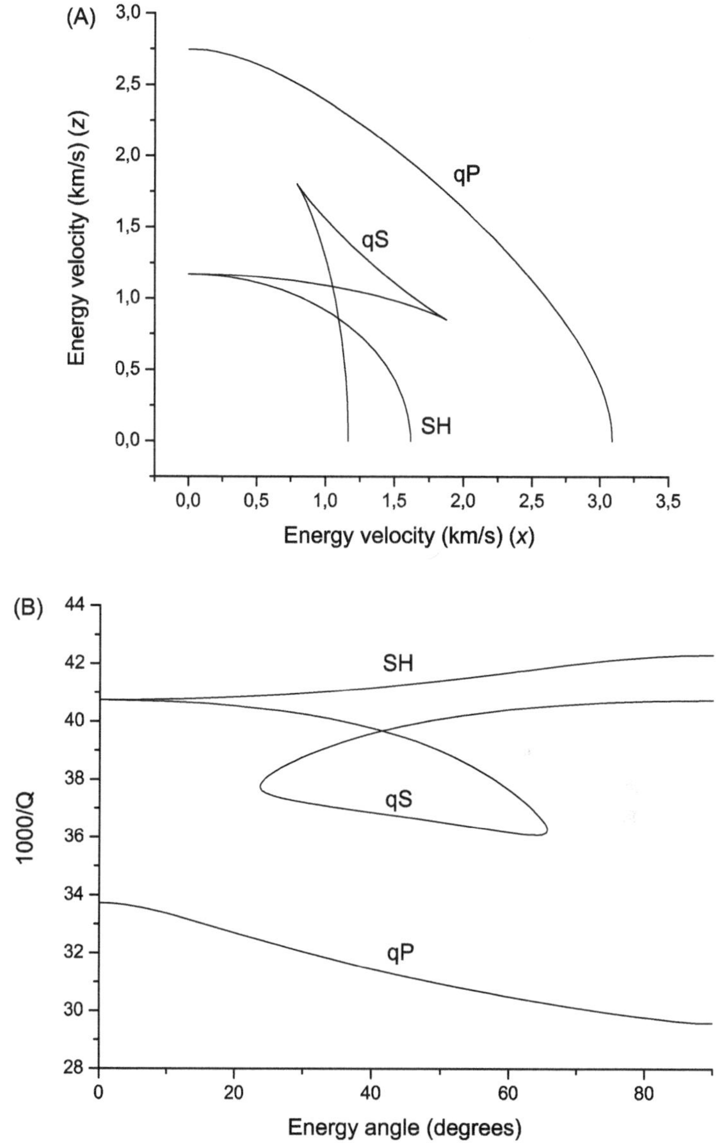

FIGURE 4.8 Polar representation of the energy velocity (A) and quality factors as a function of the energy (ray) angle (B), corresponding to the (x,z)-plane. The medium has orthorhombic symmetry and the frequency is 25 Hz.

of view, no matter how anisotropic are the single constituent layers. Even for transversely isotropic layers with tilted axes, the theory performs equally well. The theory is not limited to thin layers but the anisotropy of the single layers can be due to other causes, such as cracks and fractures.

4.3.3 Backus Averaging in Porous Media

In the theory of finely layered porous media, there is a dominant P-wave attenuation mechanism due to wave-induced fluid flow, called mesoscopic loss caused by the presence of a viscous fluid and mesoscopic-scale heterogeneities (dissimilar medium properties). The mesoscopic-scale lengths (e.g., layer thicknesses) are intended to be much larger than the grain sizes but much smaller than the wavelength of the pulse (White et al., 1975; Carcione, 2007; Section 7.10). The P-wave induces a fluid-pressure difference at mesoscopic-scale inhomogeneities generating fluid flow and slow (diffusion) Biot waves. The related relaxation peak can be located at seismic frequencies for specific values of the layer thicknesses. On the other hand, the attenuation mechanism predicted by Biot's theory has a macroscopic nature (Biot, 1956) and its peak may be located at higher frequencies (Carcione, 2007).

We present here the wave theory for an equivalent TI medium composed of plane homogeneous and isotropic layers, where the relaxation behavior can be described by a single relaxation function, based on the assumption that the fluid-flow direction is perpendicular to the layering plane and it is independent of the loading direction (Krzikalla and Müller, 2011; Carcione et al., 2013). As a consequence, besides P-wave losses, the coupling between the qP and qSV waves generates shear-wave anisotropic velocity dispersion and attenuation. This assumption can be validated by the quasi-static numerical experiments illustrated in Section 4.5.

Let us consider isotropic poroelastic layers. Using the notation of Carcione (2007), let $\mathbf{u}^{(m)}(\mathbf{x}) = (u_1^{(m)}, u_2^{(m)}, u_3^{(m)})$ and $\mathbf{w}(\mathbf{x}) = (w_1, w_2, w_3)$ be the time Fourier transform of the displacement vector of the solid (matrix) and fluid (relative to the solid) phases, respectively (if $\mathbf{u}^{(f)}$ is the fluid displacement vector, $\mathbf{w} = \phi(\mathbf{u}^{(f)} - \mathbf{u}^{(m)})$, where ϕ is the porosity). Define $\mathbf{u} = (\mathbf{u}^{(m)}, \mathbf{w})$ and let $\sigma_{ij}(\mathbf{u})$ and $p_f(\mathbf{u})$ denote the time Fourier transform of the total stress and the fluid pressure, respectively, and let $\varepsilon_{ij}(\mathbf{u}^{(m)})$ be the strain tensor of the matrix. The frequency-domain stress–strain relations of a single plane layer l in a sequence of L layers, are

$$\sigma_{ij}(\mathbf{u}) = 2\mu^{(l)}\varepsilon_{ij}(\mathbf{u}^{(m)}) + \delta_{ij}\left(\lambda_G^{(l)}\,\mathrm{div}\,\mathbf{u}^{(m)} + \alpha^{(l)}M^{(l)}\mathrm{div}\,\mathbf{w}\right), \tag{4.95}$$

$$p_f(\mathbf{u}) = -\alpha^{(l)}M^{(l)}\mathrm{div}\,\mathbf{u}^{(m)} - M^{(l)}\mathrm{div}\,\mathbf{w} \tag{4.96}$$

(Carcione, 2007; eqs. (7.33, 7.172, and 7.173)). The coefficient μ is the shear modulus of the bulk material, considered to be equal to the shear modulus of the dry matrix, μ_m. Moreover

$$\lambda_G = K_G - \frac{2}{3}\mu, \tag{4.97}$$

with K_G the bulk modulus of the saturated material (Gassmann modulus). The coefficients in equations (4.95) and (4.96) can be obtained from the relations

$$\alpha = 1 - \frac{K_m}{K_s}, \quad M = \left(\frac{\alpha - \phi}{K_s} + \frac{\phi}{K_f} \right)^{-1}, \tag{4.98}$$

$$K_G = K_m + \alpha^2 M \tag{4.99}$$

(Carcione, 2007; eqs. (7.25, 7.26, and 7.34)), where K_s, K_m, and K_f denote the bulk moduli of the solid grains, dry matrix, and fluid, respectively. The coefficient α is known as the effective stress coefficient of the bulk material. If ρ_s and ρ_f are the mass densities of the solid and fluid, respectively,

$$\rho = (1 - \phi)\rho_s + \phi\rho_f \tag{4.100}$$

denotes the mass density of the bulk material.

On the other hand, Biot's equations of motion, stated in the space-frequency domain, are

$$\begin{aligned} &\text{div}\, \boldsymbol{\sigma} = 0, \\ &\text{grad}\, p_f + \left(\frac{\eta}{\kappa} \right) \mathrm{i}\omega \mathbf{w} = 0 \end{aligned} \tag{4.101}$$

(Carcione, 2007; eqs. (7.210) and (7.211)), where η is the fluid viscosity and κ is the matrix permeability. We have ignored external sources in equation (4.101) and we have neglected the inertial (acceleration) terms that are negligible at relatively low (seismic) frequencies. Therefore, at the seismic band, the effects of wave-induced fluid flow are described by the quasi-static Biot theory, that is, stress equilibrium within the porous matrix and Darcy's flow of pore fluid.

Bakulin and Molotkov (1997) and Gelinsky and Shapiro (1997) obtained the equivalent or effective TI medium corresponding to a stack of two thin alternating porous layers of thicknesses d_1 and d_2, such that the period of the stratification is $d = d_1 + d_2$ and the proportions are $\mathrm{p}_l = d_l / d, l = 1, 2$. In the following, the number in subindices indicate coordinates in field variables and medium in material properties.

In particular, Bakulin and Molotkov (1997) derived the low-frequency limit equation of motion of the equivalent medium. Their notation (left side of the following equations) is related to ours as

$$\begin{aligned} &P = K_G + \frac{4}{3}\mu_m = E_G = E_m + \alpha^2 M, \quad E_m = K_m + \frac{4}{3}\mu_m, \\ &M' = \alpha M, \\ &R = M, \\ &F = K_G - \frac{2}{3}\mu_m, \end{aligned} \tag{4.102}$$

$$L = \mu_m,$$

$$\tilde{m}(\omega) = m + \frac{\eta}{\mathrm{i}\omega\kappa}$$

(actually M' is termed M in that work, but here M is used in equation (4.98)). The equation of motion of the equivalent medium, describing qP-qS propagation in the (x,z)-plane reads

$$\begin{aligned}
\sigma_{11} &= \bar{P}\partial_1 u_1^{(m)} + \bar{F}\partial_3 u_3^{(m)} + \bar{M}\mathrm{div}\cdot\mathbf{w},\\
\sigma_{33} &= \bar{F}\partial_1 u_1^{(m)} + \bar{C}\partial_3 u_3^{(m)} + \bar{Q}\mathrm{div}\cdot\mathbf{w},\\
\sigma_{13} &= \bar{L}(\partial_1 u_3^{(m)} + \partial_3 u_1^{(m)}),\\
-p_f &= \bar{M}\partial_1 u_1^{(m)} + \bar{Q}\partial_3 u_3^{(m)} + \bar{R}\mathrm{div}\cdot\mathbf{w},\\
\partial_1\sigma_{11} + \partial_3\sigma_{13} &= \rho_x\,\partial_{tt}^2 u_1 + \rho_{fx}\,\partial_{tt}^2 w_1,\\
\partial_1\sigma_{13} + \partial_3\sigma_{33} &= \rho_z\,\partial_{tt}^2 u_3 + \rho_{fz}\,\partial_{tt}^2 w_3,\\
-\partial_1 p_f &= \rho_{fx}\,\partial_{tt}^2 u_1^{(m)} + m_x\,\partial_{tt}^2 w_1,\\
-\partial_3 p_f &= \rho_{fz}\,\partial_{tt}^2 u_3^{(m)} + m_z\,\partial_{tt}^2 w_3,
\end{aligned} \tag{4.103}$$

where the Backus averaged properties are

$$\begin{aligned}
\rho_z &= \langle\rho\rangle, \quad \rho_x = \left\langle\frac{\rho\tilde{m} - \rho_f^2}{\tilde{m}}\right\rangle + \left\langle\frac{\rho_f}{\tilde{m}}\right\rangle^2\left\langle\frac{1}{\tilde{m}}\right\rangle^{-1},\\
\rho_{fx} &= \left\langle\frac{\rho_f}{\tilde{m}}\right\rangle\left\langle\frac{1}{\tilde{m}}\right\rangle^{-1}, \quad \rho_{fz} = \langle\rho_f\rangle, \quad m_x = \langle\tilde{m}\rangle, \quad m_z = \left\langle\frac{1}{\tilde{m}}\right\rangle^{-1},\\
\bar{P} &= \left\langle\frac{FR - M'^2}{\beta}\right\rangle\bar{F} + \left\langle\frac{2LM'}{\beta}\right\rangle\bar{M} + \left\langle\frac{D}{\beta}\right\rangle,\\
\bar{F} &= \frac{1}{\gamma}\left[\left\langle\frac{P}{\beta}\right\rangle\left\langle\frac{FR - M'^2}{\beta}\right\rangle + \left\langle\frac{M'}{\beta}\right\rangle\left\langle\frac{2LM'}{\beta}\right\rangle\right],\\
\bar{M} &= \frac{1}{\gamma}\left[\left\langle\frac{M'}{\beta}\right\rangle\left\langle\frac{FR - M'^2}{\beta}\right\rangle + \left\langle\frac{R}{\beta}\right\rangle\left\langle\frac{2LM'}{\beta}\right\rangle\right],
\end{aligned} \tag{4.104}$$

$$\bar{C} = \frac{1}{\gamma}\left\langle\frac{P}{\beta}\right\rangle, \ \bar{Q} = \frac{1}{\gamma}\left\langle\frac{M'}{\beta}\right\rangle, \ \bar{R} = \frac{1}{\gamma}\left\langle\frac{R}{\beta}\right\rangle, \ \bar{L} = \left\langle\frac{1}{L}\right\rangle^{-1},$$

$$\beta = PR - M'^2, \quad D = \begin{vmatrix} P & F & M' \\ F & P & M' \\ M' & M' & R \end{vmatrix},$$

$$\gamma = \left\langle\frac{P}{\beta}\right\rangle\left\langle\frac{R}{\beta}\right\rangle - \left\langle\frac{M'}{\beta}\right\rangle^2.$$

We have generalized the equations to the case of nonzero viscosity (see eq. $(4.102)_6$). Equations (4.103)–(4.104) describe the relaxed version of the equivalent medium with a tensorial mass density, including the Biot loss mechanism but excluding the mesoscopic loss, which is incorporated as shown below.

In terms of the undrained (wet-rock) and drained (dry-rock) stiffnesses c_{IJ}^u and $c_{IJ}^{(m)}$, respectively, and the notation for anisotropic media used here, we have

$$\begin{gathered} c_{11}^u = \bar{P}, \ c_{13}^u = \bar{F}, \ c_{33}^u = \bar{C}, \ c_{55}^u = \bar{L}, \\ M = \bar{R}, \ \alpha_1 = \bar{M}/\bar{R}, \ \alpha_3 = \bar{Q}/\bar{R}, \\ c_{IJ}^u = c_{IJ}^{(m)} + M\alpha_I\alpha_J, \quad I, J = 1, \ldots, 6 \end{gathered} \tag{4.105}$$

(Carcione, 2007; eq. (7.136)). Equation $(4.105)_3$ is similar to the stress–strain relation for saturated porous anisotropic media obtained by Brown and Korringa (1975) in terms of 4-index compliance tensors (see eq. (4.118)). The equation corresponds to the anisotropic version of Gassmann's undrained modulus K_G.

For completeness, we report the relaxed equivalent stiffnesses of Gelinsky and Shapiro (1997) [their eq. (14)] and the relation to the stiffnesses of Bakulin and Molotkov (1997)

$$c_{66}^r = B_1^* = \langle\mu_m\rangle,$$

$$c_{11}^r - 2c_{66}^r = c_{12}^r = B_2^* = 2\left\langle\frac{\lambda_m\mu_m}{E_m}\right\rangle + \left\langle\frac{\lambda_m}{E_m}\right\rangle^2\left\langle\frac{1}{E_m}\right\rangle^{-1} + \frac{B_6^{*2}}{B_8^*},$$

$$c_{13}^r = B_3^* = \left\langle\frac{\lambda_m}{E_m}\right\rangle\left\langle\frac{1}{E_m}\right\rangle^{-1} + \frac{B_6^*B_7^*}{B_8^*} = \bar{F},$$

$$c_{33}^r = B_4^* = \left\langle\frac{1}{E_m}\right\rangle^{-1} + \frac{B_7^{*2}}{B_8^*} = \left[\left\langle\frac{1}{E_m}\right\rangle - \left\langle\frac{\alpha}{E_m}\right\rangle^2\left\langle\frac{E_G}{ME_m}\right\rangle^{-1}\right]^{-1} = \bar{C},$$

$$c^r_{55} = B^*_5 = \langle \mu_m^{-1} \rangle^{-1} = \bar{L}, \tag{4.106}$$

$$B^*_6 = -B^*_8 \left(2\left\langle \frac{\alpha \mu_m}{E_m} \right\rangle + \left\langle \frac{\alpha}{E_m} \right\rangle \left\langle \frac{\lambda_m}{E_m} \right\rangle \left\langle \frac{1}{E_m} \right\rangle^{-1} \right),$$

$$B^*_7 = -B^*_8 \left\langle \frac{\alpha}{E_m} \right\rangle \left\langle \frac{1}{E_m} \right\rangle^{-1},$$

$$B^*_8 = \left[\left\langle \frac{1}{M} \right\rangle + \left\langle \frac{\alpha^2}{E_m} \right\rangle - \left\langle \frac{\alpha}{E_m} \right\rangle^2 \left\langle \frac{1}{E_m} \right\rangle^{-1} \right]^{-1} = \bar{R}.$$

Gelinsky and Shapiro (1997) considered, in addition, the case of no interlayer flow, that is, the unrelaxed regime, for which the stiffnesses are

$$c_{66} = c^r_{66},$$

$$c_{11} - 2c_{66} = c_{12} = 2\left\langle \frac{(E_G - 2\mu_m)\mu_m}{E_G} \right\rangle + \left\langle \frac{E_G - 2\mu_m}{E_G} \right\rangle^2 \left\langle \frac{1}{E_G} \right\rangle^{-1},$$

$$c_{13} = \left\langle \frac{E_G - 2\mu_m}{E_G} \right\rangle \left\langle \frac{1}{E_G} \right\rangle^{-1}, \tag{4.107}$$

$$c_{33} = \left\langle \frac{1}{E_G} \right\rangle^{-1},$$

$$c_{55} = c^r_{55}.$$

On the other hand, White et al. (1975) obtained the complex and frequency-dependent stiffness associated with the symmetry axis. This is given by

$$\bar{p}_{33} = \left[\frac{1}{c_{33}} + \frac{2(r_2 - r_1)^2}{\mathrm{i}\omega(d_1 + d_2)(I_1 + I_2)} \right]^{-1}. \tag{4.108}$$

where

$$r_l = \frac{\alpha_l M_l}{E_{Gl}}, \quad l = 1,2 \tag{4.109}$$

and

$$I_l = \frac{\eta_l}{\kappa_l a_l} \coth\left(\frac{a_l d_l}{2} \right), \qquad a_l = \sqrt{\frac{\mathrm{i}\omega \eta_l E_{Gl}}{\kappa M_l E_{ml}}}, \tag{4.110}$$

for each single layer (Carcione and Picotti, 2006; Carcione, 2007; eq. (7.400)).

Let us assume for the moment that the properties of the frame are the same in media 1 and 2. At enough low frequencies, the fluid pressure is uniform (isostress state) and the effective modulus of the pore fluid is given by Reuss's or Wood's equation (Reuss, 1929; Wood, 1955):

$$\frac{1}{K_f} = \frac{\mathrm{p}_1}{K_{f_1}} + \frac{\mathrm{p}_2}{K_{f_2}}. \tag{4.111}$$

It can be shown (e.g., Johnson, 2001) that $\bar{p}_{33}(\omega = 0)$ is equal to Gassmann's modulus K_G for a fluid whose modulus is K_f. On the other hand, at high frequencies, the pressure is not uniform but can be assumed to be constant within each phase. In such a situation, Hill's theorem (Hill, 1964) gives the high-frequency limit $\bar{p}_{33}(\omega = \infty) = c_{33}$.

The stress–strain relations of the equivalent VTIv medium (closed system) are

$$\begin{aligned}
\sigma_{11} &= \bar{p}_{11}\varepsilon_{11} + \bar{p}_{12}\varepsilon_{22} + \bar{p}_{13}\varepsilon_{33},\\
\sigma_{22} &= \bar{p}_{12}\varepsilon_{11} + \bar{p}_{11}\varepsilon_{22} + \bar{p}_{13}\,\varepsilon_{33},\\
\sigma_{33} &= \bar{p}_{13}\varepsilon_{11} + \bar{p}_{13}\varepsilon_{22} + \bar{p}_{33}\,\varepsilon_{33},\\
\sigma_{23} &= 2\bar{p}_{55}\varepsilon_{23},\\
\sigma_{13} &= 2\bar{p}_{55}\varepsilon_{13},\\
\sigma_{12} &= 2\bar{p}_{66}\varepsilon_{12},
\end{aligned} \tag{4.112}$$

where here σ_{ij} are the stress components of the equivalent medium. This viscoelastic form arises from the fact that we consider a closed (undrained) system, which has a variation of fluid content $\zeta = -\mathrm{div}\,\mathbf{w}$ equal to zero, since the fluid cannot flow in and out of the volume.

Krzikalla and Müller (2011) combined the two previous models to obtain the five complex and frequency-dependent stiffnesses of the equivalent medium. They consider a 1D character of the fluid pressure equilibration process between the poroelastic layers, assuming that the fluid-flow direction is perpendicular to the layering plane. As a consequence, the model considers one relaxation function, namely, ψ_{33}, such that $\bar{p}_{33} = \boldsymbol{F}(\partial_t \psi_{33})$, corresponding to the symmetry-axis P-wave stiffness. Therefore, knowing this relaxation function and the high and low-frequency elastic limits of the stiffness tensor, Krzikalla and Müller (2011) obtained the stiffnesses components as

$$\bar{p}_{IJ}(\omega) = c_{IJ} + \left(\frac{c_{IJ} - c_{IJ}^r}{c_{33} - c_{33}^r}\right)[\bar{p}_{33}(\omega) - c_{33}]. \tag{4.113}$$

This equation has been verified by performing harmonic finite-element experiments by Krzikalla and Müller (2011) and Carcione et al. (2011) (see Section 4.5).

The theory is valid for plane layers, since a single relaxation function cannot be used in the case of 2-D or 3-D heterogeneities. The fact that the relaxed and unrelaxed shear moduli coincide implies that there is no shear loss along the directions perpendicular and parallel to layering. The qSV wave is dispersive due to its coupling with the qP wave, but the horizontally polarized SH wave is not dispersive, since $c_{55} = c_{55}^r$ and $c_{66} = c_{66}^r$ imply $\bar{p}_{55} = c_{55}$ and $\bar{p}_{66} = c_{66}$. Moreover, it can be shown that an alternating sequence of thin layers saturated with different fluids but having the same shear modulus does not generate anisotropy.

The transition frequency separating the relaxed and unrelaxed states (i.e., the approximate location of the relaxation peak) is

$$f_0 = \frac{8\kappa M E_m}{\pi \eta d^2 E_G} \tag{4.114}$$

(Carcione, 2007; eq. (7.450)), where M, E_G, η, and d refer to each single layer. At this reference frequency, the Biot slow-wave attenuation length equals the mean layer thickness or characteristic length of the inhomogeneities. Equation (4.114) indicates that the loss mechanism moves toward the low frequencies with increasing viscosity and decreasing permeability, that is, the opposite behavior of the Biot relaxation mechanism. The loss mechanism is due to the presence of the Biot slow wave and the diffusivity constant is $D = \kappa M E_m / (\eta E_G)$ (Carcione, 2007). The critical fluid-diffusion relaxation length is $\mathcal{L} = \sqrt{D/\omega}$. The fluid pressures will be equilibrated if $\mathcal{L}$ is comparable to the period of the stratification. For smaller diffusion lengths (e.g., higher frequencies), the pressures will not be equilibrated, causing attenuation and velocity dispersion. Note that the reference frequency (4.114) is obtained for a diffusion length $\mathcal{L} = d_1/4$ in the first layer.

We consider the North-Sea Utsira formation at the Sleipner field, which contains a highly permeable sandstone and mudstone layers, whose properties are given in Table 4.2 (Carcione et al., 2006). The example considers

TABLE 4.2 Properties of the Utsira Sand

Property	Sandstone	Mudstone
Grain bulk modulus, K_s (GPa)	40	20
density, ρ_s (kg/m^3)	2600	2600
Frame bulk modulus, K_m (GPa)	1.37	7
shear modulus, μ_m (GPa)	0.82	6
porosity, ϕ	0.36	0.2
permeability, κ (D)	1.6	0.01
Brine density, ρ_w (kg/m^3)	1030	1040
viscosity, η_w (Pa s)	0.0012	0.001
bulk modulus, K_w (GPa)	2.6	2.6

alternating layers of mudstone and sandstone saturated with brine, with a thickness of 10 cm each.

The phase velocity and dissipation factor as a function of frequency along the direction perpendicular to the layering plane are shown in Figure 4.9. It can be seen a relaxation peak centered at the exploration seismic band, with a minimum quality factor of approximately 30. This shows that realistic values of

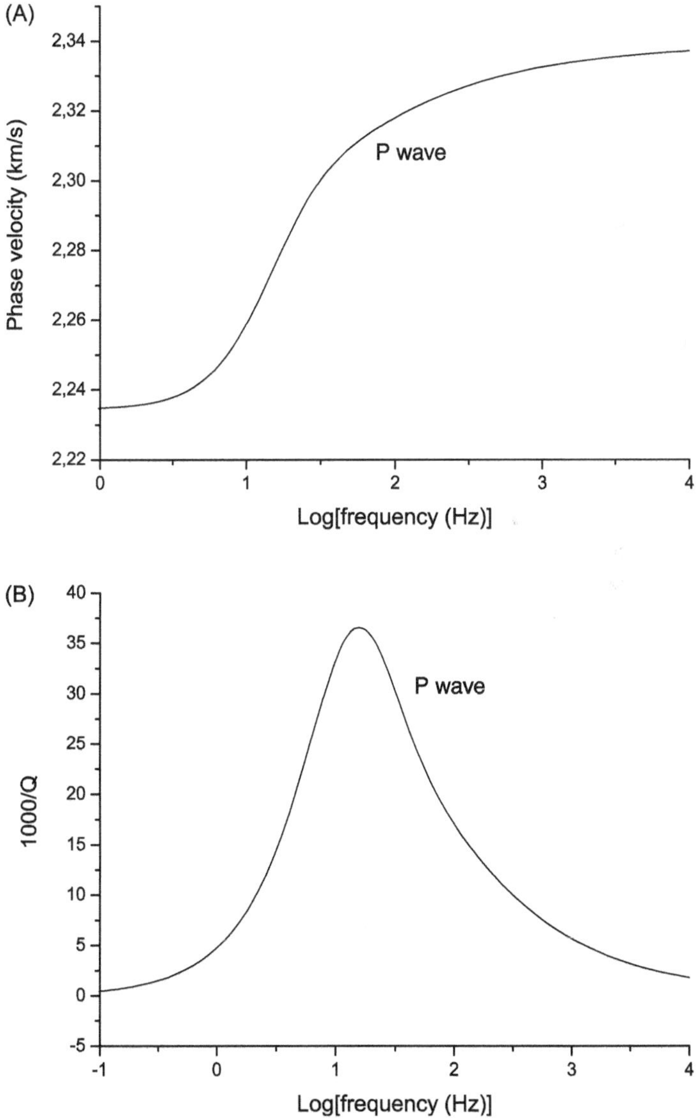

FIGURE 4.9 P-wave phase velocity (A) and dissipation factor (B) as a function of frequency in the direction perpendicular to the layering (bedding) plane. The medium consists of a sequence of mudstone and water-saturated sandstone layers of equal composition.

attenuation can be described with a physical – poroelasticity – theory without invoking phenomenological viscoelastic models, such as those based on springs and dashpots. The attenuation is higher along the direction perpendicular to the layering plane and the medium is practically lossless along this plane. Higher attenuation is associated with higher velocity dispersion.

Figure 4.10 shows the energy velocity and dissipation factor as a function of the propagation (phase) angle at a frequency of 25 Hz. The three wave modes are indicated. The velocity anisotropy of the shear modes is substantial, with a noticeable shear-wave splitting. As can be seen, the coupling between the qP

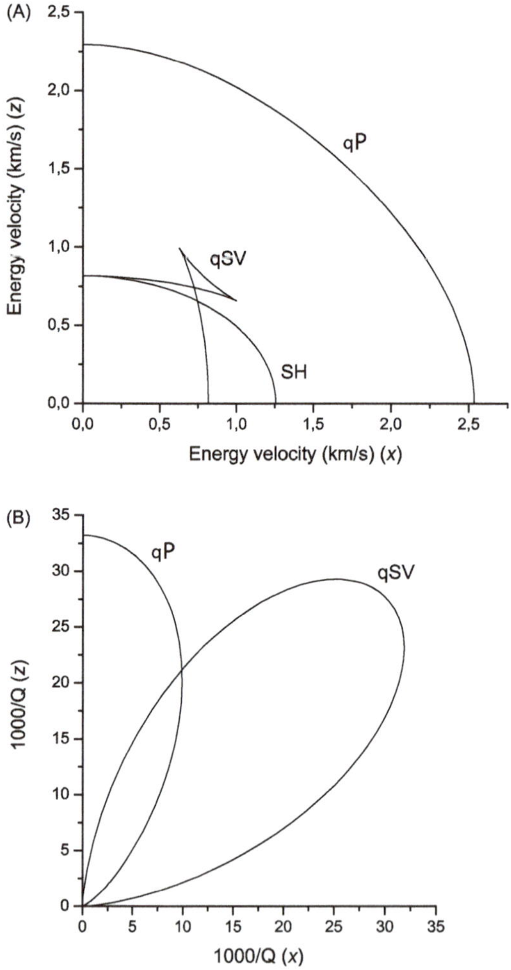

FIGURE 4.10 Polar representation of the energy velocity (A) and dissipation factor (B) corresponding to a frequency of 25 Hz. The medium consists of a sequence of mudstone and water-saturated sandstone layers.

and qSV waves generates strong shear attenuation at 45°. It can be shown that this coupling effect is mainly influenced by the contrast in dry-rock rigidity between the mudstone and the sandstone.

The mesoscopic-loss mechanism combined with Backus averaging can be implemented in the wave equation to compute synthetic seismograms of porous thinly layered formations. The seismograms can be computed with a finite-element method in the space-frequency domain. In this way, the frequency-dependent properties of the medium are modeled exactly, without the need of approximations with viscoelastic mechanical models (Gauzellino et al., 2014).

4.3.4 Gassmann Model

An equivalent medium theory to describe layering at a fine (pore) scale is based on Gassmann equation for an anisotropic frame and a solid pore infill. This theory has been developed by Ciz and Shapiro (2007), who consider lossless media and obtain the undrained compliance tensor when the pore infill and grains are anisotropic materials. Their equation reads

$$\bar{s}_{ijkl} = s^m_{ijkl} - (s^m_{ijmn} - s^s_{ijmn})[\phi(\mathbf{s}^{if} - \mathbf{s}^{\phi}) + \mathbf{s}^m - \mathbf{s}^s]^{-1}_{mnqp}(s^m_{qpkl} - s^s_{qpkl}), \quad (4.115)$$

where the s's are the components of the compliance tensor, the subscripts m, s, and if denote dry matrix (skeleton), solid grain, and pore infill, respectively, and the Einstein summation is assumed over 1, 2, and 3. The compliance tensor $\mathbf{s}^{\phi}$ is explicitly defined in Ciz and Shapiro (2007). In the case that the skeleton is made of a homogeneous material, $\mathbf{s}^{\phi} = \mathbf{s}^s$. Here, we consider that the grains and the frame are transversely isotropic. In this case, we use the following relations between the Voigt stiffnesses and compliances: $c_{11} + c_{12} = s_{33}/s$, $c_{11} - c_{12} = 1/(s_{11} - s_{12})$, $c_{13} = -s_{13}/s$, $c_{33} = (s_{11} + s_{12})/s$, $c_{55} = 1/s_{55}$, where $s = s_{33}(s_{11} + s_{12}) - 2s^2_{13}$. The equations for the inversion are obtained by interchanging all c's and s's. Note the following relations: $s_{66} = 4s_{1212}$ and $s_{55} = 4s_{1313}$, valid for all the compliance tensors, while $c_{66} = c_{1212}$ and $c_{55} = c_{1313}$. The components of the corresponding matrices in the undrained case transform in the same way. Moreover, the usual symmetry relations by interchanging the indices hold (e.g., Carcione, 2007).

In the limit of high porosities, say beyond 50%, the dry-rock elastic constants s^m_{ijkl} are zero in practice, and equation (4.115) becomes

$$\bar{s}_{ijkl} = (1 - \phi)s^s_{ijkl} + \phi s^{if}_{ijkl}, \quad (4.116)$$

that is, a generalization of the Reuss (or Wood) average.

Equation (4.115) can be inverted to obtain the dry-rock compliance tensor as a function of the undrained compliance tensor. We have

$$s^m_{ijkl} = s^s_{ijkl} + \phi(\bar{s}_{ijmn} - s^s_{ijmn})[\phi(\mathbf{s}^{if} - \mathbf{s}^{\phi}) - \bar{\mathbf{s}} + \mathbf{s}^s]^{-1}_{mnqp}(s^{if}_{qpkl} - s^{\phi}_{qpkl}). \quad (4.117)$$

This equation can be used to obtain the dry-rock compliance tensor by using calibration data (seismic, well, or laboratory data). Note that the Backus $\bar{p}_{55}$ and $\bar{p}_{66}$ in (4.68) are Reuss and Voigt averages, respectively (consider lossless media to compare). Hence, according to equation (4.116), the first stiffness is obtained from Gassmann's relations by assuming zero dry-rock stiffnesses. This means that an inversion of Backus generated stiffnesses by using equation (4.117) gives $c_{55}^m = 0$, because Backus's equation for $\bar{c}_{55}$ is a Reuss average as equation (4.116), that is satisfied when the dry-rock stiffness is zero. Equation (4.115) can easily be generalized to the anelastic case by invoking the correspondence principle.

A particular case of equation (4.115) is the stress–strain relation obtained by Brown and Korringa (1975), when the pore-infill is a fluid

$$\bar{s}_{ijkl} = s_{ijkl}^m - \frac{(s_{ijnn}^m - s_{ijnn}^s)(s_{klnn}^m - s_{klnn}^s)}{(s_{mmnn}^m - C_s) + \phi(C_f - C_s)}, \tag{4.118}$$

where C_f is the compressibility of the fluid and $C_s = s_{mmnn}^s$ (Mavko et al., 2009).

Gassmann equation requires the knowledge of the dry-rock elastic constants. Krief et al. (1990) proposed a simple heuristic equation

$$K_m = K_s(1-\phi)^{A/(1-\phi)} \quad \text{and} \quad \mu_m = K_m \mu_s / K_s, \tag{4.119}$$

where A is a constant that depends on the type of rock (the second expression in eq. (4.119) is assumed here). The porosity dependence is consistent with the concept of critical porosity, since the moduli should be small above a certain value of the porosity (usually between 0.4 and 0.6). The properties of the dry matrix can be described by an anisotropic version of the Krief model:

$$\begin{aligned}
c_{11}^m &= c_{11}^s(1-\phi)^{A/(1-\phi)},\\
c_{66}^m &= c_{66}^s(1-\phi)^{A/(1-\phi)},\\
c_{13}^m &= c_{13}^s(1-\phi)^{A/(1-\phi)},\\
c_{33}^m &= c_{33}^s(1-\phi)^{B/(1-\phi)},\\
c_{55}^m &= c_{55}^s(1-\phi)^{B/(1-\phi)},
\end{aligned} \tag{4.120}$$

where A and B are constants. The use of two constants is somehow equivalent to vary the Krief exponent as a function of the propagation (phase) angle, since c_{11}^m and c_{66}^m describe the velocities along the stratification, and c_{33}^m and c_{55}^m along the perpendicular direction. Equation (4.120) reduces to equation (4.119) for $A = B$ in the isotropic case. Another possibility is to obtain the dry-rock elastic

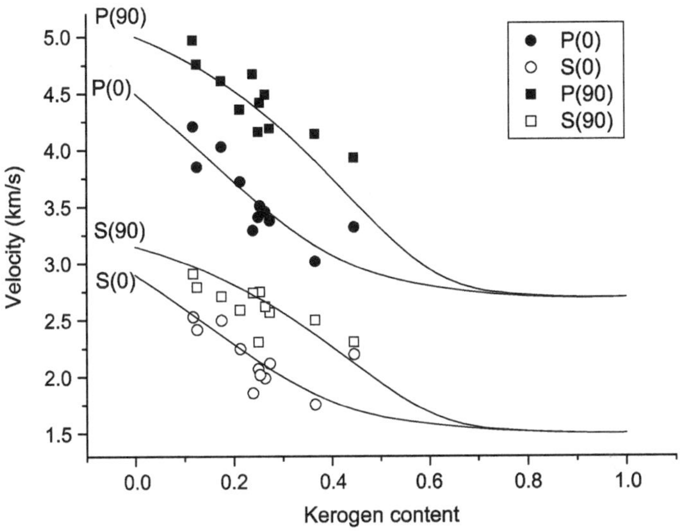

FIGURE 4.11 Gassmann velocities as a function of kerogen content for Bakken shales. The dry-rock moduli have been obtained by the generalization of Krief equations to the anisotropic case. The symbols correspond to the experimental data reported by Vernik and Nur (1992).

constants from wet-rock data by using the inverse Gassmann relation (4.117) (Carcione et al., 2011).

4.3.4.1 *Source rocks*

We consider the Bakken shales (Vernik and Nur, 1992). The P and S-wave velocities parallel to the bedding plane (90) and perpendicular to the bedding plane (0) as a function of kerogen content are shown in Figure 4.11. We use the Krief/Gassmann equations, where $A = 1.8$ and $B = 8$.

More examples can be found in Carcione et al. (2011), where the results of Backus averaging are compared to those obtained with Gassmann equation.

4.4 FRACTURES

The behavior of wave propagation at fractures, cracks, and any partially welded interface can be defined through appropriate boundary conditions. We first define these boundary conditions for a single fracture and analyze its reflection and transmission coefficients and then obtain the wave characteristics of a dense set of fractures in solid and porous media.

4.4.1 Scattering at a Single Fracture

In this section, we define the boundary condition and obtain the reflection and transmission coefficients of a single fracture in a solid medium by considering

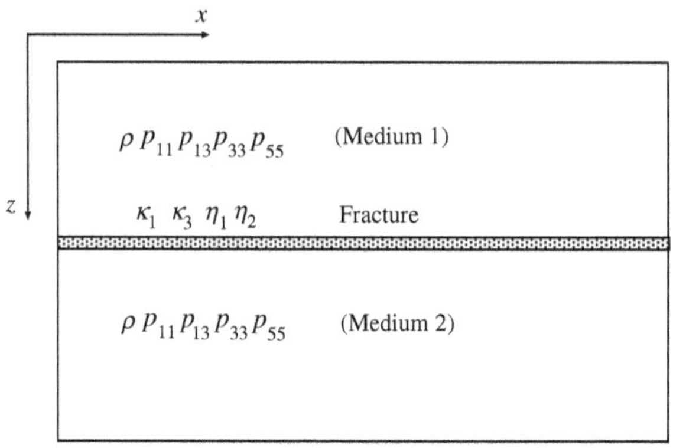

FIGURE 4.12 Planar fracture embedded in transversely isotropic media of dissimilar density and stiffness constants. κ_1 and κ_3 are specific stiffnesses and η_1 and η_3 are specific viscosities of the fracture.

displacement and particle-velocity discontinuities at the interface. The fracture is embedded in a VTIv medium. The imperfect bonding is described by four parameters, namely, the normal and tangential specific stiffnesses and corresponding viscosities (Carcione and Picotti, 2012).

4.4.1.1 Boundary Conditions and Stress–Strain Relations

A planar fracture separates two elastic VTIv media and the nonideal characteristics of the interface are modeled through the boundary conditions, that is, continuity of stress and discontinuity of the displacement and particle-velocity fields across the interface. We consider the (x,z)-plane, and refer to the upper and lower media with the labels 1 and 2, respectively, with z increasing toward the lower medium (see Figure 4.12).

The boundary conditions for a wave impinging on the fracture ($z = 0$) are

$$
\begin{aligned}
\kappa_1[u_1] + \eta_1[v_1] &= \sigma_{13}, \\
\kappa_3[u_3] + \eta_3[v_3] &= \sigma_{33}, \\
[\sigma_{13}] &= 0, \\
[\sigma_{33}] &= 0
\end{aligned}
\qquad (4.121)
$$

(Schoenberg, 1980; Schoenberg and Douma, 1983; Carcione, 1998; Carcione, 2007, Section 3.11.1), where u_1 and u_3 are the displacement components, v_1 and v_3 are the particle-velocity components, σ_{13} and σ_{33} are the stress components, κ_1 and κ_3 are specific stiffnesses, and η_1 and η_3 are specific viscosities. They have dimensions of stiffness and viscosity per unit length, respectively.

Moreover, the brackets denote discontinuities across the interface, such that for a field variable a, it is $[a] = a_2 - a_1$.

The model simulates the fracture by a zero width layer of distributed spring-dashpots. It can be shown that relaxation-like functions of Maxwell or Kelvin–Voigt type govern the tangential and normal coupling properties of the crack. The interface exhibits time dependent mechanical properties through the relaxation functions, and, as in a viscoelastic material, this implies energy dissipation. A displacement discontinuity ($\kappa_3 \neq 0$) yields a change of phase, while a discontinuity in the particle velocity ($\eta_3 \neq 0$) implies an energy loss at the interface (Carcione, 1996, 1998, 2007); $\kappa_i = 0$, $i = 1(x), 3(z)$ gives the particle-velocity discontinuity model and $\eta_i = 0$ gives the displacement discontinuity model. On the other hand, if $\kappa_i \to \infty$ or $\eta_i \to \infty$, the model gives the welded interface.

The characteristics of the medium are completed with the constitutive relations

$$
\begin{aligned}
\sigma_{11} &= p_{11}\,\partial_1 u_1 + p_{13}\,\partial_3 u_3, \\
\sigma_{33} &= p_{13}\,\partial_1 u_1 + p_{33}\,\partial_3 u_3, \\
\sigma_{13} &= p_{55}(\partial_1 u_3 + \partial_3 u_1)
\end{aligned}
\tag{4.122}
$$

(Carcione, 2007, Chapter 1, eq. 1.43), where p_{IJ} are the stiffness components (complex and frequency dependent) of the background medium.

4.4.1.2 Propagation Characteristics

A general plane-wave solution for the displacement field $\mathbf{u} = (u_1, u_3)$ is

$$
\mathbf{u} = \mathbf{u}_0 \exp[\mathrm{i}\omega(t - s_1 x - s_3 z)], \tag{4.123}
$$

where s_1 and s_3 are the components of the slowness vector and $\mathbf{u}_0$ is a constant complex vector.

On the basis of the kernel (4.123), the dispersion equation of a TIV medium can be obtained from equation $(4.36)_2$ by setting $p_{46} = p_{35} = p_{15} = 0$ and considering that $p_{44} = p_{55}$,

$$
(p_{11}s_1^2 + p_{55}s_3^2 - \rho)(p_{33}s_3^2 + p_{55}s_1^2 - \rho) - (p_{13} + p_{55})^2 s_1^2 s_3^2 = 0 \tag{4.124}
$$

(Carcione, 2007, Chapter 6, eq. 6.113). Equation (4.124) has two solutions corresponding to the quasi-compressional (qP) and quasi-shear (qS) waves.

To distinguish between down and up propagating waves, the slowness relation equation (4.124) is solved for s_3, given the horizontal slowness s_1 (Carcione, 1997). This yields

$$
s_3 = \pm\frac{1}{\sqrt{2}}\sqrt{K_1 \mp \mathrm{pv}\sqrt{K_1^2 - 4K_2K_3}}, \tag{4.125}
$$

where

$$K_1 = \rho\left(\frac{1}{p_{55}} + \frac{1}{p_{33}}\right) + \frac{1}{p_{55}}\left[\frac{p_{13}}{p_{33}}(p_{13} + 2p_{55}) - p_{11}\right]s_1^2,$$

$$K_2 = \frac{1}{p_{33}}(p_{11}s_1^2 - \rho), \quad K_3 = s_1^2 - \frac{\rho}{p_{55}},$$

and "pv" denotes the principal value. The signs in s_3 correspond to

(+,−) downward propagating qP wave
(+,+) downward propagating qS wave
(−,−) upward propagating qP wave
(−,+) upward propagating qS wave.

The plane-wave eigenvectors (polarizations) belonging to a particular eigenvalue can be obtained from the qP-qS Kelvin-Christoffel equation (Carcione, 2007, Chapter 6, eq. 6.115) as

$$\mathbf{u}_0 = \bar{u}_0\begin{pmatrix}\beta \\ \xi\end{pmatrix}, \tag{4.126}$$

where $\bar{u}_0$ is the plane-wave amplitude and

$$\begin{aligned}\beta &= \text{pv}\sqrt{\frac{p_{55}s_1^2 + p_{33}s_3^2 - \rho}{p_{11}s_1^2 + p_{33}s_3^2 + p_{55}(s_1^2 + s_3^2) - 2\rho}},\\ \xi &= \pm\text{pv}\sqrt{\frac{p_{11}s_1^2 + p_{55}s_3^2 - \rho}{p_{11}s_1^2 + p_{33}s_3^2 + p_{55}(s_1^2 + s_3^2) - 2\rho}},\end{aligned} \tag{4.127}$$

where the + and − signs correspond to the qP and qS waves, respectively. From equations (4.123) and (4.127), the particle-velocity field can be written as

$$\mathbf{v} = \mathrm{i}\omega\mathbf{u} = \mathrm{i}\omega\bar{u}_0\begin{pmatrix}\beta \\ \xi\end{pmatrix}\exp[\mathrm{i}\omega(t - s_1x - s_3z)]. \tag{4.128}$$

Substituting the plane wave (4.128) into the stress–strain relation (4.122) yields

$$\begin{aligned}\sigma_{11} &= -\mathrm{i}\omega\bar{u}_0(\beta p_{11}s_1 + \xi p_{13}s_3) \equiv X,\\ \sigma_{33} &= -\mathrm{i}\omega\bar{u}_0(\beta p_{13}s_1 + \xi p_{33}s_3) \equiv Z,\\ \sigma_{13} &= -\mathrm{i}\omega\bar{u}_0 p_{55}(\xi s_1 + \beta s_3) \equiv W.\end{aligned} \tag{4.129}$$

The slowness components are given by

$$s_1 = \frac{\sin\theta}{v(\theta)}, \quad s_3 = \frac{\cos\theta}{v(\theta)} \tag{4.130}$$

where θ is the phase propagation angle, measured with respect to the z-axis, and

$$v = \frac{1}{s} = \frac{1}{\sqrt{s_1^2 + s_3^2}} \tag{4.131}$$

is the complex velocity that can be obtained from the slowness relation (4.124). Hence, we have

$$\begin{aligned} \rho v^2 &= \frac{1}{2}(p_{55} + p_{11}\sin^2\theta + p_{33}\cos^2\theta \pm A), \\ A &= \sqrt{[(p_{11} - p_{55})\sin^2\theta + (p_{55} - p_{33})\cos^2\theta]^2 + (p_{13} + p_{55})^2 \sin^2 2\theta}, \end{aligned} \tag{4.132}$$

where the + sign corresponds to the qP wave, and the − sign to the qS wave. Equation (4.132) is equivalent to equations $(4.55)_2$ and $(4.55)_3$.

4.4.1.3 Reflection and Transmission Coefficients

In the following, the symbols P and S indicate the qP and qS waves, respectively. Moreover, the subscripts I, R, and T denote the incident, reflected, and transmitted waves, respectively. Using symmetry properties to define the polarization of the reflected waves, the displacements for a qP wave incident from above the interface are given by

$$\mathbf{u}_1 = \mathbf{u}_{\mathrm{P}_I} + \mathbf{u}_{\mathrm{P}_R} + \mathbf{u}_{\mathrm{S}_R}, \tag{4.133}$$

$$\mathbf{u}_2 = \mathbf{u}_{\mathrm{P}_T} + \mathbf{u}_{\mathrm{S}_T}, \tag{4.134}$$

where

$$\begin{aligned} \mathbf{u}_{\mathrm{P}_I} &= (\beta_{\mathrm{P}_1}, \xi_{\mathrm{P}_1})\exp[\mathrm{i}\omega(t - s_1 x - s_{3\mathrm{P}_1} z)], \\ \mathbf{u}_{\mathrm{P}_R} &= R_{\mathrm{PP}}(\beta_{\mathrm{P}_1}, -\xi_{\mathrm{P}_1})\exp[\mathrm{i}\omega(t - s_1 x + s_{3\mathrm{P}_1} z)], \\ \mathbf{u}_{\mathrm{S}_R} &= R_{\mathrm{PS}}(\beta_{\mathrm{S}_1}, -\xi_{\mathrm{S}_1})\exp[\mathrm{i}\omega(t - s_1 x + s_{3\mathrm{S}_1} z)], \\ \mathbf{u}_{\mathrm{P}_T} &= T_{\mathrm{PP}}(\beta_{\mathrm{P}_2}, \xi_{\mathrm{P}_2})\exp[\mathrm{i}\omega(t - s_1 x - s_{3\mathrm{P}_2} z)], \\ \mathbf{u}_{\mathrm{S}_T} &= T_{\mathrm{PS}}(\beta_{\mathrm{S}_2}, \xi_{\mathrm{S}_2})\exp[\mathrm{i}\omega(t - s_1 x - s_{3\mathrm{S}_2} z)]. \end{aligned} \tag{4.135}$$

Application of Snell's law (see Section 4.2.3) implies the continuity of the horizontal slowness s_1. The boundary conditions do not influence the emergence angles of the transmitted and reflected waves. The vertical slownesses s_{3P} and s_{3S}, as well as β_P, β_S, ξ_P, and ξ_S, follow, respectively the $(+,-)$ and $(+,+)$ sign sets given in equation (4.125). The choice $\bar{u}_0 = 1$ implies no loss of generality.

The boundary conditions (4.121) are

$$[u_1] = c_1\sigma_{13}, \; [u_3] = c_3\sigma_{33}, \; [\sigma_{13}] = 0, \; [\sigma_{33}] = 0, \tag{4.136}$$

where

$$c_i = \frac{1}{\kappa_i + i\omega\eta_i}, \quad i = 1(x), 3(z) \tag{4.137}$$

is a complex compliance per unit length characterizing the fracture.

Using the equations for the displacements and stresses, the boundary conditions generate the following matrix equation for the reflection and transmission coefficients:

$$\begin{pmatrix} \beta_{P_1} - c_1 W_{P_1} & \beta_{S_1} - c_1 W_{S_1} & -\beta_{P_2} & -\beta_{S_2} \\ \xi_{P_1} - c_3 Z_{P_1} & \xi_{S_1} - c_3 Z_{S_1} & \xi_{P_2} & \xi_{S_2} \\ Z_{P_1} & Z_{S_1} & -Z_{P_2} & -Z_{S_2} \\ W_{P_1} & W_{S_1} & W_{P_2} & W_{S_2} \end{pmatrix} \cdot \begin{pmatrix} R_{PP} \\ R_{PS} \\ T_{PP} \\ T_{PS} \end{pmatrix} = \begin{pmatrix} -\beta_{P_1} - c_1 W_{P_1} \\ \xi_{P_1} + c_3 Z_{P_1} \\ -Z_{P_1} \\ W_{P_1} \end{pmatrix}, \tag{4.138}$$

where W and Z are given by equation (4.129).

The steps to compute the reflection and transmission coefficients are the following:

1. The horizontal slowness s_1 is the independent parameter. It is the same for all the waves (Snell's law). For an incident wave, the independent variable becomes the incidence angle θ, and s_1 is obtained from equation (4.130).
2. Compute s_{3P_1}, s_{3S_1}, s_{3P_2}, and s_{3S_2} from equation (4.125), where the first sign is positive. For an incident wave, s_{3P_1} can be calculated either from equation (4.125) or from equation (4.130).
3. Compute β_{P_1}, β_{S_1}, β_{P_2}, β_{S_2}, ξ_{P_1}, ξ_{S_1}, ξ_{P_2}, and ξ_{S_2} from equation (4.127).
4. Compute W_{P_1}, W_{S_1}, W_{P_2}, and W_{S_2} and Z_{P_1}, Z_{S_1}, Z_{P_2}, and Z_{S_2} from equation (4.129).
5. Compute the reflection and transmission coefficients by numerically solving equation (4.138).

Equation (4.138) gives the results of Carcione (1996) in the isotropic lossless case, (when $c_{33} = c_{11}$ and $c_{13} = c_{11} - 2c_{55}$) and those of Carcione (1997) in the case of a welded interface ($c_1 = c_3 = 0$). Carcione (1996) obtained the potential amplitude coefficients, which are related to the displacement amplitude coefficients by a conversion factor. In the case of isotropy and similar upper and lower media, the conversion factor from one type of coefficient to the other is 1 for PP coefficients and c_S/c_P for PS coefficients, where c_P and c_S denote the P-wave and S-wave velocities (Aki and Richards, 1980, p. 139). When the background medium is losses and $c_1 = c_3 = 0$, equation (4.138) yields the reflection and transmission coefficients of a welded interface as in Wright (1987).

At normal incidence and similar upper and lower media, we obtain

$$R_{PP} = \left(\frac{i\omega_P}{\omega} - \frac{2\eta_3}{I_P} - 1 \right)^{-1} = \left(\frac{2i}{\omega I_P c_3} - 1 \right)^{-1} \tag{4.139}$$

and

$$T_{PP} = 1 + R_{PP} \tag{4.140}$$

(Carcione, 1998), where $I_P = \sqrt{\rho p_{33}}$ and $\omega_P = 2\kappa_3 / I_P$ is a complex frequency. For weakly attenuating media and lossless media, $p_{33} \simeq c_{33}$, ω_P defines the transition frequency from an apparently perfect interface to the apparently decoupled one. If $\kappa_3 = 0$, it is $\omega_P = 0$ and the particle-velocity discontinuity model is obtained. On the other hand, when $\eta_3 = 0$, the theory gives the displacement discontinuity model. A discontinuity in the particle velocity implies energy dissipation at the interface (Carcione, 1996, 1998). Moreover, if $\eta_3 \to 0$ and $\kappa_3 \to 0$, $R_{PP} \to -1$ and $T_{PP} \to 0$, and the free surface condition is obtained, when $\kappa_3 \to \infty$ or $\eta_3 \to \infty$, $R_{PP} \to 0$ and $T_{PP} \to 1$, giving the solution for a welded contact.

The coefficients are shown here as a function of the incidence ray angle ψ, which defines the direction of the energy-flux vector of the incidence wave. The ray angle can be obtained as

$$\tan\psi = \frac{\beta_{P_1} X_{P_1} + \xi_{P_1} W_{P_1}}{\beta_{P_1} W_{P_1} + \xi_{P_1} Z_{P_1}} \tag{4.141}$$

(Carcione, 2007).

4.4.1.4 *Energy Balance and Loss*

We assume the background medium to be lossless to analyze the attenuation due to the fracture alone. In a completely welded interface, the normal component of the time-averaged energy flux is continuous across the plane separating the two media. This is a consequence of the boundary conditions that impose continuity of normal stress and particle velocity. If the viscosities are nonzero, there is energy dissipation at the interface as shown in the following. Since the media are elastic, the interference fluxes between different waves vanish and only the fluxes corresponding to each single plane wave need be considered (Carcione, 2007, Section 6.2.6). Denoting by F the vertical component of the energy flux, we have

$$
\begin{aligned}
-2F_{\mathrm{P}_I} &= \mathrm{Re}(\sigma_{13\mathrm{P}_I} v_{1\mathrm{P}_I}^* + \sigma_{33\mathrm{P}_I} v_{3\mathrm{P}_I}^*),\\
-2F_{\mathrm{P}_R} &= \mathrm{Re}(\sigma_{13\mathrm{P}_R} v_{1\mathrm{P}_R}^* + \sigma_{33\mathrm{P}_R} v_{3\mathrm{P}_R}^*),\\
-2F_{\mathrm{S}_R} &= \mathrm{Re}(\sigma_{13\mathrm{S}_R} v_{1\mathrm{S}_R}^* + \sigma_{33\mathrm{S}_R} v_{3\mathrm{S}_R}^*),\\
-2F_{\mathrm{P}_T} &= \mathrm{Re}(\sigma_{13\mathrm{P}_T} v_{1\mathrm{P}_T}^* + \sigma_{33\mathrm{P}_T} v_{3\mathrm{P}_T}^*),\\
-2F_{\mathrm{S}_T} &= \mathrm{Re}(\sigma_{13\mathrm{S}_T} v_{1\mathrm{S}_T}^* + \sigma_{33\mathrm{S}_T} v_{3\mathrm{S}_T}^*)
\end{aligned}
\tag{4.142}
$$

(Carcione, 2007). As shown by Carcione (1997, 2007), further algebra implies that the fluxes given in the preceding equations are proportional to the real parts of

$$
\begin{aligned}
F_{\mathrm{P}_I} &\propto \beta_{\mathrm{P}_1}^* W_{\mathrm{P}_1} + \xi_{\mathrm{P}_1}^* Z_{\mathrm{P}_1},\\
F_{\mathrm{P}_R} &\propto -(\beta_{\mathrm{P}_1}^* W_{\mathrm{P}_1} + \xi_{\mathrm{P}_1}^* Z_{\mathrm{P}_1}) | R_{\mathrm{PP}} |^2,\\
F_{\mathrm{S}_R} &\propto -(\beta_{\mathrm{S}_1}^* W_{\mathrm{S}_1} + \xi_{\mathrm{S}_1}^* Z_{\mathrm{S}_1}) | R_{\mathrm{PS}} |^2,\\
F_{\mathrm{P}_T} &\propto (\beta_{\mathrm{P}_2}^* W_{\mathrm{P}_2} + \xi_{\mathrm{P}_2}^* Z_{\mathrm{P}_2}) | T_{\mathrm{PP}} |^2,\\
F_{\mathrm{S}_T} &\propto (\beta_{\mathrm{S}_2}^* W_{\mathrm{S}_2} + \xi_{\mathrm{S}_2}^* Z_{\mathrm{S}_2}) | T_{\mathrm{PS}} |^2,
\end{aligned}
\tag{4.143}
$$

where the proportionality factor is $\frac{\mathrm{i}}{2}\omega$. We define the energy reflection and transmission coefficients as

$$
ER_{\mathrm{PP}} = \sqrt{\frac{F_{\mathrm{P}_R}}{F_{\mathrm{P}_I}}},\quad ER_{\mathrm{PS}} = \sqrt{\frac{F_{\mathrm{S}_R}}{F_{\mathrm{P}_I}}},\quad ET_{\mathrm{PP}} = \sqrt{\frac{F_{\mathrm{P}_T}}{F_{\mathrm{P}_I}}},\quad ET_{\mathrm{PS}} = \sqrt{\frac{F_{\mathrm{S}_T}}{F_{\mathrm{P}_I}}}
\tag{4.144}
$$

The energy loss at the interface is obtained by subtracting the energies of the reflected and transmitted waves from the energy of the incident wave. The normalized dissipated energy is

$$
E_{\mathrm{loss}} = 1 - ER_{\mathrm{PP}}^2 - ER_{\mathrm{PS}}^2 - ET_{\mathrm{PP}}^2 - ET_{\mathrm{PS}}^2.
\tag{4.145}
$$

At normal incidence, we have $E_{\mathrm{loss}} = 1 - | R_{\mathrm{PP}} |^2 - | T_{\mathrm{PP}} |^2$. Substituting equations (4.139) and (4.140), the energy loss becomes

$$
E_{\mathrm{loss}} = \frac{4\eta_3 / I_{\mathrm{P}}}{(1 + 2\eta_3 / I_{\mathrm{P}})^2 + (\omega_{\mathrm{P}} / \omega)^2}.
\tag{4.146}
$$

The maximum loss is obtained for $\eta_3 = (I_{\mathrm{P}} / 2)\sqrt{1 + (\omega_{\mathrm{P}} / \omega)^2}$, where we assumed a lossess background (ω_{P} is a real quantity).

We consider a planar fracture in a lossless background medium whose properties are c_{11} = 38 GPa, c_{13} = 5 GPa, c_{33} = 12 GPa, c_{55} = 3 GPa, and

ρ = 2300 kg/m^3. The fracture has the following parameters: $\kappa_1 = \pi f_0 I_S$, $\kappa_3 = \pi f_0 I_P$, $\eta_1 = I_S / 50$ and $\eta_3 = I_P / 50$, where f_0 = 100 Hz is a reference frequency, $I_P = \sqrt{\rho c_{33}}$ and $I_S = \sqrt{\rho c_{55}}$. The fracture stiffness values per unit length are κ_1 = 0.5 GPa/m and κ_3 = 1.1 $\overline{1}$ GPa/m. Figure 4.13 shows the

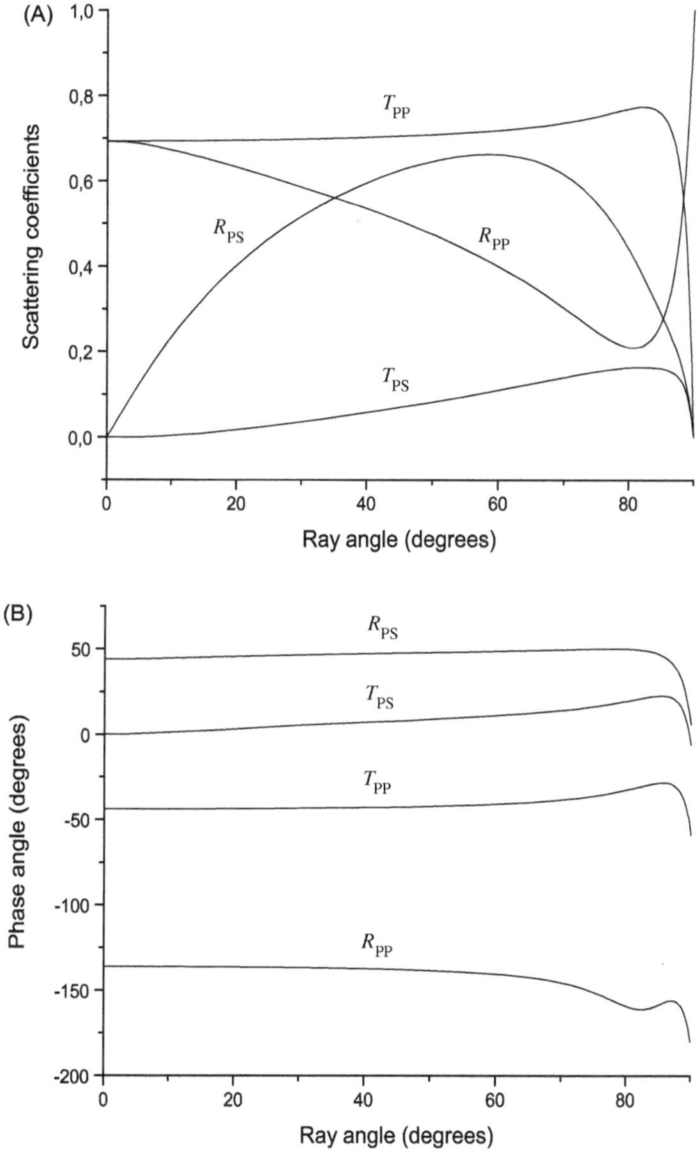

FIGURE 4.13 Absolute value (A) and phase (B) of the reflection and transmission coefficients as a function of the ray angle for an incident P wave and f_0 = 100 Hz.

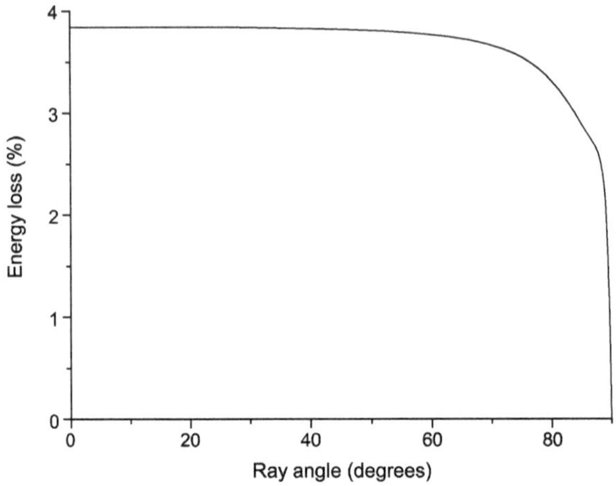

FIGURE 4.14 Normalized energy loss as a function of the ray angle for f_0 = 100 Hz.

absolute value (A) and phase (B) of the coefficients as a function of the ray angle for f_0 = 100 Hz. The energy loss is displayed in Figure 4.14, indicating that almost 4% of the energy is lost at the fracture at most angles of incidence.

4.4.2 Multiple Dense Fractures

In a dense set of planar fractures the average separation between fractures is much smaller than the seismic wavelength. As in the case of thin layers, the medium behaves as an equivalent TI medium. Energy loss is in this case considerable due to the dense nature of fracturing.

4.4.2.1 Solid Medium

As in the case of a single fracture, the model is based on the discontinuity of the displacement and particle-velocity fields across the interface. We consider a vector equation similar to equation (4.121)

$$\boldsymbol{\kappa} \cdot [\mathbf{u}] + \boldsymbol{\eta} \cdot [\mathbf{v}] = \boldsymbol{\sigma} \cdot \hat{\mathbf{n}}, \tag{4.147}$$

where $\mathbf{u}$ and $\mathbf{v}$ are the displacement and particle-velocity vectors, respectively, $\boldsymbol{\sigma}$ is the 3 × 3 stress tensor, $\hat{\mathbf{n}}$ is the unit normal to the fracture, $\boldsymbol{\kappa}$ is the specific stiffness matrix, and $\boldsymbol{\eta}$ is the specific viscosity matrix (both of dimension 3 × 3). The particle velocity is given by $\mathbf{v} = \partial_t \mathbf{u}$. In the Fourier domain ($\mathbf{v} = \mathrm{i}\omega\mathbf{u}$), equation (4.147) becomes

$$[\mathbf{u}] = \mathbf{Z} \cdot (\boldsymbol{\sigma} \cdot \hat{\mathbf{n}}), \qquad \mathbf{Z} = (\boldsymbol{\kappa} + \mathrm{i}\omega\,\boldsymbol{\eta})^{-1} \tag{4.148}$$

where **Z** is a fracture compliance matrix, whose dimension is length/stress. This approach is equivalent to the linear-slip model introduced by Schoenberg (1980). Equation (4.148), with $\eta = 0$, is given in Coates and Schoenberg (1995) and Schoenberg and Helbig (1997). Three models have been studied by Liu et al. (2000) to obtain the expression of **Z** for different fracture models. Two of the models only provide the real part of **Z**. The third model describes the fracture as a thin and soft viscoelastic layer embedded in an isotropic elastic background medium, where **Z** can be obtained in terms of the thickness, Lamé constants, and viscosity of the soft material infill. The resulting equivalent medium has VTI symmetry. Similarly, Carcione (1996) described the infill by a viscous fluid.

The compliance matrix **Z** of the set of fractures is the diagonal matrix

$$\mathbf{Z} = \begin{pmatrix} Z_1 & 0 & 0 \\ 0 & Z_2 & 0 \\ 0 & 0 & Z_3 \end{pmatrix} = \begin{pmatrix} (\kappa_1 + \mathrm{i}\omega\eta_1)^{-1} & 0 & 0 \\ 0 & (\kappa_2 + \mathrm{i}\omega\eta_2)^{-1} & 0 \\ 0 & 0 & (\kappa_3 + \mathrm{i}\omega\eta_3)^{-1} \end{pmatrix}, \tag{4.149}$$

where Z_1 is the normal compliance, Z_2 is the horizontal tangential compliance, and Z_3 is the vertical tangential compliance. The fact that $Z_2 \neq Z_3$ means that the texture of the fracture surface has different roughnesses vertically and horizontally, while the fact that there are no off-diagonal components means that, across the fractures, the normal motion is uncoupled from the tangential motion.

A common situation in the Earth's crust is to have a finely layered medium and vertical fractures. Figure 4.15 shows such a case, where $q = 2$ sets of long vertical fractures are embedded in a TI medium with a vertical symmetry. This background medium is the long-wavelength equivalent of the finely layered medium, according to Backus (1962).

The mechanical (viscoelastic) representation of the fracture boundary condition by a Kelvin–Voigt model is illustrated in the figure, according to the stress–displacement relation $\boldsymbol{\sigma} \cdot \hat{\mathbf{n}} = (\kappa + \mathrm{i}\omega\eta) \cdot [\mathbf{u}]$, where the model simulates the fracture by a zero width layer of distributed spring-dashpots. The quantity $(\kappa + \mathrm{i}\omega\eta)$ is the complex modulus per unit length of the Kelvin–Voigt element (e.g., Carcione, 2007).

A displacement discontinuity yields compliance, while a discontinuity in the particle velocity implies energy loss at the interface (Carcione, 1996, 1998, 2007); $\kappa = 0$ gives the particle-velocity discontinuity model and $\eta = 0$ gives the displacement discontinuity model. On the contary, if $\kappa \to \infty$ or $\eta \to \infty$, the model gives a welded interface.

4.4.2.2 Monoclinic Equivalent Medium

Let us consider a background VTIv medium. In the frequency domain, the stress–strain relation is

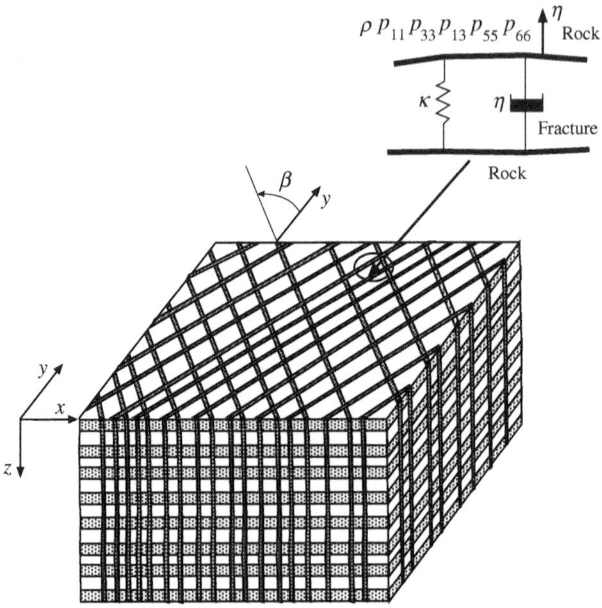

FIGURE 4.15 Two sets of vertical fractures embedded in a TI medium. Transverse isotropy is due to fine layering. If the two sets are orthogonal, the equivalent medium is orthorhombic, otherwise the symmetry is monoclinic with a horizontal single plane of symmetry. The boundary condition at the fracture describes an imperfect bonding in terms of the specific stiffness κ and specific viscosity η. Angle β is measured from the y-axis towards the strike direction.

$$\begin{pmatrix} \sigma_1 \\ \sigma_2 \\ \sigma_3 \\ \sigma_4 \\ \sigma_5 \\ \sigma_6 \end{pmatrix} = \begin{pmatrix} \sigma_{11} \\ \sigma_{22} \\ \sigma_{33} \\ \sigma_{23} \\ \sigma_{13} \\ \sigma_{12} \end{pmatrix} = \begin{pmatrix} p_{11} & p_{12} & p_{13} & 0 & 0 & 0 \\ p_{12} & p_{11} & p_{13} & 0 & 0 & 0 \\ p_{13} & p_{13} & p_{33} & 0 & 0 & 0 \\ 0 & 0 & 0 & p_{55} & 0 & 0 \\ 0 & 0 & 0 & 0 & p_{55} & 0 \\ 0 & 0 & 0 & 0 & 0 & p_{66} \end{pmatrix} \begin{pmatrix} e_1 \\ e_2 \\ e_3 \\ e_4 \\ e_5 \\ e_6 \end{pmatrix} \equiv \mathbf{P} \begin{pmatrix} e_1 \\ e_2 \\ e_3 \\ e_4 \\ e_5 \\ e_6 \end{pmatrix}, \tag{4.150}$$

where

$$2p_{66} = p_{11} - p_{12}.$$

The background medium can be described by using one of the three viscoelastic models of anisotropic anelasticity proposed in Chapter 4 of Carcione (2007). The stiffness constants p_{IJ} are complex and frequency dependent. Schoenberg and Muir (1989) show how to combine arbitrary sets of elastic thin layers and find their long-wavelength-equivalent medium properties. The method is also applicable when a set of layers is infinitely thin and compliant, as it is the fracture set shown in Figure 4.15. Nichols et al. (1989) and Hood (1991) simplify the procedure to include fractures by using compliance matrices instead of stiffness matrices. As shown below, the compliance matrix of the equivalent medium is found by addition of expanded 6 × 6 versions of the fracture compliance matrices to the compliance matrix of the background medium.

Let us consider many sets of vertical fractures, each set indicated by the index q, so that the fracture normal of a given set makes an angle β with the x-axis (see Figure 4.15). Here, we avoid the fracture index (q) for simplicity. The expanded fracture compliance matrix of each set is

$$\mathbf{S}_f = \begin{pmatrix} s_{11} & s_{12} & 0 & 0 & 0 & s_{16} \\ s_{12} & s_{22} & 0 & 0 & 0 & s_{26} \\ 0 & 0 & 0 & 0 & 0 & 0 \\ 0 & 0 & 0 & s_{44} & s_{45} & 0 \\ 0 & 0 & 0 & s_{45} & s_{55} & 0 \\ s_{16} & s_{26} & 0 & 0 & 0 & s_{66} \end{pmatrix}, \tag{4.151}$$

where

$$\begin{aligned} s_{11} &= \frac{3Z_N + Z_H}{8} + \frac{Z_N}{2}\cos 2\beta + \frac{Z_N - Z_H}{8}\cos 4\beta, \\ s_{12} &= \frac{Z_N - Z_H}{8}(1 - \cos 4\beta), \\ s_{16} &= \frac{Z_N \sin 2\beta}{2} + \frac{Z_N - Z_H}{4}\sin 4\beta, \\ s_{22} &= \frac{3Z_N + Z_H}{8} - \frac{Z_N}{2}\cos 2\beta + \frac{Z_N - Z_H}{8}\cos 4\beta, \\ s_{26} &= \frac{Z_N \sin 2\beta}{2} - \frac{Z_N - Z_H}{4}\sin 4\beta, \\ s_{44} &= \frac{Z_V(1 - \cos 2\beta)}{2}, \\ s_{45} &= \frac{Z_V \sin 2\beta}{2}, \\ s_{55} &= \frac{Z_V(1 + \cos 2\beta)}{2}, \\ s_{66} &= \frac{Z_N + Z_H}{2} - \frac{Z_N - Z_H}{2}\cos 4\beta \end{aligned} \tag{4.152}$$

(Schoenberg et al., 1999). The computation of the compliance matrix (4.151) can be derived by carrying out the matrix multiplication given by Nichols et al. (1989) or directly using fourth-rank tensor notation.

We have introduced

$$Z_N = \frac{Z_1}{d}, \quad Z_H = \frac{Z_2}{d}, \quad Z_V = \frac{Z_3}{d}, \tag{4.153}$$

where d is a characteristic length, such that these quantities have dimensions of compliance, and

$$\begin{aligned} \kappa_N &= d\kappa_1, \quad \kappa_H = d\kappa_2, \quad \kappa_V = d\kappa_3, \\ \eta_N &= d\eta_1, \quad \eta_H = d\eta_2, \quad \eta_V = d\eta_3, \end{aligned} \tag{4.154}$$

such that these quantities have dimensions of stiffness and viscosity, respectively. The quantity d, called h in Grechka et al. (2003) (see their eq. 3), is the average fracture spacing, which has to be much smaller than the dominant wavelength of the pulse. It is

$$Z_N = \frac{1}{\kappa_N + i\omega\eta_N}, \quad Z_H = \frac{1}{\kappa_H + i\omega\eta_H}, \quad Z_V = \frac{1}{\kappa_V + i\omega\eta_V}. \tag{4.155}$$

Then, the complex and frequency-dependent stiffness matrix of the equivalent medium is given by

$$\bar{\mathbf{P}} = \left(\mathbf{P}^{-1} + \sum_q \mathbf{S}_f^{(q)} \right)^{-1}, \tag{4.156}$$

where the sum is over the q sets of fractures. The equivalent homogeneous anisotropic medium (in the long wavelength limit) is a monoclinic medium with a horizontal mirror plane of symmetry. Its complex and frequency-dependent stiffness matrix has the form

$$\bar{\mathbf{P}} = \begin{pmatrix} \bar{p}_{11} & \bar{p}_{12} & \bar{p}_{13} & 0 & 0 & \bar{p}_{16} \\ \bar{p}_{12} & \bar{p}_{22} & \bar{p}_{23} & 0 & 0 & \bar{p}_{26} \\ \bar{p}_{13} & \bar{p}_{23} & \bar{p}_{33} & 0 & 0 & \bar{p}_{36} \\ 0 & 0 & 0 & \bar{p}_{44} & \bar{p}_{45} & 0 \\ 0 & 0 & 0 & \bar{p}_{45} & \bar{p}_{55} & 0 \\ \bar{p}_{16} & \bar{p}_{26} & \bar{p}_{36} & 0 & 0 & \bar{p}_{66} \end{pmatrix}. \tag{4.157}$$

At zero frequency, we obtain a lossless medium, where $Z_N = 1/\kappa_N, Z_H = 1/\kappa_H$, and $Z_V = 1/\kappa_V$, and the medium remains monoclinic. At "infinite" frequency, the fracture stiffnesses vanish ($\mathbf{S}_f^{(q)} \to 0$) and we obtain the properties of the background (lossless) medium. By "infinite" frequency, we mean frequencies such that the wavelength is much larger than the distance between single fractures, that is, the long-wavelength approximation. It can be shown that the wave velocities at zero frequency are smaller than the wave velocities at infinite frequency, that is, the medium is more compliant.

4.4.2.3 Orthorhombic Equivalent Medium

Assume two fracture sets with $\beta = 0$ (fracture strike along the y-direction) and $\beta = \pi/2$ (fracture strike points along the x-direction). The fracture stiffness matrices are given by

$$\mathbf{S}_f^{(1)} = \begin{pmatrix} Z_N^{(1)} & 0 & 0 & 0 & 0 & 0 \\ 0 & 0 & 0 & 0 & 0 & 0 \\ 0 & 0 & 0 & 0 & 0 & 0 \\ 0 & 0 & 0 & 0 & 0 & 0 \\ 0 & 0 & 0 & 0 & Z_V^{(1)} & 0 \\ 0 & 0 & 0 & 0 & 0 & Z_H^{(1)} \end{pmatrix} \tag{4.158}$$

and

$$\mathbf{S}_f^{(2)} = \begin{pmatrix} 0 & 0 & 0 & 0 & 0 & 0 \\ 0 & Z_N^{(2)} & 0 & 0 & 0 & 0 \\ 0 & 0 & 0 & 0 & 0 & 0 \\ 0 & 0 & 0 & Z_V^{(2)} & 0 & 0 \\ 0 & 0 & 0 & 0 & 0 & 0 \\ 0 & 0 & 0 & 0 & 0 & Z_H^{(2)} \end{pmatrix}, \tag{4.159}$$

respectively.

The evaluation of equation (4.156) requires simple operations with 3×3 matrices. We obtain

$$\begin{aligned}
\bar{p}_{11} &= \frac{p_{11} + Z_N^{(2)}(p_{11}^2 - p_{12}^2)}{1 + Z_N^{(1)}Z_N^{(2)}(p_{11}^2 - p_{12}^2) + p_{11}(Z_N^{(1)} + Z_N^{(2)})}, \\
\bar{p}_{12} &= \frac{p_{12}}{1 + Z_N^{(1)}Z_N^{(2)}(p_{11}^2 - p_{12}^2) + p_{11}(Z_N^{(1)} + Z_N^{(2)})}, \\
\bar{p}_{13} &= \frac{p_{13}[1 + Z_N^{(2)}(p_{11} - p_{12})]}{1 + Z_N^{(1)}Z_N^{(2)}(p_{11}^2 - p_{12}^2) + p_{11}(Z_N^{(1)} + Z_N^{(2)})}, \\
\bar{p}_{22} &= \frac{p_{11} + Z_N^{(1)}(p_{11}^2 - p_{12}^2)}{1 + Z_N^{(1)}Z_N^{(2)}(p_{11}^2 - p_{12}^2) + p_{11}(Z_N^{(1)} + Z_N^{(2)})}, \\
\bar{p}_{23} &= \frac{p_{13}[1 + Z_N^{(1)}(p_{11} - p_{12})]}{1 + Z_N^{(1)}Z_N^{(2)}(p_{11}^2 - p_{12}^2) + p_{11}(Z_N^{(1)} + Z_N^{(2)})}, \\
\bar{p}_{33} &= \frac{p_{33} + (Z_N^{(1)} + Z_N^{(2)})(p_{11}p_{33} - p_{13}^2) + Z_N^{(1)}Z_N^{(2)}(p_{11} - p_{12})[p_{33}(p_{11} + p_{12}) - 2p_{13}^2]}{1 + Z_N^{(1)}Z_N^{(2)}(p_{11}^2 - p_{12}^2) + p_{11}(Z_N^{(1)} + Z_N^{(2)})}, \\
\bar{p}_{44} &= \frac{p_{55}}{1 + p_{55}Z_V^{(2)}}, \\
\bar{p}_{55} &= \frac{p_{55}}{1 + p_{55}Z_V^{(1)}}, \\
\bar{p}_{66} &= \frac{p_{66}}{1 + p_{66}(Z_H^{(1)} + Z_H^{(2)})}.
\end{aligned} \tag{4.160}$$

Since the two sets are orthogonal, the equivalent medium has orthorhombic symmetry. In the case of a single fracture set, for example, $\beta = 0$, equation (4.160) gives the stiffness matrix of Schoenberg and Helbig (1997):

$$\bar{\mathbf{P}} = \begin{pmatrix} p_{11}(1-\delta_N) & p_{12}(1-\delta_N) & p_{13}(1-\delta_N) & 0 & 0 & 0 \\ p_{12}(1-\delta_N) & p_{11}(1-\delta_N p_{12}^2/p_{11}^2) & p_{13}(1-\delta_N p_{12}/p_{11}) & 0 & 0 & 0 \\ p_{13}(1-\delta_N) & p_{13}(1-\delta_N p_{12}/p_{11}) & p_{33}[1-\delta_N p_{13}^2/(p_{11}p_{33})] & 0 & 0 & 0 \\ 0 & 0 & 0 & p_{55} & 0 & 0 \\ 0 & 0 & 0 & 0 & p_{55}(1-\delta_V) & 0 \\ 0 & 0 & 0 & 0 & 0 & p_{66}(1-\delta_H) \end{pmatrix}, \tag{4.161}$$

where

$$\delta_N = [1+1/(Z_N p_{11})]^{-1}, \quad \delta_H = [1+1/(Z_H p_{66})]^{-1}, \quad \delta_V = [1+1/(Z_V p_{55})]^{-1}. \tag{4.162}$$

4.4.2.4 HTIv Equivalent Media

If the background medium is isotropic ($p_{11} = p_{12} + 2p_{55}$, $p_{12} = p_{13}$, $p_{55} = p_{66}$) and the fracture set is rotationally invariant, we have $Z_H = Z_V \equiv Z_T$, and the equivalent medium is TI with a horizontal symmetry axis (HTIv), whose stiffness matrix is

$$\bar{\mathbf{P}} = \begin{pmatrix} p_{11}c_N & p_{12}c_N & c_{12}c_N & 0 & 0 & 0 \\ p_{12}c_N & p_{11} - p_{12}^2 Z_N c_N & p_{12} - p_{12}^2 Z_N c_N & 0 & 0 & 0 \\ p_{12}c_N & p_{12} - p_{12}^2 Z_N c_N & p_{11} - p_{12}^2 Z_N c_N & 0 & 0 & 0 \\ 0 & 0 & 0 & p_{55} & 0 & 0 \\ 0 & 0 & 0 & 0 & p_{55}c_T & 0 \\ 0 & 0 & 0 & 0 & 0 & p_{55}c_T \end{pmatrix}, \tag{4.163}$$

where

$$c_N = (1+p_{11}Z_N)^{-1} \quad \text{and} \quad c_T = (1+p_{55}Z_T)^{-1}. \tag{4.164}$$

The stiffness matrix in Coates and Schoenberg (1995) is equivalent to equation (4.163) with a rotation of $\pi/2$ around the y-axis and considering the lossless case ($\eta = 0$). If $\eta \neq 0$, equation (4.163) is equivalent to the medium studied by Chichinina et al. (2009b).

The fractured medium defined by equation (4.163) can be obtained from Backus averaging of a periodic medium composed of two isotropic constituents with proportions p_i, and P-wave and S-wave moduli, respectively, given by $E_i = \lambda_i + 2\mu_i$ and μ_i, $i = 1, 2$. Backus's equivalent stiffness constants are given by equation (4.64)

$$
\begin{aligned}
\overline{p}_{11} &= [E_1 E_2 + 4\mathrm{p}_1\mathrm{p}_2(\mu_1 - \mu_2)(\lambda_1 + \mu_1 - \lambda_2 - \mu_2)]D,\\
\overline{p}_{12} &= [\lambda_1\lambda_2 + 2(\lambda_1\mathrm{p}_1 + \lambda_2\mathrm{p}_2)(\mu_2\mathrm{p}_1 + \mu_1\mathrm{p}_2)]D,\\
\overline{p}_{13} &= (\lambda_1\mathrm{p}_1 E_2 + \lambda_2\mathrm{p}_2 E_1)D,\\
\overline{p}_{33} &= E_1 E_2 D,\\
\overline{p}_{55} &= \mu_1\mu_2(\mathrm{p}_1\mu_2 + \mathrm{p}_2\mu_1)^{-1},\\
\overline{p}_{66} &= \mathrm{p}_1\mu_1 + \mathrm{p}_2\mu_2,\\
D &= (\mathrm{p}_1 E_2 + \mathrm{p}_2 E_1)^{-1}.
\end{aligned}
\tag{4.165}
$$

Equation (4.163) is then obtained by taking the limits $\mathrm{p}_1 \to 1$ and $\mathrm{p}_2 \to 0$ (or $\mathrm{p}_2 \ll \mathrm{p}_1$) and setting $E_1 = p_{11}$ and $\mu_1 = p_{55}$ (background medium) and $E_2 = \mathrm{p}_2/Z_N$ and $\mu_2 = \mathrm{p}_2/Z_H$ (fracture). This is shown by Schoenberg (1983) in the lossless case.

Expression of the phase velocities, attenuation factor, energy velocities, and quality factors are obtained from the equations given in Section 4.1.5 by replacing the p_{IJ} with the $\overline{p}_{IJ}$. The quality-factor expressions for each symmetry plane are similar to the equations obtained for the symmetry plane of the monoclinic medium. Along pure mode directions, we have equation (4.54). The attenuation has a maximum for a given value of the specific viscosity. Let us consider, for instance, a lossless background medium, the 11-component, the P-wave, and equation (4.160). A calculation yields

$$
Q_{11}^{-1} = \frac{c_{11}\omega\eta_N}{\kappa_N(\kappa_N + c_{11}) + \omega^2\eta_N^2}. \tag{4.166}
$$

This function has a maximum at

$$
\omega\eta_N = \sqrt{\kappa_N(\kappa_N + c_{11})} \approx \kappa_N \tag{4.167}
$$

if $\kappa_N \gg c_{11}$. On the other hand,

$$
Q_{JJ}^{-1} = \frac{(c_{1J}^2 / c_{JJ})\omega\eta_N}{(\kappa_N + c_{11})(\kappa_N + c_{11} - c_{1J}^2 / c_{JJ}) + \omega^2\eta_N^2}, \quad J = 2,3, \tag{4.168}
$$

with a maximum at

$$
\omega\eta_N = \sqrt{(\kappa_N + c_{11})(\kappa_N + c_{11} - c_{1J}^2 / c_{JJ})}. \tag{4.169}
$$

The attenuation of the shear waves (the 44, 55, and 66 components) have similar expressions. Two different values of the viscosity may give the same value of the quality factor, although the phase velocities differ.

For the analysis of the HTI medium, one can use the exact velocity expressions or the following approximations:

$$
\begin{aligned}
v_1^2 &= (c_{55}/\rho)(1-\delta_H \sin^2\theta), \\
v_2^2 &= (c_{55}/\rho)(1-\delta_H \cos^2 2\theta - (c_{55}/c_{11})\delta_N \sin^2 2\theta), \\
v_3^2 &= (c_{11}/\rho)[1-(c_{55}/c_{11})\delta_H \sin^2 2\theta - (1-2(c_{55}/c_{11})\delta_N \cos^2\theta)^2],
\end{aligned}
\tag{4.170}
$$

where θ is the angle between the wavenumber vector and the vertical direction.

Similar approximations have been obtained by Schoenberg and Douma (1988) (their eq. 26) for VTI media (horizontal fractures). Schoenberg and Douma (1988) have written those expressions in terms of the coefficients $\delta_N/(1-\delta_N)$ and $\delta_H/(1-\delta_H)$, instead of δ_N and δ_H, respectively, assuming $\delta_N \ll 1$ and $\delta_H \ll 1$. Without imposing these conditions, more accurate expressions are obtained by using δ_N and δ_H, as in Chichinina et al. (2009a, 2009b).

We consider two sets of fractures with $\beta_1 = 20^\circ$ and $\beta_2 = 65^\circ$ embedded in a TI lossless background medium representing a typical shale (numbers in MPa)

$$
\mathbf{C} = \rho \begin{pmatrix} 10 & 4 & 2.5 & 0 & 0 & 0 \\ 4 & 10 & 2.5 & 0 & 0 & 0 \\ 2.5 & 2.5 & 6 & 0 & 0 & 0 \\ 0 & 0 & 0 & 2 & 0 & 0 \\ 0 & 0 & 0 & 0 & 2 & 0 \\ 0 & 0 & 0 & 0 & 0 & 3 \end{pmatrix}, \tag{4.171}
$$

where ρ = 2300 kg/m^3, for example, c_{11} = 23 GPa. We assume a frequency $f = \omega/(2\pi)$ = 50 Hz. The fracture stiffnesses of the first set are $\kappa_N^{(1)} = 9c_{11}$, $\kappa_H^{(1)} = \frac{8}{9}c_{66}$, and $k_V^{(1)} = 4c_{55}$. Moreover, we set $\kappa_N^{(2)} = b\kappa_N^{(1)}$, $\kappa_H^{(2)} = b\kappa_H^{(1)}$, and $k_V^{(2)} = bk_V^{(1)}$, with $b = 0.5$. The fracture viscosities are assumed to be $\eta_N^{(q)} = a\kappa_N^{(q)}$, $\eta_H^{(q)} = a\kappa_H^{(q)}$, and $\eta_V^{(q)} = ak_V^{(q)}$, where a = 1 ms, for both the fracture sets.

The fracture parameters corresponding to the first example are considered. The stiffness-matrix components, given by equation (4.156), are

$$
\begin{aligned}
&\bar{p}_{11} = (18.05, 1.1), \quad \bar{p}_{12} = (8.98, 0.29), \quad \bar{p}_{13} = (4.83, 0.23), \quad \bar{p}_{16} = (-1.07, 0.13), \\
&\bar{p}_{22} = (17.27, 1.26), \quad \bar{p}_{23} = (4.69, 0.26), \quad \bar{p}_{26} = (-0.15, 0.05), \quad \bar{p}_{33} = (13.44, 0.09), \\
&\bar{p}_{36} = (-0.22, 0.03), \quad \bar{p}_{44} = (3.37, 0.25), \quad \bar{p}_{45} = (-0.67, 0.11), \quad \bar{p}_{55} = (3.70, 0.19), \\
&\bar{p}_{66} = (4.53, 0.46),
\end{aligned}
\tag{4.172}
$$

in GPa. Figure 4.16 shows the low-frequency limit energy velocities (A) and dissipation factors (B) in the (x,y) symmetry plane as a function of the phase (propagation) angle. The qS wave shows the maximum attenuation.

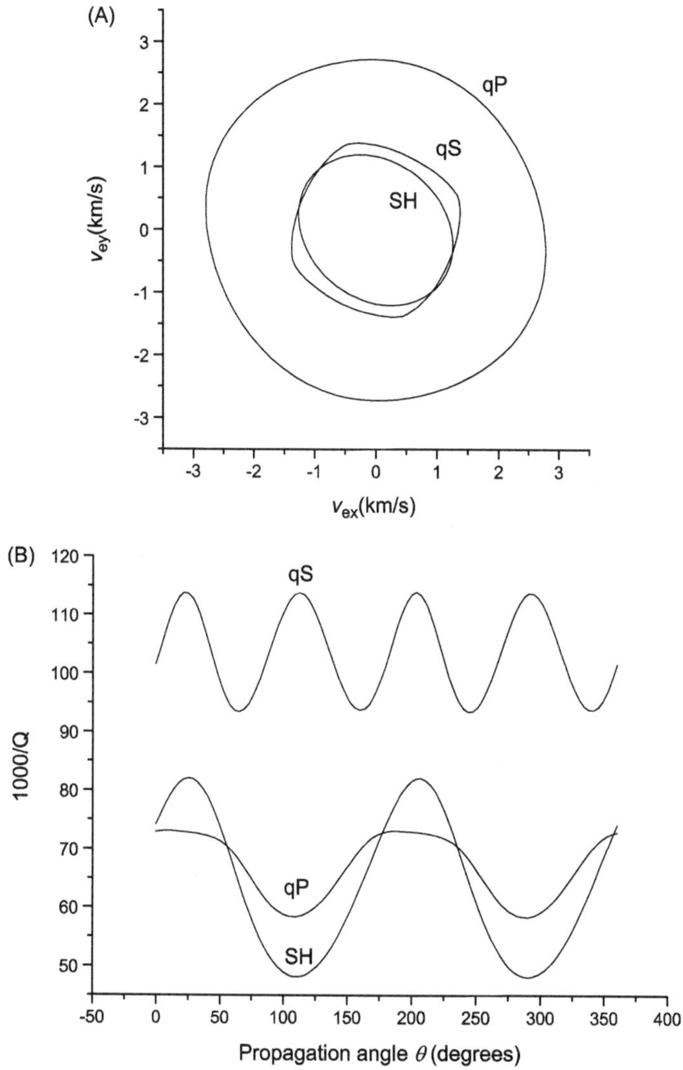

FIGURE 4.16 Low-frequency limit energy velocities (A) and dissipation factors (B) in the (x, y) symmetry plane as a function of the phase (propagation) angle. The equivalent medium has monoclinic symmetry.

4.4.2.5 Porous Medium

The theory of wave propagation in a dense fractured porous medium is a particular case of the fine layering case studied in Section 4.3.3 when one of the layers is infinitely thin and soft. This is the case of an highly permeable fracture. We consider two alternating media and equations (4.104), and let medium 2 be a very thin and soft layer. Following Bakulin and Molotkov (1997), we assume that all the stiffnesses of this layer are proportional to p_2 that is assumed to be small, that is, $P_2 = \mathrm{p}_2 P^*$, $M'_2 = \mathrm{p}_2 M'^*$, $R_2 = \mathrm{p}_2 R^*$, $F_2 = \mathrm{p}_2 F^*$, and $L_2 = \mathrm{p}_2 L^*$, where the starred quantities correspond to a stiff medium. Then, we obtain

$$\bar{P} = a_1 \bar{F} + b_1 \bar{M} + \frac{D_1}{\beta_1}, \quad a_1 = \frac{F_1 R_1 - M_1'^2}{\beta_1}, \quad b_1 = \frac{2 L_1 M_1'}{\beta_1},$$

$$\bar{F} = a_1 \bar{C} + b_1 \bar{Q}, \quad \bar{M} = a_1 \bar{Q} + b_1 \bar{R},$$

$$\bar{C} = \frac{1}{\gamma}\left(\frac{P_1}{\beta_1} + \frac{P^*}{\beta^*}\right), \quad \bar{Q} = \frac{1}{\gamma}\left(\frac{M_1'}{\beta_1} + \frac{M'^*}{\beta^*}\right),$$

$$\bar{R} = \frac{1}{\gamma}\left(\frac{R_1}{\beta_1} + \frac{R^*}{\beta^*}\right), \quad \bar{L} = \left(\frac{1}{L_1} + \frac{1}{L^*}\right)^{-1}, \tag{4.173}$$

$$\gamma = \left(\frac{P_1}{\beta_1} + \frac{P^*}{\beta^*}\right)\left(\frac{R_1}{\beta_1} + \frac{R^*}{\beta^*}\right) - \left(\frac{M_1'}{\beta_1} + \frac{M'^*}{\beta^*}\right)^2,$$

$$\beta = PR - M'^2.$$

In this case, the density coefficients (see eq. (4.104)) become those of the background medium

$$\rho_z = \rho_1, \quad \rho_x = \rho_1, \quad \rho_{fx} = \rho_{f1}, \quad \rho_{fz} = \rho_{f1}, \quad m_x = \tilde{m}_1, \quad m_z = \tilde{m}_1. \tag{4.174}$$

When the background medium is porous and saturated with fluids, the mesoscopic loss mechanism plays an important role in the physics of wave propagation. According to Bakulin and Molotkov (1997), the boundary conditions at the fracture interfaces are

$$[\sigma_{33}] = [\sigma_{13}] = [p_f] = 0, \quad \begin{bmatrix} u_3 \\ w_3 \\ u_1 \end{bmatrix} = \begin{pmatrix} \eta_{11} & \eta_{12} & 0 \\ \eta_{12} & \eta_{22} & 0 \\ 0 & 0 & \eta_{33} \end{pmatrix} \begin{pmatrix} \sigma_{33} \\ -p_f \\ \sigma_{13} \end{pmatrix}, \tag{4.175}$$

where

$$\eta_{11} = \frac{d_2 R_2}{\beta_2}, \quad \eta_{12} = -\frac{d_2 M_2'}{\beta_2}, \quad \eta_{22} = \frac{d_2 P_2}{\beta_2}, \quad \eta_{33} = \frac{d_2}{L_2} \tag{4.176}$$

are compliances per unit length. Defining the dimensionless fracture compliances

$$E_{11} = \frac{\beta_1 R_2}{\beta_2 R_1}\mathrm{p}_2, \quad E_{12} = -\frac{\beta_1 M_2'}{\beta_2 M_1'}\mathrm{p}_2, \quad E_{22} = \frac{\beta_1 P_2}{\beta_2 P_1}\mathrm{p}_2, \quad E_{33} = \frac{L_1}{L_2}\mathrm{p}_2, \tag{4.177}$$

we obtain the following stiffnesses from equation (4.173):

$$\bar{P} = a_1\bar{F} + b_1\bar{M} + \frac{D_1}{\beta_1}, \quad a_1 = \frac{F_1R_1 - M_1'^2}{\beta_1}, \quad b_1 = \frac{2L_1M_1'}{\beta_1},$$

$$\bar{F} = a_1\bar{C} + b_1\bar{Q},$$

$$\bar{M} = a_1\bar{Q} + b_1\bar{R},$$

$$\bar{C} = \frac{P_1}{\gamma\beta_1}(1+E_{22}), \quad \bar{Q} = \frac{M_1'}{\gamma\beta_1}(1-E_{12}), \tag{4.178}$$

$$\bar{R} = \frac{R_1}{\gamma\beta_1}(1+E_{11}), \quad \bar{L} = \frac{L_1}{1+E_{33}},$$

$$\beta_1^2\gamma = P_1R_1(1+E_{11})(1+E_{22}) - M_1'^2(1-E_{12})^2.$$

On the other hand, the open-fracture boundary conditions obtained by Nakagawa and Schoenberg (1997) (their eq. 53) are

$$[\sigma_{33}] = [\sigma_{13}] = [p_f] = 0,$$

$$[u_1] = \eta_T\sigma_{13},$$

$$[u_3] = \eta_{ND}(\sigma_{33} + \alpha_2 p_f), \tag{4.179}$$

$$[w_3] = -\alpha_2\eta_{ND}(\sigma_{33} + p_f/B_2),$$

where B_2 is the Skempton coefficient of the fractured material

$$B_2 = \frac{\alpha_2 M_2}{P_2}. \tag{4.180}$$

(The impermeable fracture boundary conditions (their eq. 54) implies $[u_3] = \eta_{ND}(1-\alpha_2B_2)\sigma_{33} = (d_2/P_2)\sigma_{33}$.)

Comparing equations (4.175) to (4.179), we have

$$\eta_{11} = \eta_{N_D}, \quad \eta_{12} = -\alpha_2\eta_{N_D}, \quad \eta_{22} = \frac{\alpha_2\eta_{N_D}}{B_2}, \quad \eta_{33} = \eta_T. \tag{4.181}$$

These equations are verified since $\eta_{ND} = d_2/E_{m2}$ and $\beta/R = \alpha\beta/M' = E_m$. Then, both boundary conditions are equivalent.

The dry-rock fracture compliances η_{11} and η_{33} are equivalent to the compliances Z_N and Z_T defined in Section 4.4.2, respectively, with the viscosity terms equal to zero. In this case, we have

$$\eta_{11} = Z_N d \quad \text{and} \quad \eta_{33} = Z_T d, \tag{4.182}$$

such that

$$E_{11}=\frac{Z_N\beta_1}{R_1},\quad E_{12}=-\frac{Z_N\beta_1\alpha_2}{M_1'}\simeq-\frac{Z_N\beta_1}{M_1'},\quad E_{22}=\frac{Z_N\beta_1P_2}{P_1M_2'}\simeq\frac{Z_N\beta_1}{P_1},\quad E_{33}=Z_TL_1, \tag{4.183}$$

since $\alpha_2\simeq 1$ and $P_2\simeq M_2'$ ($E_{G2}\simeq\alpha_2M_2$).

On the other hand, if $d_2\to 0$, $d_1\simeq d$, $E_{G2}\simeq M_2$, $K_{E2}\simeq E_{m2}$, $\alpha_2\simeq 1$ and $\coth(a_2d_2)\simeq 1/(a_2d_2)$, equation (4.108) becomes

$$\bar{p}_{33}(\omega)=\left[\frac{1}{E_{G1}}+\frac{((\alpha_1M_1/E_{G1})-1)^2}{dK_{E1}\vartheta_1\cot(\vartheta_1d)+Z_N^{-1}}\right]^{-1}, \tag{4.184}$$

where

$$\vartheta_1=\frac{1}{2}\sqrt{\frac{\omega\eta_1}{\mathrm{i}\kappa_1K_{E1}}},\quad K_{E1}=\frac{M_1E_{m1}}{E_{G1}} \tag{4.185}$$

and

$$Z_N=\frac{d_2}{dE_{m2}}=\frac{\mathrm{p}_2}{E_{m2}}=\frac{\eta_{11}}{d} \tag{4.186}$$

is the normal compliance of the fracture. Brajanovski et al. (2005) obtained equation (4.184) using the opposite sign convention to express the properties in the Fourier domain, that is, ω has to be replaced by $-\omega$ in their equation 15 (there is an omission in this equation; using their notation, M_b/C_b should be replaced by L_bM_b/C_b in the denominator).

The low-frequency limit (relaxed) stiffness components of the equivalent medium obtained by Brajanovski et al. (2005) are

$$c_{11}^r=\frac{c_{13}^{r\,2}}{c_{33}^r}+4\left[(1-\gamma)\mu_{m1}+\frac{\alpha_1^2\gamma^2K_{E1}}{1+Z_NK_{E1}}\right],$$

$$c_{13}^r=c_{33}^r\left[1-2\gamma+2\alpha_1\gamma\frac{M_1}{E_{G1}}\frac{\alpha_1+Z_NE_{m1}}{1+Z_NK_{E1}}\right],$$

$$c_{33}^r=E_{G1}\left[1+\frac{Z_N(\alpha_1M_1-E_{G1})^2}{E_{G1}(1+Z_NK_{E1})}\right]^{-1}, \tag{4.187}$$

$$c_{55}^r=\left(\mu_{m1}^{-1}+Z_T\right)^{-1},$$

$$c_{66}^r=\mu_{m1},$$

where $\gamma=\mu_{m1}/E_{m1}$ and

$$Z_T = \frac{d_2}{d\mu_{m2}} = \frac{\mathrm{p}_2}{\mu_{m2}} = \frac{\eta_{33}}{d} \tag{4.188}$$

is the shear compliance of the fracture (Brajanovski et al., 2005). Equations (4.187) are exactly equations (4.178) if the coefficients (4.183) are used in the latter.

Equation (4.107) yields the high-frequency limit (unrelaxed) stiffness components

$$\begin{aligned} c_{11} &= c_{33} = E_{G1}, \\ c_{13} &= c_{11} - 2\mu_{m1}, \\ c_{55} &= c_{55}^r, \\ c_{66} &= \mu_{m1}. \end{aligned} \tag{4.189}$$

Carcione et al. (2013) obtain the complex and frequency-dependent stiffnesses of the equivalent fractured medium combining equation (4.113) with equations (4.184)–(4.189).

An alternative theory, considering idealized pore shapes, has been proposed by Chapman (2003, 2009), who describes attenuation and dispersion resulting from local flow in equant-porosity media permeated by fracture sets. His predictions are similar to those of the mesoscopic-loss models. In both cases, the relative fluid–solid motion is perpendicular to the mesoscale heterogeneity. Along similar lines, the T-matrix model presented by Jakobsen and Chapman (2009) allows for nondilute concentrations of inclusions characterized by different shapes, orientations, and spatial distributions. Ali and Jakobsen (2013) use this model to obtain information about the fracture aperture, density and orientation from frequency-dependent seismic amplitude variations with angle and azimuth.

The example considers an oil-saturated sandstone with the following properties: $K_s = 37$ GPa, $\mu_s = 44$ GPa, $\rho_s = 2650$ kg/m^3, $K_f = 2$ GPa, $\rho_f = 870$ kg/m^3, and $\eta = 0.3$ cP. The dry-rock bulk and shear moduli are given by the Krief model

$$\frac{K_m}{K_s} = \frac{\mu_m}{\mu_s} = (1-\phi)^{3/(1-\phi)}. \tag{4.190}$$

Porosity and permeability are related by an equation derived by Carcione et al. (2000)

$$\kappa = \frac{r_g^2 \phi^3}{45(1-\phi)^2}, \tag{4.191}$$

where $r_g = 20$ μm denotes the average radius of the grains.

We consider the fracture stiffnesses $1/Z_N = 121$ GPa and $1/Z_T = 14$ GPa, and assume $\phi = 0.25$ and $d = 0.2$ m. We use equation (4.113) to obtain the complex stiffnesses at 35 Hz, where the peak quality factor has a value of 26. We obtain

$$\begin{aligned} \bar{p}_{11} &= (33.68, 0.001), \\ \bar{p}_{13} &= (5.75, 0.034), \\ \bar{p}_{33} &= (30.71, 1.17), \\ \bar{p}_{55} &= 6.96, \\ \bar{p}_{66} &= 13.92 \end{aligned} \tag{4.192}$$

in GPa. The energy velocities and dissipation factors are represented in Figure 4.17A and B, respectively. There is a noticeable shear-wave splitting and the coupling between the qP and qS waves generates strong shear attenuation

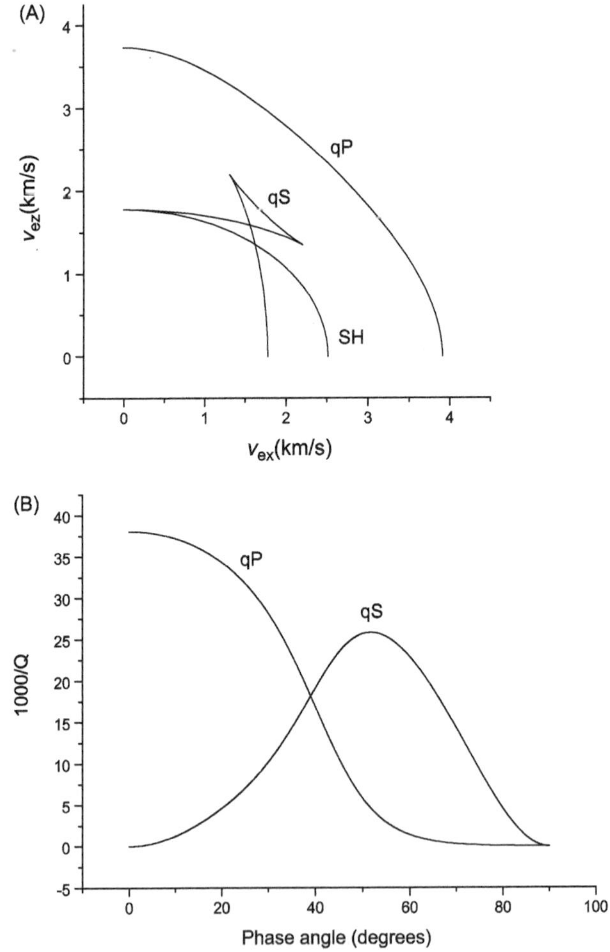

FIGURE 4.17 Energy velocity (A) and dissipation factor (B) for an oil saturated fractured medium with $(\Delta N, \Delta T) = (0.2, 0.5)$. The frequency is 35 Hz.

($Q \approx 40$) at approximately 50°, with magnitudes comparable to those of the qP wave. The SH wave is lossless. These curves have the same characteristics of those obtained for finely layered media (see Figure 4.10).

4.5 NUMERICAL HARMONIC EXPERIMENTS

The stiffness components of the equivalent medium (layered or fractured) can be computed by performing a set of quasistatic numerical harmonic experiments (Santos et al., 2012; Carcione et al., 2011). The modeling technique can be based on finite-element (FE) or finite-differences methods, and the equations of motion are solved in the space-frequency domain.

4.5.1 Solid Medium

Let $\mathbf{u}$, $\varepsilon_{ij}(\mathbf{u})$ and $\sigma_{ij}(\mathbf{u})$ denote the displacement vector, strain components, and stress components of each single isotropic medium in the frequency domain and consider the (x,z)-plane. We assume isotropic single layers but the method can be used for anisotropic media as well. The stress–strain relation is

$$\sigma_{ij} = \lambda\vartheta\delta_{ij} + 2\mu\varepsilon_{ij}, \quad \vartheta = \varepsilon_{ii} = u_{i,i}, \tag{4.193}$$

where λ and μ are given, for instance, by equation (4.62). The frequency-domain equation of motion in the (x,z)-plane is

$$\rho\omega^2\mathbf{u}(x,z,\omega) + \text{div}\,\boldsymbol{\sigma}\,(x,z,\omega) = 0, \tag{4.194}$$

where we have assumed no body forces.

Let us consider a rectangular sample (see Figure 4.18). The boundary of the sample is $\Gamma = \Gamma^L \cup \Gamma^B \cup \Gamma^R \cup \Gamma^T$, where

$$\begin{aligned} \Gamma^L &= \{(x,z) \in \Gamma : x = 0\}, \quad \Gamma^R = \{(x,z) \in \Gamma : x = L\}, \\ \Gamma^B &= \{(x,z) \in \Gamma : z = 0\}, \quad \Gamma^T = \{(x,z) \in \Gamma : z = L\}, \end{aligned} \tag{4.195}$$

where L is the size of the sample, Γ^L, Γ^R, Γ^B, and Γ^T are the left, right, bottom, and top boundaries of the sample, respectively. Denote by $\hat{\mathbf{n}}$ the unit outer normal on Γ and let $\hat{\mathbf{m}}$ be a unit tangent on Γ so that $\{\hat{\mathbf{n}}, \hat{\mathbf{m}}\}$ is an orthonormal system on Γ.

The sample is subjected to time-harmonic compressions $\Delta P \exp(i\omega t)$, where P denotes pressure, and time-harmonic tangential forces $\Delta G \exp(i\omega t)$, where G is the shear stress (see Figure 4.18). (Here, Δ denotes a variation of the field variable.) In the following, we establish the boundary conditions to be used at the sides of the sample to obtain each stiffness component. Then, we solve equation (4.194) with those conditions.

i) $\bar{p}_{33}$: The boundary conditions are

$$\begin{aligned} (\boldsymbol{\sigma} \cdot \hat{\mathbf{n}}) \cdot \hat{\mathbf{n}} &= -\Delta P, \quad (x,z) \in \Gamma^T, \\ (\boldsymbol{\sigma} \cdot \hat{\mathbf{n}}) \cdot \hat{\mathbf{m}} &= 0, \qquad\;\; (x,z) \in \Gamma, \\ \mathbf{u} \cdot \hat{\mathbf{n}} &= 0, \qquad\;\; (x,z) \in \Gamma^L \cup \Gamma^R \cup \Gamma^B. \end{aligned} \tag{4.196}$$

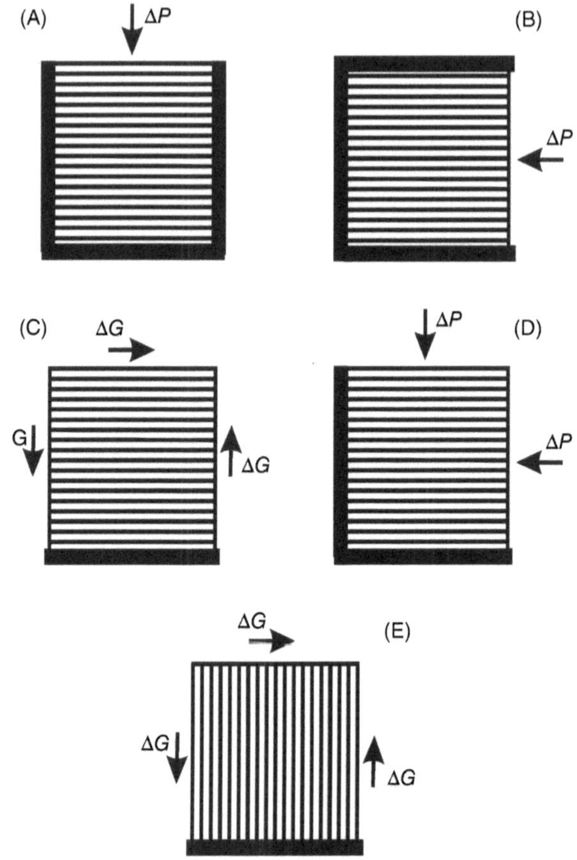

FIGURE 4.18 Oscillatory tests performed to obtain $\overline{p}_{33}$ (A), $\overline{p}_{11}$ (B), $\overline{p}_{55}$ (C), $\overline{p}_{13}$ (D), and $\overline{p}_{66}$ (E). The orientation of the layers and the directions of the applied stresses are indicated. The thick black lines at the edges indicate rigid boundary conditions (zero displacements).

In this experiment, $\varepsilon_{11} = \varepsilon_{22} = 0$ in the stress–strain relations (4.2). Denoting by Ω the original volume of the sample and by $\Delta\Omega(\omega)$ its (complex) oscillatory volume change, we note that

$$\frac{\Delta\Omega(\omega)}{\Omega} = -\frac{\Delta P}{\overline{p}_{33}(\omega)}, \tag{4.197}$$

valid in the quasistatic case. After solving equation (4.194) with the boundary conditions (4.196), we obtain the average vertical displacement $u_3^T(\omega)$ at the boundary Γ^T from the vertical displacements $u_3(x, L, \omega)$ measured on Γ^T. Then, for each frequency ω, the volume change produced by the compressibility test can be approximated by $\Delta\Omega(\omega) \approx L u_3^T(\omega)$, which yields $\overline{p}_{33}(\omega)$ using the relation (4.197).

ii) $\bar{p}_{11}$: The boundary conditions are:

$$\begin{aligned} (\sigma \cdot \hat{\mathbf{n}}) \cdot \hat{\mathbf{n}} &= -\Delta P, & (x,z) &\in \Gamma^R, \\ (\sigma \cdot \hat{\mathbf{n}}) \cdot \hat{\mathbf{m}} &= 0, & (x,z) &\in \Gamma, \\ \mathbf{u} \cdot \hat{\mathbf{n}} &= 0, & (x,z) &\in \Gamma^L \cup \Gamma^B \cup \Gamma^T. \end{aligned} \tag{4.198}$$

In this experiment, $\varepsilon_{33} = \varepsilon_{22} = 0$, and it determines $\bar{p}_{11}$ in the same way indicated for $\bar{p}_{33}$.

iii) $\bar{p}_{55}$: The boundary conditions are:

$$\begin{aligned} \sigma \cdot \hat{\mathbf{m}} &= \mathbf{g}, & (x,z) &\in \Gamma^T \cup \Gamma^L \cup \Gamma^R, \\ \mathbf{u} &= 0, & (x,z) &\in \Gamma^B, \end{aligned} \tag{4.199}$$

where

$$\mathbf{g} = \begin{cases} (0, \Delta G), & (x,z) \in \Gamma^R, \\ (0, -\Delta G), & (x,z) \in \Gamma^L, \\ (\Delta G, 0), & (x,z) \in \Gamma^T. \end{cases} \tag{4.200}$$

The change in shape of the sample gives $\bar{p}_{55}(\omega)$ from the relation

$$\tan[\beta(\omega)] = \frac{\Delta G}{\bar{p}_{55}(\omega)}, \tag{4.201}$$

where $\beta(\omega)$ is the angle between the original positions of the lateral boundaries and the location after applying the shear stresses.

The horizontal displacements $u_1(x, L, \omega)$ at the top boundary Γ^T are used to obtain, for each frequency, an average horizontal displacement $u_1^T(\omega)$ at the boundary Γ^T. This average value gives $\tan[\beta(\omega)] \approx u_1^T(\omega) / L$, which from equation (4.201) yields $\bar{p}_{55}(\omega)$.

iv) $\bar{p}_{66}$: Since this stiffness is associated with shear waves traveling in the (x,y)-plane, we rotate the sample by 90° and apply the shear test as indicated for $\bar{p}_{55}$.

v) $\bar{p}_{13}$: The boundary conditions are

$$\begin{aligned} (\sigma \cdot \hat{\mathbf{n}}) \cdot \hat{\mathbf{n}} &= -\Delta P, & (x,z) &\in \Gamma^R \cup \Gamma^T, \\ (\sigma \cdot \hat{\mathbf{n}}) \cdot \hat{\mathbf{m}} &= 0, & (x,z) &\in \Gamma, \\ \mathbf{u} \cdot \hat{\mathbf{n}} &= 0, & (x,z) &\in \Gamma^L \cup \Gamma^L \cup \Gamma^B. \end{aligned} \tag{4.202}$$

Thus, in this experiment $\varepsilon_{22} = 0$, and from the stress–strain relations (4.2) we obtain

$$\begin{aligned} \sigma_{11} &= \bar{p}_{11}\varepsilon_{11} + \bar{p}_{13}\varepsilon_{33}, \\ \sigma_{33} &= \bar{p}_{13}\varepsilon_{11} + \bar{p}_{33}\varepsilon_{33}, \end{aligned} \tag{4.203}$$

where ε_{11} and ε_{33} are the strain components at the right lateral side and top side of the sample, respectively. Then, from equation (4.203) and using $\sigma_{11} = \sigma_{33} = -\Delta P$, we obtain

$$\bar{p}_{13}(\omega) = \frac{\bar{p}_{11}\varepsilon_{11} - \bar{p}_{33}\varepsilon_{33}}{\varepsilon_{11} - \varepsilon_{33}}. \tag{4.204}$$

To estimate the equivalent complex stiffnesses, we use a FE procedure to approximate the solution of the equations of motion under the boundary conditions described above. We use bilinear functions to approximate the solid displacement vector, while to approximate the fluid displacement a closed subspace of the vector part of the Raviart–Thomas–Nedelec space of zero order is employed (Raviart and Thomas, 1975; Nedelec, 1980). The arguments given in Santos et al. (2009) to derive a-priori error estimates for the isotropic case can be extended to these FE problems to show that the corresponding errors measured in the energy norm, is of the order of the size of the computational mesh.

4.5.2 Porous Medium

We assume an undrained composite medium, for which the variation of fluid content $\zeta = -\text{div}\,\mathbf{w}$ is zero. This approach provides the complex velocities of the fast wave modes. To obtain also the complex velocity of the slow Biot wave, one needs to consider the stiffness coefficients related to the variation fluid content and fluid pressure, where the stiffness matrix has a 7×7 dimension (Carcione 2007, Section 7.14). This calculation requires to know the complex and frequency-dependent version of the coefficients B_6^*, B_7^*, and B_8^* (see eq. (4.106)).

In order to determine the coefficients $\bar{p}_{IJ}$ we proceed as follows. We solve equation (4.101) by considering the boundaries defined in equation (4.195) and, as above, the sample is subjected to time-harmonic compressions $\Delta P \exp(i\omega t)$ and tangential forces $\Delta G \exp(i\omega t)$ (see Figure 4.18). We obtain the stiffness components as follows.

i) $\bar{p}_{33}$: We solve equation (4.101) with the following boundary conditions

$$\begin{aligned}
&[\sigma(\mathbf{u})\cdot\hat{\mathbf{n}}]\cdot\hat{\mathbf{n}} = -\Delta P, \quad (x,z)\in\Gamma^T,\\
&[\sigma(\mathbf{u})\cdot\hat{\mathbf{n}}]\cdot\hat{\mathbf{m}} = 0, \quad (x,z)\in\Gamma^T,\\
&[\sigma(\mathbf{u})\cdot\hat{\mathbf{n}}]\cdot\hat{\mathbf{m}} = 0, \quad (x,z)\in\Gamma^L\cup\Gamma^R,\\
&\mathbf{u}^{(m)}\cdot\hat{\mathbf{n}} = 0, \quad (x,z)\in\Gamma^L\cup\Gamma^R,\\
&\mathbf{u}^{(m)} = 0, \quad (x,z)\in\Gamma^B,\\
&\mathbf{w}\cdot\hat{\mathbf{n}} = 0, \quad (x,z)\in\Gamma.
\end{aligned} \tag{4.205}$$

In this experiment, $\varepsilon_{11}[\mathbf{u}^{(m)}] = \varepsilon_{22}[\mathbf{u}^{(m)}] = \zeta = 0$. We obtain $\overline{p}_{33}$ as in the solid case by using the relation (4.197).

ii) $\overline{p}_{11}$: The boundary conditions are:

$$
\begin{aligned}
&[\sigma(\mathbf{u})\cdot\hat{\mathbf{n}}]\cdot\hat{\mathbf{n}} = -\Delta P, \quad (x,z)\in\Gamma^R,\\
&[\sigma(\mathbf{u})\cdot\hat{\mathbf{n}}]\cdot\hat{\mathbf{m}} = 0, \quad (x,z)\in\Gamma^R,\\
&[\sigma(\mathbf{u})\cdot\hat{\mathbf{n}}]\cdot\hat{\mathbf{m}} = 0, \quad (x,z)\in\Gamma^B\cup\Gamma^T,\\
&\mathbf{u}^{(m)}\cdot\hat{\mathbf{n}} = 0, \quad (x,z)\in\Gamma^B\cup\Gamma^T,\\
&\mathbf{u}^{(m)} = 0, \quad (x,z)\in\Gamma^L,\\
&\mathbf{w}\cdot\hat{\mathbf{n}} = 0, \quad (x,z)\in\Gamma.
\end{aligned}
\tag{4.206}
$$

In this experiment, $\varepsilon_{33}[\mathbf{u}^{(m)}] = \varepsilon_{22}[\mathbf{u}^{(m)}] = \zeta = 0$, and from equation (4.101), this experiment determines $\overline{p}_{11}$ in the same way indicated for $\overline{p}_{33}$.

iii) $\overline{p}_{55}$: The boundary conditions are:

$$
\begin{aligned}
&-\sigma(\mathbf{u})\cdot\hat{\mathbf{n}} = \mathbf{g}, \quad (x,z)\in\Gamma^T\cup\Gamma^L\cup\Gamma^R,\\
&\mathbf{u}^{(m)} = 0, \quad (x,z)\in\Gamma^B,\\
&\mathbf{w}\cdot\hat{\mathbf{n}} = 0, \quad (x,z)\in\Gamma,
\end{aligned}
\tag{4.207}
$$

where **g** is given in equation (4.200). The change in shape of the rock sample allow us to compute $\overline{p}_{55}(\omega)$ by using the relation (4.201).

iv) $\overline{p}_{66}$: Since this stiffness is associated with shear waves traveling in the (x,y)-plane, we take the layered sample, rotate it 90° and apply the shear test as indicated for $\overline{p}_{55}$.

v) $\overline{p}_{13}$: The boundary conditions are

$$
\begin{aligned}
&[\sigma(\mathbf{u})\cdot\hat{\mathbf{n}}]\cdot\hat{\mathbf{n}} = -\Delta P, \quad (x,z)\in\Gamma^R\cup\Gamma^T,\\
&[\sigma(\mathbf{u})\cdot\hat{\mathbf{n}}]\cdot\hat{\mathbf{m}} = 0, \quad (x,z)\in\Gamma,\\
&\mathbf{u}^{(m)}\cdot\hat{\mathbf{n}} = 0, \quad (x,z)\in\Gamma^L\cup\Gamma^B,\\
&\mathbf{w}\cdot\hat{\mathbf{n}} = 0, \quad (x,z)\in\Gamma.
\end{aligned}
\tag{4.208}
$$

Thus, in this experiment $\varepsilon_{22} = \zeta = 0$, and from equation (4.112) we obtain

$$
\begin{aligned}
\sigma_{11} &= \overline{p}_{11}\varepsilon_{11} + \overline{p}_{13}\varepsilon_{33},\\
\sigma_{33} &= \overline{p}_{13}\varepsilon_{11} + \overline{p}_{33}\varepsilon_{33},
\end{aligned}
\tag{4.209}
$$

where ε_{11} and ε_{33} are the strain components at the right lateral side and top side of the sample, respectively. Then, using $\sigma_{11} = \sigma_{33} = -\Delta P$, we obtain $\bar{p}_{13}$ as

$$\bar{p}_{13} = \frac{\bar{p}_{11}\varepsilon_{11} - \bar{p}_{33}\varepsilon_{33}}{\varepsilon_{11} - \varepsilon_{33}}. \tag{4.210}$$

Figure 4.19 shows the energy velocity (A) and dissipation factor as a function of the propagation angle (B) at a frequency of 50 Hz. The three wave modes are

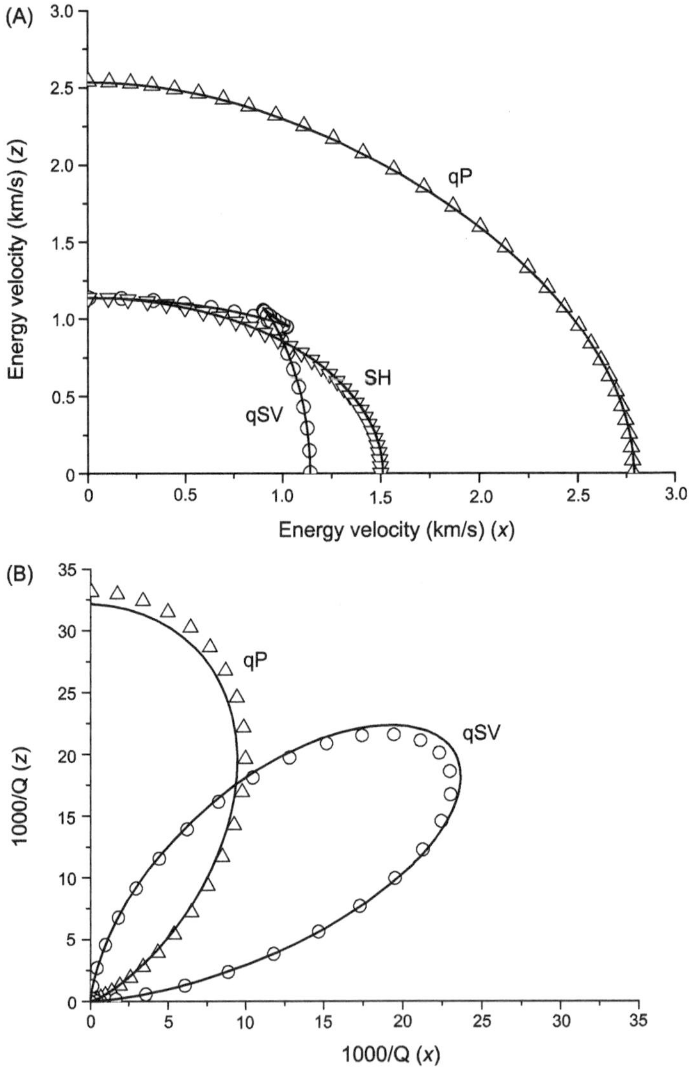

FIGURE 4.19 Polar representation of the energy velocity (A) and dissipation factor $[(\mathbf{1000}/Q)(\sin\theta, \cos\theta)]$ (B) corresponding to a frequency of 50 Hz. The medium consists of a sequence of mudstone and water-saturated sandstone layers of 5 and 1 cm, respectively. This example is taken from Carcione et al. (2011).

indicated in the plots. The symbols correspond to the numerical evaluation with the FE method. The velocity anisotropy of the shear modes is substantial, with a noticeable shear-wave splitting. As can be seen, the coupling between the qP and qSV waves generates strong shear attenuation at 45°. This coupling effect is mainly influenced by the contrast in dry-rock rigidity between the mudstone and the sandstone.

The FE variational formulation used to obtain the numerical results is explained by Carcione et al. (2010) and Santos et al. (2011). Similar numerical experiments have been applied to obtain the equivalent-medium stiffnesses of finely layered viscoelastic media (Picotti et al., 2010), porous media (Masson and Pride, 2007; Carcione et al., 2011; Quintal et al., 2011) and fractured media (Carcione et al., 2012). Quintal et al. (2012) used the method to analyze the influence of wave-induced fluid flow on the attenuation of S-waves.

REFERENCES

Aki, K., Richards, P.G., 1980. Quantitative Seismology: Theory and Methods. W.H. Freeman and Co., New York.

Ali, A., Jakobsen, M., 2013. Anisotropic permeability in fractured reservoirs from frequency-dependent seismic amplitude versus angle and azimuth data. Geophys. Prosp., DOI: 10.1111/1365-2478.12084.

Auld, B.A., 1990. Acoustic fields and waves in solids, Vol I, Krieger Publ. Co, Malabar, Florida.

Backus, G.E., 1962. Long-wave elastic anisotropy produced by horizontal layering. J. Geophys. Res. 7, 4427–4440.

Bakulin, A., Molotkov, L., 1997. Poroelastic medium with fractures as limiting case of stratified poroelastic medium with thin and soft Biot layers. 67th Annu. Int. Mtg. Soc. Expl. Geophys. Expanded Abstracts, 1001–1004.

Biot, M.A., 1956. Theory of propagation of elastic waves in a fluid-saturated porous solid. I. Low-frequency range. J. Acoust. Soc. Am. 28, 168–178.

Blitz, J., Simpson, G., 1996. Ultrasonic Methods of Non-Destructive Testing. Chapman & Hall Pub., London.

Brown, R., Korringa, J., 1975. On the dependence of the elastic properties of a porous rock on the compressibility of the pore fluid. Geophysics 40, 608–616.

Bruggeman, D.A.G., 1937. Berechnung der verschiedenen physikalischen Konstanten von heterogenen Substanzen, 3: die elastischen Konstanten de quasiisotropen Mischkörper aus isotropen Substanzen. Annalen der Physik 20, 160–178.

Carcione, J.M., 1992. Anisotropic Q and velocity dispersion of finely layered media. Geophys. Prosp. 40, 761–783.

Carcione, J.M., 1996. Elastodynamics of a non-ideal interface: application to crack and fracture scattering. J. Geophys. Res. 101 (b12), 28177–28188.

Carcione, J.M., 1998. Scattering of elastic waves by a plane crack of finite width in a transversely isotropic medium. Int. J. Numer. Anal. Methods Geomech. 22, 263–275.

Carcione, J.M., 2000. A model for seismic velocity and attenuation in petroleum source rocks. Geophysics 65, 1080–1092.

Carcione, J.M., 2007. Wave Fields in Real Media. Theory and Numerical Simulation of Wave Propagation in Anisotropic, Anelastic, Porous and Electromagnetic Media. Elsevier, Amsterdam (Second edition, extended and revised).

Carcione, J.M., Cavallini, F., 1994. A rheological model for anelastic anisotropic media with applications to seismic wave propagation. Geophys. J. Int. 119, 338–348.

Carcione, J.M., Cavallini, F., 1995. Attenuation and quality factors surfaces in anisotropic-viscoelastic media. Mech. Mater. 19, 311–327.

Carcione, J.M., Cavallini, F., Helbig, K., 1998. Anisotropic attenuation and material symmetry. Acustica 84, 495–502.

Carcione, J.M., Gurevich, B., Cavallini, F., 2000. A generalized Biot-Gassmann model for the acoustic properties of shaley sandstones. Geophys. Prosp. 48, 539–557.

Carcione, J.M., Gurevich, B., Santos, J.E., Picotti, S., 2013. Angular and frequency dependent wave velocity and attenuation in fractured porous media. Pure Appl. Geophys. 170, 1673–1683.

Carcione, J.M., Helbig, K., Helle, H.B., 2003. Effects of pressure and saturating fluid on wave velocity and attenuation of anisotropic rocks. Int. J. Rock Mech. Min. Sci. 40, 389–403.

Carcione, J.M., Helle, H.B., Avseth, P., 2011. Source-rock seismic-velocity models: Gassmann versus Backus. Geophysics 76, N37–N45.

Carcione, J.M., Helle, H., Zhao, T., 1998. The effects of attenuation and anisotropy on reflection amplitude versus offset. Geophysics 63, 1652–1658.

Carcione, J.M., Kosloff, D., Behle, A., 1991. Long wave anisotropy in stratified media: a numerical test. Geophysics 56, 245–254.

Carcione, J.M., Morency, C., Santos, J.E., 2010. Computational poroelasticity – a review. Geophysics 75, A229–A243.

Carcione, J.M., Picotti, S., 2006. P-wave seismic attenuation by slow-wave diffusion: effects of inhomogeneous rock properties. Geophysics 71, O1–O8.

Carcione, J.M., Picotti, S., 2012. Reflection and transmission coefficients of a fracture in transversely isotropic media. Stud. Geophys. Geod. 56, 307–322.

Carcione, J.M., Picotti, S., Cavallini, F., Santos, J.E., 2012. Numerical test of Schoenberg-Muir averaging theory. Geophysics 77, 27–35.

Carcione, J.M., Picotti, S., Gei, D., Rossi, G., 2006. Physics and seismic modeling for monitoring CO_2 storage. Pure Appl. Geophys. 163, 175–207.

Carcione, J.M., Picotti, S., Santos, J.E., 2012. Numerical experiments of fracture-induced velocity and attenuation anisotropy. Geophys. J. Int. 191, 1179–1191.

Carcione, J.M., Poletto, F., 2013. Seismic rheological model and reflection coefficients of the brittle-ductile transition. Pure Appl. Geophys. 170, 2021–2035.

Carcione, J.M., Santos, J.E., Picotti, S., 2011. Anisotropic poroelasticity and wave-induced fluid flow. Harmonic finite-element simulations. Geophys. J. Int. 186, 1245–1254.

Carcione, J.M., Santos, J.E., Picotti, S., 2012. Fracture-induced anisotropic attenuation. Rock Mech. Rock Eng. 45, 929–942.

Chapman, M., 2003. Frequency dependent anisotropy due to mesoscale fractures in the presence of equant porosity. Geophys. Prosp. 51, 369–379.

Chapman, M., 2009. Modeling the effect of multiple fracture sets of mesoscale fractures in porous rock on frequency-dependent anisotropy. Geophysics 74, D97–D103.

Chichinina, T.I., Obolentseva, I.R., Ronquillo-Jarillo, G., 2009a. Anisotropy of seismic attenuation in fractured media: theory and ultrasonic experiment. Trans. Porous Media 79, 1–14.

Chichinina, T.I., Obolentseva, I.R., Gik, L., Bobrov, B., Ronquillo-Jarillo, G., 2009b. Attenuation anisotropy in the linear-slip model: interpretation of physical modeling data. Geophysics 74, WB165–WB176.

Christensen, R.M., 1982. Theory of Viscoelasticity. Academic Press Inc., New York.

Coates, R.T., Schoenberg, M., 1995. Finite-difference modelling of faults and fractures. Geophysics 60, 1514–1526.

Daub, E.G., Carlson, J.M., 2010. Friction, fracture, and earthquakes. Annu. Rev. Condens. Mat. Phys. 1, 397–418.

Galvin, R.J., Gurevich, B., 2009. Effective properties of a poroelastic medium containing a distribution of aligned cracks. J. Geophys. Res. 114, B07305, doi:10.1029/2008JB006032.

Gauzellino, P., Carcione, J.M., Santos, J.E., and Picotti, S., 2014, A rheological equation for anisotropic-anelastic media and simulation of synthetic seismograms: Wave Motion, in press.

Gelinsky, S., Shapiro, S.A., 1997. Poroelastic Backus-averaging for anisotropic, layered fluid and gas saturated sediments. Geophysics 62, 1867–1878.

Golden, J.M., Graham, G.A.C., 1988. Boundary Value Problems in Linear Viscoelasticity. Springer-Verlag, Berlin.

Grechka, V.A., Bakulin, A., Tsvankin, I., 2003. Seismic characterization of vertical fractures described as general linear-slip interfaces. Geophys. Prosp. 51, 117–130.

Gurevich, B., 2003. Elastic properties of saturated porous rocks with aligned fractures. J. Appl. Geophys. 54, 203–218.

Gurevich, B., Brajanovski, M., Galvin, R.J., Müller, T.M., Toms-Stewart, J., 2009. P-wave dispersion and attenuation in fractured and porous reservoirs–poroelasticity approach. Geophys. Prosp. 57, 225–237.

Helbig, K., 1984. Anisotropy and dispersion in periodically layered media. Geophysics 49, 364–373.

Helbig, K., 1999. Layer-induced elastic anisotropy – Part 1: forward relations between constituent parameters and compound medium parameters. Revista Brasileira de Geofisica 16, 103–112.

Hertz, H., 1882. Ueber die Berührung fester elastischer Körper. J. Reine Angewandte Mathematik 92, 156–171.

Hill, R., 1964. Theory of mechanical properties of fibre-strengthened materials. J. Mech. Phys. Solids 11, 357–372.

Hood, J.A., 1991. A simple method for decomposing fracture-induced anisotropy. Geophysics 56, 1275–1279.

Hudson, J.A., Crampin, S., 1991. On: "A calculus for finely layered anisotropic media" by M. Schoenberg and F. Muir. Geophysics 54, 581–589, Geophysics 56, 572–575.

Jakobsen, M., Chapman, M., 2009. Unified theory of global flow and squirt flow in cracked porous media. Geophysics 74, WA65–WA76.

Johnson, D.L., 2001. Theory of frequency dependent acoustics in patchy-saturated porous media. J. Acoust. Soc. Am 110, 682–694.

Johnson, K.L., 2003. Contact Mechanics. Cambridge University Press, Cambridge.

Krief, M., Garat, J., Stellingwerff, J., Ventre, J., 1990. A petrophysical interpretation using the velocities of P and S waves (full waveform sonic). Log Anal. 31, 355–369.

Krzikalla, F., Müller, T., 2011. Anisotropic P-SV-wave dispersion and attenuation due to inter-layer flow in thinly layered porous rocks. Geophysics 76, WA135–WA145.

Kuster, G.T., Toksöz, M.N., 1974. Velocity and attenuation of seismic waves in two-phase media: part I. Theor. Formul. Geophys. 39, 587–606.

Li, Y., Cheng, D.H., Töksoz, M.N., 1998. Seismic monitoring of the growth of a hydraulic fracture zone at Fenton Hill, New Mexico. Geophysics 63, 120–131.

Liu, E., Hudson, J.A., Pointer, T., 2000. Equivalent medium representation of fractured rock. J. Geophys. Res. 105, 2981–3000.

Liu, E., Martinez, A., 2013. Seismic Fracture Characterization: Concepts and Practical Applications. Eagle Pub., Amsterdam.

Masson, Y.J., Pride, S.R., 2007. Poroelastic finite difference modeling of seismic attenuation and dispersion due to mesoscopic-scale heterogeneity. J. Geophy. Res. 112, B03204, doi:10.1029/2006JB004592.

Mavko, G., Mukerji, T., Dvorkin, J., 2009. The Rock Physics Handbook. Cambridge University Press, Cambridge.

McCain, Jr., W.D., 1984. The Properties of Petroleum Fluids. PennWell Books, Tulsa, OK.

Meissner, R., Strehlau, J., 1982. Limits of stresses in continental crusts and their relation to the depth-frequency distribution of shallow earthquakes. Tectonics 1, 73–89.

Melia, P.J., Carlson, R.L., 1984. An experimental test of P-wave anisotropy in stratified media. Geophysics 49, 374–378.

Mindlin, R.D., 1960. Waves and Vibrations in Isotropic Elastic Planes. In: Goodier, J.W., Hoff, W.J. (Eds.), Structural Mechanics. Pergamon Press, New York, p. 199.

Mukherjee, S., Paulino, G.H., 2003. The elastic-viscoelastic correspondence principle for functionally graded materials, revisited. J. Appl. Mech. 70, 359–363.

Müller, T., Gurevich, B., Lebedev, M., 2010. Seismic wave attenuation and dispersion resulting from wave-induced flow in porous rocks – a review. Geophysics 75, A147–A164.

Nakagawa, S., Schoenberg, M.A., 2007. Poroelastic modeling of seismic boundary conditions across a fracture. J. Acoust. Soc. Am. 122, 831–847.

Newmark, N.M., Siess, C.P., Viest, I.M., 1951. Test and analysis of composite beams with incomplete interaction. Proc. Soc. Experim. Stress Anal. 9, 75–92.

Nedelec, J.C., 1980. Mixed finite elements in R^3. Numer. Math 35, 315–341.

Nichols, D., Muir, F., Schoenberg, M., 1989. Elastic properties of rocks with multiple sets of fractures. 63rd Annu. Int. Mtg., Soc. Expl. Geophys., Extended Abstracts, 471–474.

Picotti, S., Carcione, J.M., Santos, J.E., and Gei, D., 2010, Q-anisotropy in finely-layered media: Geophys. Res. Lett. 37, L06302, doi:10.1029/2009GL042046. Correction: 38, L10307, doi:10.1029/2011GL047613.

Picotti, S., Santos, J.E., Carcione, J.M., 2012. Oscillatory numerical experiments in finely layered anisotropic viscoelastic media. Comput. Geosci. 43, 83–89.

Postma, G.W., 1955. Wave propagation in a stratified medium. Geophysics 20, 780–806.

Pride, S.R., Berryman, J.G., Harris, J.M., 2004. Seismic attenuation due to wave-induced flow. J. Geophys. Res 109, 1–19, B01201.

Pinna, G., Carcione, J.M., Poletto, F., 2011. Kerogen to oil conversion in source rocks. Pore-pressure build-up and effects on seismic velocities. J. Appl. Geophys. 74, 229–235.

Pyrak-Nolte, L.J., Myer, L.R., Cook, N.G.W., 1990. Transmission of seismic waves across single natural fractures. J. Geophys. Res. 95, 8617–8638.

Quintal, B., Steeb, H., Frehner, M., Schmalholz, S.M., 2011. Quasi-static finite-element modeling of seismic attenuation and dispersion due to wave-induced fluid flow in poroelastic media. J. Geophys. Res. 116, B01201, doi: 10.1029/2010JB007475.

Quintal, B., Steeb, H., Frehner, M., Schmalholz, S.M., Saenger, E.H., 2012. Pore fluid effects on S-wave attenuation caused by wave-induced fluid flow. Geophysics 77, L13–L23.

Raviart, P.A., Thomas, J.M., 1975. Mixed finite element method for 2nd order elliptic problems. Mathematical Aspects of the Finite Element Methods, Lecture Notes of Mathematics vol. 606 Springer, Berlin Heidelberg.

Reuss, A., 1929. Berechnung der Fliessgrenzen von Mischkristallen auf Grund der Plastizitätsbedingung für Einkristalle. Z. Angew. Math. Mech. 9, 49–58.

Riznichenko, Y.V., 1949. Seismic quasi-anisotropy. Bull. Acad. Sci. USSR Geograph. Geophys. Serv. 13, 518–544.

Santos, J.E., Carcione, J.M., Picotti, S., 2011. Viscoelastic-stiffness tensor of anisotropic media from oscillatory numerical experiments. Comput. Methods Appl. Mech. Eng. 200, 896–904.

Santos, J.E., Picotti, S., Carcione, J.M., 2012. Evaluation of the stiffness tensor of a fractured medium with harmonic experiments. Comput. Methods Appl. Mech. Eng. 247–248, 130–145.

Santos, J.E., Rubino, J.G., Ravazzoli, C.L., 2009. A numerical upscaling procedure to estimate effective plane wave and shear moduli in heterogeneous fluid-saturated poroelastic media. Comput. Methods Appl. Mech. Eng. 198, 2067–2077.

Schoenberg, M., 1980. Elastic wave behavior across linear slip interfaces. J. Acous. Soc. Am. 68, 1516–1521.

Schoenberg, M., 1983. Reflection of elastic waves from periodically stratified media with interfacial slip. Geophys. Prosp. 31, 265–292.

Schoenberg, M., Dean, S., Sayers, C., 1999. Azimuth-dependent tuning of seismic waves reflected from fractured reservoirs. Geophysics 64, 1160–1171.

Schoenberg, M., Douma, J., 1988. Elastic wave propagation in media with parallel fractures and aligned cracks. Geophys. Prosp. 36, 571–590.

Schoenberg, M., Helbig, K., 1997. Orthorhombic media: modeling elastic wave behavior in a vertically fractured earth. Geophysics 62, 1954–1974.

Schoenberg, M., Muir, F., 1989. A calculus for finely layered media. Geophysics 54, 581–589.

Schoenberg, M., Muir, F., 1991. Reply by authors to J. A. Hudson and S. Crampin. Geophysics 56, 575–576.

Scholz, C., 2002. The Mechanics of Earthquakes and Faulting, 2nd edition. Cambridge University Press, Cambridge.

Sidler, R., Carcione, J.M., Holliger, K., 2008. On the evaluation of the plane-wave reflection coefficients in anelastic media. Geophys. J. Int. 175, 94–102.

Steen, O., Sverdrup, E., Hanssen, T.H., 1998. Predicting the distribution of small faults in a hydrocarbon reservoir by combining outcrop, seismic and well data. Geol. Soc. Special Publicat. 147, 27–50.

Thomsen, L., 1986. Weak elastic anisotropy. Geophysics 51, 1954–1966.

Vernik, L., 1994. Hydrocarbon-generation-induced microcraking of source rocks. Geophysics 59, 555–563.

Vernik, L., Nur, A., 1992. Ultrasonic velocity and anisotropy of hydrocarbon source rocks. Geophysics 57, 727–735.

White, J.E., Mikhaylova, N.G., Lyakhovitskiy, F.M., 1975. Low-frequency seismic waves in fluid saturated layered rocks. Izvestija Academy of Sciences USSR Phys. Solid Earth 11, 654–659.

Wood, A.W., 1955. A Textbook of Sound. MacMillan Co., New York.

Wright, J., 1987. The effects of transverse isotropy on reflection amplitude versus offset. Geophysics 52, 564–567.

Xu, T.F., Pruess, K., 2001. Modeling multiphase non-isothermal fluid flow and reactive geochemical transport in variably saturated fractured rocks: 1. Methodology. Am. J. Sci. 301, 16–33.

Zhang, J., 2005. Elastic wave modeling in fractured media with an explicit approach. Geophysics 70, T75–T85.

Zhang, J., Gao, H., 2009. Elastic wave modelling in 3-D fractured media: an explicit approach. Geophys. J. Int. 177, 1233–1241.

Chapter 5

Characteristics of Seismic Wave Propagation in Viscoelastic Anisotropic Fractured Reservoirs

Xianzheng Zhao[1], Xuming Bai[2], Qizhen Du[3]
[1]*Huabei Oilfield Branch, China National Petroleum Corp., Renqiu, Hebei, China;*
[2]*Bureau of Geophysical Prospecting INC., China National Petroleum Corporation, Zhuozhou, Hebei, China;*
[3]*School of Geosciences, China University of Petroleum (East China), Qingdao, China*

Chapter Outline

Seismic Exploration of Hydrocarbons in Heterogeneous Reservoirs. http://dx.doi.org/10.1016/B978-0-12-420151-4.00005-2

5.1 INTRODUCTION

As an important type of hydrocarbon reservoirs, fractured reservoirs formed by fracture systems can serve as both hydrocarbon reserve spaces and percolation paths for oil and gas. The exploration and development of oil/gas in fractured reservoirs have become one of the focuses of applied geophysical research.

Based on the long-wavelength approximation and scattering theory analysis, Hudson (1980, 1981, 1986) presented an effective theoretical model to describe wave characteristics in the elastic medium containing coin-shaped thin ellipsoidal fractures/inclusions. The theory is applicable in cases of thin ellipsoidal fractures with low distribution density and small aspect ratio. In this theory, fractures can be aligned or in random arrangement and can be void or filled with fluid or solids. According to the scattering theory, high-order interactions may be present between scattering objects. When the original Hudson theory was proposed, it only took into account the second-order scattering effects. Schoenberg (1980) (Schoenberg and Douma, 1988; Schoenberg and Sayers, 1995) neglected the shape and details of spatial distribution of fractures, and treated the medium containing parallel fractures as the infinitely thin and soft layers under the long-wavelength approximation. When seismic waves propagate through the fractures, the displacement tensor is discontinuous and the stress tensor is continuous at the fracture surface. Each tensor is linearly dependent on the other. The newly established theory is called linear slip theory. In this theory, the effective medium theory has relatively simpler forms, requires fewer undetermined parameters, and can describe fracture effects more accurately. The effect of fractures is described in the two excess weaknesses, and the equivalent compliance tensor of the fractured medium is expressed as the sum of solid compliance tensor and fracture compliance tensor. Linear slip theory has laid the foundation for the research on effective fractured medium theory. For porous rocks with parallel fractures, Thomsen (1995) studied the influence of presence of fractures on the wave nature of effective medium for a fractured reservoir with unique fracture sizes based on weak anisotropy (Thomsen, 1986). However, this theory only considers the fracture-induced anisotropy and does not consider the effects of viscoelasticity of fractures and the fluid in fractures on seismic wavefields.

A lot of work has been done for solid viscoelasticity theory and its application. The Kelvin–Voigt model and Maxwell model are commonly used in discussing the stress–strain relationships in viscoelastic medium. In order to better describe the characteristics of most viscoelastic media, a standard linear solid (SLS) model (Aki and Richards, 1980) has been developed. SLS simultaneously considers the sudden changes in strains and residual strains and the characteristics of creeps. Casula and Carcione (1992) compared the attenuation models of Kelvin–Voigt body, Maxwell body, standard linear body, and the two extended standard linear bodies for velocity dispersion and attenuation changes. The Kelvin–Voigt model, which is a linear viscoelasticity model, is often used to describe the viscoelasticity of solid rocks (Shan, 2007; Jia, 2011; Fan and Teng, 1988; Nie and Yang, 2008; Nie et al., 2008, 2012) in typical seismic band. The main reason is that the theoretical form of Kelvin–Voigt model is simple, implementation is straightforward, and the elastic wave attenuation increases with frequency. Although high-frequency components have excessive attenuation in this model, it is able to describe the low-frequency (0–100 Hz) attenuation behavior in the studies of seismic exploration (Fan and Teng, 1988). Seismic waves received in the location far away from sources have relatively weak high-frequency components and strong low-frequency components. Fan and Teng (1988) showed that, with the increase of propagation distance in earth medium, seismic waveforms recorded deviate from the source wavelet and the amplitude spectrum gradually shifts to the low-frequency end. Therefore, earth medium can be treated as Kelvin–Voigt viscoelastic body in seismic band. The computation based on Kelvin–Voigt is simple and no need to have additional memory during forward modeling. Carcione et al. (2004) used this model for 3D modeling of viscoelastic waves.

Researchers have done many studies on the forward modeling of waves in viscoelastic anisotropic medium. Carcione et al. (1988) and Carcione (1993) systematically worked on forward modeling of wave propagation in viscoelastic anisotropic media. Since the time-domain forward modeling of waves in viscoelastic medium cannot be done directly, memory variables were introduced to avoid the convolution operation between stress and strain. For simulating seismic wave dispersion and attenuation in real medium, Robertsson et al. (1994) developed a finite-difference algorithm on the basis of the viscoelasticity model to simulate the waves in 2D and 3D space, respectively. Later, Stekl and Pratt (1998) simulated the waves in viscoelastic medium in frequency domain. Dong and McMechan (1995) conducted wavefield numerical simulations in viscoelastic anisotropic medium using a 3D staggered-grid finite-difference technique and obtained the multicomponent multiazimuth modeling results.

By realizing the issues of stability and accuracy of standard staggered grid (SSG) (Virieux, 1984, 1986) method in wave simulation of viscoelastic anisotropic medium, researchers started to use the rotated staggered grid (RSG) in forward modeling (Saenger et al., 2000). Saegner and Bohlen (2002, 2004) simulated seismic waves in the viscoelastic medium and the anisotropic medium with a RSG finite-difference algorithm. Chu et al. (2009) analyzed the numerical dispersion and stability conditions of the high-order convolution finite-difference operator and compare it with the traditional finite-difference scheme. They found that the convolution finite-difference operator have better dispersion features and higher computation efficiency and make it a good choice for RSG. Forward modeling in 3D anisotropic medium was performed using this operator based on RSG finite-difference method. Du et al. (2011) started from the first-order velocity–stress equations in TTI medium and numerically simulated the three-component elastic waves in 2D TTI medium using free surface boundary condition and RSG high-order finite-difference algorithm. Yan and Liu (2012) derived the three-component velocity–stress equations of a 2D viscoelastic TTI medium based on Carcione's viscoelastic anisotropic theory and modeled the wave fields with a RSG finite-difference algorithm.

In this chapter, an effective medium model that is suitable for analyzing wave phenomena in the fractured reservoirs is formulated based on previous studies. Three-dimensional seismic forward modeling is implemented by applying a RSG finite-difference algorithm, and the effect of the types of rocks and pore fluid and the fracture development patterns on seismic wave propagation is analyzed.

5.2 EFFECTIVE MEDIUM MODEL OF VISCOELASTIC ANISOTROPIC FRACTURED RESERVOIRS

Assuming that viscoelasticity of fractured reservoirs is mainly caused by rock minerals, the Kelvin–Voigt model of viscoelasticity is applicable to solid contents. In the long-wavelength approximation, anisotropy caused by the parallel fractures can be described by Schoenberg's linear slip theory. The relationships between the stiffness parameters and the microscopic fracture properties in a viscoelastic fractured medium can then be established by means of the studies related to Hudson's theory. From this a viscoelastic anisotropic effective medium model that is suitable for modeling seismic wave propagation in fractured reservoirs is developed.

5.2.1 Characterization of Rock Mineral Viscoelasticity

Based on the homogeneous elastic isotropic model of solid rocks, Kelvin–Voigt viscoelasticity is introduced to build a viscoelastic mineral mixture.

In real earth medium, the mechanical properties of rocks are described by the generalized Hooke's Law (Figure 5.1). For the 1D Kelvin–Voigt model, that is, $\sigma = C \cdot \varepsilon$, where σ is the stress, ε is the strain, and C is the rock coefficient of a system containing of a spring and a damper in parallel. For viscoelastic medium, C is a combined tensor of elasticity coefficients and viscoelasticity coefficients. The correct expression of $\boldsymbol{C}$ quantitatively

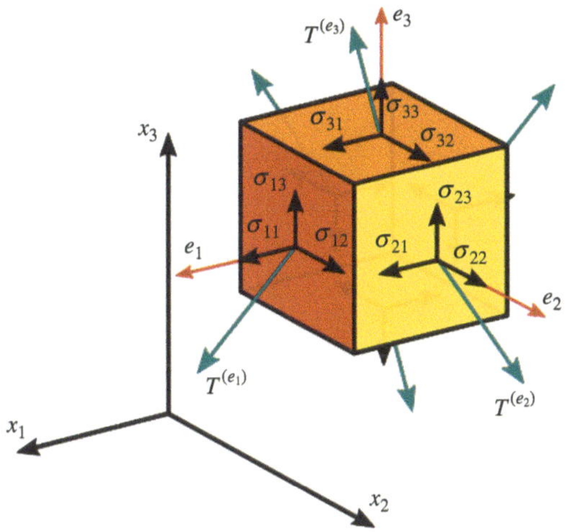

FIGURE 5.1 Schematic diagram of the stress analysis on a elastic body.

reflects the "viscoelastic anisotropy" nature of a medium with fractures saturated with different fluids.

According to the generalized 3D Hooke's Law with Kelvin–Voigt viscoelasticity model, the constitutive relation of a homogeneous isotropic viscoelastic medium can be expressed as follows:

$$\sigma = \sigma_E + \sigma_V = \left(C^E + i\omega \cdot C^V\right)\varepsilon \tag{5.1}$$

where σ_E is the elastic stress, σ_V is the viscous stress, ε is the strain, and $C^E + i\omega C^V$ is the overall viscoelastic coefficient tensor of the medium. C^E denotes the elastic tensor, C^V denotes the viscous tensor, i is the imaginary unit, and ω is the angular frequency.

For elastic homogeneous isotropic medium, the stiffness matrix is expressed as

$$C^E = \begin{bmatrix} M^e & \lambda^e & \lambda^e & 0 & 0 & 0 \\ \lambda^e & M^e & \lambda^e & 0 & 0 & 0 \\ \lambda^e & \lambda^e & M^e & 0 & 0 & 0 \\ 0 & 0 & 0 & \mu^e & 0 & 0 \\ 0 & 0 & 0 & 0 & \mu^e & 0 \\ 0 & 0 & 0 & 0 & 0 & \mu^e \end{bmatrix} \tag{5.2}$$

where the superscript e denotes the elasticity. $M^e = \lambda^e + 2\mu^e$, where λ^e and μ^e denote Lame constants of the elastic material.

Using Kelvin–Voigt viscoelasticity model and including linear viscosity, the viscosity tensor is expressed as

$$C^V = \begin{bmatrix} M^v & \lambda^v & \lambda^v & 0 & 0 & 0 \\ \lambda^v & M^v & \lambda^v & 0 & 0 & 0 \\ \lambda^v & \lambda^v & M^v & 0 & 0 & 0 \\ 0 & 0 & 0 & \mu^v & 0 & 0 \\ 0 & 0 & 0 & 0 & \mu^v & 0 \\ 0 & 0 & 0 & 0 & 0 & \mu^v \end{bmatrix} \tag{5.3}$$

where the superscript v denotes the viscosity. $M^v = \lambda^v + 2\mu^v$. λ^v and μ^v denote the viscosity coefficients of the medium, which are related to the Lame constants and P-/S-wave quality factors

$$\begin{aligned} Q_p &= \frac{\lambda^e + 2\mu^e}{\lambda^v + 2\mu^v}\frac{1}{\omega_c} \\ Q_s &= \frac{\mu^e}{\mu^v}\frac{1}{\omega_c} \end{aligned} \tag{5.4}$$

where ω_c denotes the characteristic frequency corresponding to the specific Q_P and Q_S.

λ^v and μ^v can be written as

$$\begin{aligned} \lambda^v &= \frac{\lambda^e + 2\mu^e}{Q_p\omega_c} - \frac{2\mu^e}{Q_s\omega_c} \\ \mu^v &= \frac{\mu^e}{Q_s\omega_c} \end{aligned} \tag{5.5}$$

The stiffness matrix of the viscoelastic isotropic homogeneous rock skeleton is expressed as follows:

$$\mathbf{C} = C^E + i\omega \cdot C^V = \begin{bmatrix} M & \lambda & \lambda & 0 & 0 & 0 \\ \lambda & M & \lambda & 0 & 0 & 0 \\ \lambda & \lambda & M & 0 & 0 & 0 \\ 0 & 0 & 0 & \mu & 0 & 0 \\ 0 & 0 & 0 & 0 & \mu & 0 \\ 0 & 0 & 0 & 0 & 0 & \mu \end{bmatrix} \tag{5.6}$$

where $\lambda = \lambda^e + i\omega\lambda^v$, $\mu = \mu^e + i\omega\mu^v$, and $M = \lambda + 2\mu$.

5.2.2 Characterization of Fracture-Induced Anisotropy

While building the model, we assume that the viscoelastic fractured medium is composed of a viscoelastic isotropic homogeneous host solid and the inserted parallel fractures are saturated with one type of fluid. In order to build an effective medium model to describe such type of rocks, we use Schoenberg's linear slip theory to include the anisotropy that is induced by parallel fractures imbedded in a homogeneous host rock.

Linear slip theory neglects fracture shapes and microstructures. The effect of parallel fractures is reflected on the normal weakness Δ_N and the tangential weakness Δ_T. Therefore, after including anisotropy of the parallel fractures on the basis of host rock, the stiffness matrix of a viscoelastic anisotropic (HTI) medium is expressed as

$$C_{\mathrm{HTI}} = \begin{bmatrix} M(1-\Delta_N) & \lambda(1-\Delta_N) & \lambda(1-\Delta_N) & 0 & 0 & 0 \\ \lambda(1-\Delta_N) & M(1-\xi^2\Delta_N) & \lambda(1-\xi\Delta_N) & 0 & 0 & 0 \\ \lambda(1-\Delta_N) & \lambda(1-\xi\Delta_N) & M(1-\xi^2\Delta_N) & 0 & 0 & 0 \\ 0 & 0 & 0 & \mu & 0 & 0 \\ 0 & 0 & 0 & 0 & \mu(1-\Delta_T) & 0 \\ 0 & 0 & 0 & 0 & 0 & \mu(1-\Delta_T) \end{bmatrix} \tag{5.7}$$

where $\xi = \lambda / M$.

Thus, the Lame constants λ^E and μ^E and the viscosity coefficients λ^V and μ^V of host solid as well as the fracture-induced normal weakness Δ_N and tangential weakness Δ_T are essential for characterization of the anisotropic media with viscoelastic fractures. Lame constants λ^E and μ^E are related to the P-/S-wave velocities and density of the elastic isotropic solid. Viscosity coefficients λ^V and μ^V are related to the P-/S-wave quality factors of the viscoelastic solid. The relationships between fracture weakness parameters Δ_N and Δ_T and the fracture microscopic parameters and fluid properties are determined by using Hudson's fracture theory.

5.2.3 Characterization of the Elasticity Coefficients in a Viscoelastic Medium With Vertical Fractures

In order to relate the coefficients of fracture weakness and the microscopic structure parameters, we need to build the connection between Schoenberg's linear slip theory and Hudson's fracture theory. We make long-wavelength approximation and assume the embedded fractures are parallel. The physical quantities describing fracture-induced anisotropy in the two theories are related. The relationships between the fracture weakness Δ_N and Δ_T of the Schoenberg's linear slip theory and the fracture response parameters U_{11} and U_{33} of the Hudson's theory can be expressed as (Schoenberg and Douma, 1988):

$$\begin{aligned} \Delta_N &= \frac{\lambda + 2\mu}{\mu} U_{33} e \\ \Delta_T &= U_{11} e \end{aligned} \tag{5.8}$$

Assuming that the rock contains only one type of fluid and no fluid flow occurs in the fractures, the normal weakness Δ_N and tangential weakness Δ_T can be expressed as

$$\Delta_N = \frac{4e}{3g(1-g)\left(1+\dfrac{K_f}{\pi\alpha\mu(1-g)}\right)}$$
$$\Delta_T = \frac{16e}{3(3-2g)} \tag{5.9}$$

where e is fracture density, α is fracture aspect ratio, and K_f is bulk modulus of fracture fluid ($g = \mu/M$).

By substituting the expressions of Δ_N and Δ_T in (5.9) into formula (5.7), the equivalent stiffness matrix of a vertically fractured viscoelastic medium can be expressed as a function of solid's P-/S-wave velocities, P-/S-wave quality factors, fracture density, aspect ratio, and fluid parameters. It is obvious that the anisotropy of the fractured medium increases with the increase of fracture density and aspect ratio, or the decrease of fluid bulk modulus.

During forward modeling, the presence of $i\omega$-related complex operations in the stiffness matrix of HTI medium, TTI medium, and the more complex anisotropic medium leads to complicated finite-difference algorithm. The stiffness matrix needs to be simplified during numerical simulation.

For $\boldsymbol{C}_{\text{HTI}}$, the term outside the parentheses of each stiffness element is $i\omega$-related first derivative. In numerical simulation with velocity–stress equations, it is easy to implement by transforming the terms into time-related first derivatives. However, $i\omega$-related expressions inside the parentheses are complex and will increase numerical dispersions. Therefore, in this work, we mainly considered how to simplify the terms in the parentheses to eliminate the effects of $i\omega$.

Δ_N and Δ_T are expanded as follows:

$$\begin{aligned}
\Delta_N &= \frac{4e\pi\alpha M^2}{3\left(\pi\alpha\mu(\lambda+\mu)+K_f M\right)} \\
&= \frac{4e\pi\alpha\left(\lambda^e+2\mu^e+i\omega\left(\lambda^v+2\mu^v\right)\right)^2}{3\left(\pi\alpha\left(\mu^e+i\omega\mu^v\right)\left(\lambda^e+2\mu^e+i\omega\left(\lambda^v+2\mu^v\right)\right)+\left(\lambda^e+2\mu^e+i\omega\left(\lambda^v+2\mu^v\right)K_f\right)\right.} \\
\Delta_T &= \frac{16eM}{3(3\lambda+4\mu)} \\
&= \frac{16e\left(\lambda^e+2\mu^e+i\omega\left(\lambda^v+2\mu^v\right)\right)}{3\left(3\lambda^e+4\mu^e+i\omega\left(3\lambda^v+4\mu^v\right)\right)}
\end{aligned} \tag{5.10}$$

In the stiffness matrix of the HTI medium, the terms involving Δ_N and Δ_T include

$$\begin{aligned}
\xi\Delta_N &= \frac{4e\pi\alpha\lambda M}{3\left(\pi\alpha\mu(\lambda+\mu)+K_f M\right)} \\
&= \frac{4e\pi\alpha\left(\lambda^e+i\omega\lambda^v\right)\left(\lambda^e+2\mu^e+i\omega\left(\lambda^v+2\mu^v\right)\right)}{3\left(\pi\alpha\left(\mu^e+i\omega\mu^v\right)\left(\lambda^e+\mu^e+i\omega\left(\lambda^v+\mu^v\right)\right)+\left(\lambda^e+2\mu^e+i\omega\left(\lambda^v+2\mu^v\right)\right)K_f\right)} \\
\xi^2\Delta_N &= \frac{4e\pi\alpha\lambda^2}{3\left(\pi\alpha\mu(\lambda+\mu)+K_f M\right)} \\
&= \frac{4e\pi\alpha\left(\lambda^e+i\omega\lambda^v\right)^2}{3\left(\pi\alpha\left(\mu^e+i\omega\mu^v\right)\left(\lambda^e+\mu^e+i\omega\left(\lambda^\mu+\mu^v\right)\right)+\left(\lambda^e+2\mu^e+i\omega\left(\lambda^\mu+2\mu^v\right)\right)K_f\right)}
\end{aligned} \tag{5.11}$$

We use the simplification of Δ_N as an example. Its numerator and denominator are both expanded as the $i\omega$-related polynomials

$$\begin{aligned}
\Delta_N^{N\mu} &= 4e\pi\alpha(\lambda^{e2}+4\lambda^e\mu^e+4\mu^{e2}) \\
&\quad + 4e\pi\alpha\left[2i\lambda^e\left(\lambda^v+2\mu^v\right)+4i\mu^e\left(\lambda^v+2\mu^v\right)\right]\omega \\
&\quad - 4e\pi\alpha\left(\lambda^v+2\mu^v\right)^2\omega^2 \\
\Delta_N^{De} &= 3\left(\pi\alpha\lambda^e\mu^e+\pi\alpha\mu^{e2}+\lambda^e K_f+2\mu^e K_f\right) \\
&\quad + 3\omega\left[i\pi\alpha\lambda^e\mu^v+i\pi\alpha\mu^e\mu^v+i\pi\alpha\mu^e\left(\lambda^v+\mu^v\right)+i\left(\lambda^v+2\mu^v\right)K_f\right] \\
&\quad - 3\pi\alpha\mu^v\left(\lambda^v+\mu^v\right)\omega^2
\end{aligned} \tag{5.12}$$

According to the definition of viscosity coefficients, we have

$$\begin{aligned}
\lambda^v+2\mu^v &= \frac{\lambda^e+2\mu^e}{Q_p\omega_c} \\
\mu^v &= \frac{\mu^e}{Q_s\omega_c}
\end{aligned} \tag{5.13}$$

Therefore,

$$\begin{aligned}
\Delta_N^{N\mu} &= 4e\pi\alpha(\lambda^e+2\mu^e)^2 \\
&\quad + 8e\pi\alpha(\lambda^e+2\mu^e)^2 i\omega / Q_p\omega_c \\
&\quad - 4e\pi\alpha(\lambda^e+2\mu^e)^2\omega^2 / Q_P^2\omega_c^2 \\
\Delta_N^{De} &= 3\left[\left(\pi\alpha\mu^e+K_f\right)\left(\lambda^e+2\mu^e\right)-\pi\alpha\mu^{e2}\right] \\
&\quad + 3i\omega\left[(\pi\alpha\mu^e+K_f)(\lambda^e+2\mu^e)/Q_p w_c+\pi\alpha\lambda^e\mu^e/Q_s\omega_c\right] \\
&\quad - 3\omega^2\left[\pi\alpha\mu^e\left(\lambda^e+2\mu^e\right)/Q_P Q_s\omega_c^2-\pi\alpha\mu^{e2}/Q_s^2\omega_c^2\right]
\end{aligned} \tag{5.14}$$

In the numerator and denominator of ΔN, the corresponding coefficient of $i\omega$-related items is $1/Q_P\ \omega_c$ (or $1/Q_S\ \omega_c$). Therefore, by considering the typical values of $Q_P\ \omega_c$ or $Q_S\ \omega_c$ (assuming Q_P are equal to Q_S, with a value in range of 10–100, and $\omega_c = 2\pi f$ ($f = 30$ Hz)), the coefficients of $i\omega$ are all in the magnitude of 10^{-4}–10^{-5}. Thus, the $i\omega$-related first-order and second-order items in Δ_N can be approximately omitted, so $\Delta_N{}^{sp}$ is simplified as

$$\Delta_N^{sp} = \frac{4e\pi\alpha M^{e2}}{3\left(\pi\alpha\mu^e\left(\lambda^e+\mu^e\right)+K_f M^e\right)}. \tag{5.15}$$

Similarly, in the effective stiffness matrix C^{HTI} of a viscoelastic anisotropic medium, the $i\omega$-related items $\xi_{\Delta N}$, $\xi^2\Delta_N$, and Δ_T in the parentheses can be simplified to contain Lame constants of the elastic medium without containing $i\omega$-related items. Therefore, the simplified normal and tangential weaknesses can be written as

$$\begin{aligned}\Delta_N^{sp} &= \frac{4e\pi\alpha M^{e2}}{3\left(\pi\alpha\mu^e\left(\lambda^e+\mu^e\right)+K_f M^e\right)}\\ \Delta_T^{sp} &= \frac{16eM^e}{3\left(3\lambda^e+4\mu^e\right)}\end{aligned} \tag{5.16}$$

and $\xi^{sp} = \lambda^e/M^e = 1-2g^{sp}$, $g^{sp} = \mu^e/M^e$.

The stiffness matrix of a viscoelastic HTI medium can be expressed in the form of a complex matrix

$$C_{\mathrm{HTI}} = C_{\mathrm{HTI}}^R + i\omega C_{\mathrm{HTI}}^I \tag{5.17}$$

where

$$\begin{aligned}C_{\mathrm{HTI}}^R &= \begin{bmatrix} M^e(1-\Delta_N^{sp}) & \lambda^e(1-\Delta_N^{sp}) & \lambda^e(1-\Delta_N^{sp}) & 0 & 0 & 0\\ \lambda^e(1-\Delta_N^{sp}) & M^e(1-\xi^{sp2}\Delta_N^{sp}) & \lambda^e(1-\xi^{sp}\Delta_N^{sp}) & 0 & 0 & 0\\ \lambda^e(1-\Delta_N^{sp}) & \lambda^e(1-\xi^{sp}\Delta_N^{sp}) & M^e(1-\xi^{sp2}\Delta_N^{sp}) & 0 & 0 & 0\\ 0 & 0 & 0 & \mu^e & 0 & 0\\ 0 & 0 & 0 & 0 & \mu^e(1-\Delta_T^{sp}) & 0\\ 0 & 0 & 0 & 0 & 0 & \mu^e(1-\Delta_T^{sp}) \end{bmatrix}\\ C_{\mathrm{HTI}}^I &= \begin{bmatrix} M^v(1-\Delta_N^{sp}) & \lambda^v(1-\Delta_N^{sp}) & \lambda^v(1-\Delta_N^{sp}) & 0 & 0 & 0\\ \lambda^v(1-\Delta_N^{sp}) & M^v(1-\xi^{sp2}\Delta_N^{sp}) & \lambda^v(1-\xi^{sp}\Delta_N^{sp}) & 0 & 0 & 0\\ \lambda^v(1-\Delta_N^{sp}) & \lambda^v(1-\xi^{sp}\Delta_N^{sp}) & M^v(1-\xi^{sp2}\Delta_N^{sp}) & 0 & 0 & 0\\ 0 & 0 & 0 & \mu^e & 0 & 0\\ 0 & 0 & 0 & 0 & \mu^v(1-\Delta_T^{sp}) & 0\\ 0 & 0 & 0 & 0 & 0 & \mu^v(1-\Delta_T^{sp}) \end{bmatrix}\end{aligned} \tag{5.18}$$

5.2.4 Characterization of the Elasticity Coefficients of a Viscoelastic Medium With Tilted Fractures

In real earth medium, the patterns of the developed fractures are very complex. Most fractures are not vertically but with certain dips and azimuths. In this case, the fractured medium is equivalent to a viscoelastic TTI medium and its stiffness matrix in Cartesian coordinate is no longer to the same as the HTI medium's constitutive equations in the constitutive coordinate. For TTI medium with a tilted axis of symmetry, there is an angle between its constitutive coordinates and the observation coordinates. θ is the polarization angle, which is the angle between the symmetry axis of TTI medium in the plane *OXZ* and the axis Z of the observation coordinates. ϕ is the angle between the symmetry axis of TTI medium in the plane *OXY* and the axis *X* (Figure 5.2).

Using bond transformation, the elasticity coefficients of TTI medium with any azimuths and dips can be transformed from the constitutive coordinates to the Cartesian coordinates. Subsequent wave equation derivation and solution can be facilitated. Bond transformation matrix includes the polarization transformation matrix M_θ and the orientation transformation matrix M_ϕ

$$M_\theta = \begin{bmatrix} \cos^2\theta & 0 & \sin^2\theta & 0 & -\sin 2\theta & 0 \\ 0 & 1 & 0 & 0 & 0 & 0 \\ \sin^2\theta & 0 & \cos^2\theta & 0 & \sin 2\theta & 0 \\ 0 & 0 & 0 & \cos\theta & 0 & \sin\theta \\ \frac{1}{2}\sin 2\theta & 0 & -\frac{1}{2}\sin 2\theta & 0 & \cos 2\theta & 0 \\ 0 & 0 & 0 & -\sin\theta & 0 & \cos\theta \end{bmatrix}$$
$$M_\phi = \begin{bmatrix} \cos^2\phi & \sin^2\phi & 0 & 0 & 0 & -\sin 2\phi \\ \sin^2\phi & \cos^2\phi & 0 & 0 & 0 & \sin 2\phi \\ 0 & 0 & 1 & 0 & 0 & 0 \\ 0 & 0 & 0 & \cos\phi & \sin\phi & 0 \\ 0 & 0 & 0 & -\cos\phi & \cos\phi & 0 \\ \frac{1}{2}\sin 2\phi & -\frac{1}{2}\sin 2\phi & 0 & 0 & 0 & \cos 2\phi \end{bmatrix} \quad (5.19)$$

In Cartesian observation system, the elastic coefficient matrix of a TTI medium is

$$C_{\mathrm{TTI}} = M_\phi M_\theta C_{\mathrm{HTI}} M_\theta^T M_\varphi^T \quad (5.20)$$

where the superscript *T* denotes the matrix transpose.

We have,

$$C_{\mathrm{TTI}} = C_{\mathrm{TTI}}^R + i\omega C_{\mathrm{TTI}}^I \quad (5.21)$$

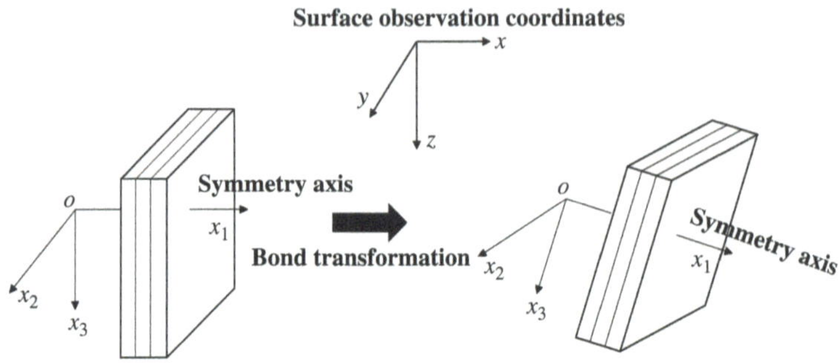

FIGURE 5.2 Schematic diagrams of fractures with different occurrences.

where

$$
\begin{aligned}
C_{\mathrm{TTI}}^{R} &= M_{\phi} M_{\theta} C_{\mathrm{HTI}}^{R} M_{\theta}^{T} M_{\varphi}^{T} \\
C_{\mathrm{TTI}}^{I} &= M_{\phi} M_{\theta} C_{\mathrm{HTI}}^{I} M_{\theta}^{T} M_{\varphi}^{T}
\end{aligned}
\tag{5.22}
$$

5.3 NUMERICAL SIMULATION OF WAVEFIELD IN VISCOELASTIC ANISOTROPIC FRACTURED MEDIUM

In this chapter, we first derive the first-order velocity–stress wave equations to describe seismic wave propagation in viscoelastic anisotropic fractured medium. We use time and space domain RSG finite-difference algorithm and the corresponding boundary conditions for numerical simulation of seismic wavefield in fractured reservoirs.

5.3.1 Derivation of Viscoelastic Anisotropic Velocity–Stress Equations

The general form of viscoelastic coefficient matrix in a viscoelastic TTI fractured medium is

$$
C_{\mathrm{TTI}} = \begin{bmatrix}
C_{11} & C_{12} & C_{13} & C_{14} & C_{15} & C_{16} \\
C_{12} & C_{22} & C_{23} & C_{24} & C_{25} & C_{26} \\
C_{13} & C_{23} & C_{33} & C_{34} & C_{35} & C_{36} \\
C_{14} & C_{24} & C_{34} & C_{44} & C_{45} & C_{46} \\
C_{15} & C_{25} & C_{35} & C_{45} & C_{55} & C_{56} \\
C_{16} & C_{26} & C_{36} & C_{46} & C_{56} & C_{66}
\end{bmatrix}
\tag{5.23}
$$

Based on equation (5.21), it can be transformed into the frequency domain. Each element of the matrix can be expressed as the sum of one term with no time partial-derivative and the other term with time partial-derivative,

$$C_{ij} = C_{ij}^{R} + l_t C_{ij}^{I} \tag{5.24}$$

where the superscripts R and I denote the real and imaginary parts, respectively. The subscripts i and j range from 1 to 6. l_t denotes time partial derivative.

The constitutive equations in a 3D viscoelastic anisotropic medium with tilted fractures are

$$\sigma_{\text{TTI}} = C_{\text{TTI}} \varepsilon_{\text{TTI}} \tag{5.25}$$

The constitutive equations in a viscoelastic TTI fractured medium are

$$\begin{aligned}
\sigma_{xx} &= \left(C_{11}^{R} + l_t C_{11}^{I}\right) e_{xx} + \left(C_{12}^{R} + l_t C_{12}^{I}\right) e_{yy} + \left(C_{13}^{R} + l_t C_{13}^{I}\right) e_{zz} \\
&\quad + \left(C_{14}^{R} + l_t C_{14}^{I}\right) e_{yz} + \left(C_{15}^{R} + l_t C_{15}^{I}\right) e_{zx} + \left(C_{16}^{R} + l_t C_{16}^{I}\right) e_{xy} \\
\sigma_{yy} &= \left(C_{12}^{R} + l_t C_{12}^{I}\right) e_{xx} + \left(C_{22}^{R} + l_t C_{22}^{I}\right) e_{yy} + \left(C_{23}^{R} + l_t C_{23}^{I}\right) e_{zz} \\
&\quad + \left(C_{24}^{R} + l_t C_{24}^{I}\right) e_{yz} + \left(C_{25}^{R} + l_t C_{25}^{I}\right) e_{zx} + \left(C_{26}^{R} + l_t C_{26}^{I}\right) e_{xy} \\
\sigma_{zz} &= \left(C_{13}^{R} + l_t C_{13}^{I}\right) e_{xx} + \left(C_{23}^{R} + l_t C_{23}^{I}\right) e_{yy} + \left(C_{33}^{R} + l_t C_{33}^{I}\right) e_{zz} \\
&\quad + \left(C_{34}^{R} + l_t C_{34}^{I}\right) e_{yz} + \left(C_{35}^{R} + l_t C_{35}^{I}\right) e_{zx} + \left(C_{36}^{R} + l_t C_{36}^{I}\right) e_{xy} \\
\sigma_{yz} &= \left(C_{14}^{R} + l_t C_{14}^{I}\right) e_{xx} + \left(C_{24}^{R} + l_t C_{24}^{I}\right) e_{yy} + \left(C_{34}^{R} + l_t C_{34}^{I}\right) e_{zz} \\
&\quad + \left(C_{44}^{R} + l_t C_{44}^{I}\right) e_{yz} + \left(C_{45}^{R} + l_t C_{45}^{I}\right) e_{zx} + \left(C_{46}^{R} + l_t C_{46}^{I}\right) e_{xy} \\
\sigma_{zx} &= \left(C_{15}^{R} + l_t C_{15}^{I}\right) e_{xx} + \left(C_{25}^{R} + l_t C_{25}^{I}\right) e_{yy} + \left(C_{35}^{R} + l_t C_{35}^{I}\right) e_{zz} \\
&\quad + \left(C_{45}^{R} + l_t C_{45}^{I}\right) e_{yz} + \left(C_{55}^{R} + l_t C_{55}^{I}\right) e_{zx} + \left(C_{56}^{R} + l_t C_{56}^{I}\right) e_{xy} \\
\sigma_{xy} &= \left(C_{16}^{R} + l_t C_{16}^{I}\right) e_{xx} + \left(C_{26}^{R} + l_t C_{26}^{I}\right) e_{yy} + \left(C_{36}^{R} + l_t C_{36}^{I}\right) e_{zz} \\
&\quad + \left(C_{46}^{R} + l_t C_{46}^{I}\right) e_{yz} + \left(C_{56}^{R} + l_t C_{56}^{I}\right) e_{zx} + \left(C_{66}^{R} + l_t C_{66}^{I}\right) e_{xy}
\end{aligned} \tag{5.26}$$

The geometric equations are expressed as

$$\begin{aligned}
e_{xx} &= \frac{\partial u_x}{\partial x} \\
e_{yy} &= \frac{\partial u_y}{\partial y} \\
e_{zz} &= \frac{\partial u_z}{\partial z} \\
e_{yz} &= \frac{\partial u_y}{\partial z} + \frac{\partial u_z}{\partial y} \\
e_{zx} &= \frac{\partial u_z}{\partial x} + \frac{\partial u_x}{\partial z} \\
e_{xy} &= \frac{\partial u_x}{\partial y} + \frac{\partial u_y}{\partial x}
\end{aligned} \tag{5.27}$$

By neglecting body force terms, the equilibrium equations of motion can be written as

$$\begin{aligned}
\rho\frac{\partial^2 u_x}{\partial t^2} &= \frac{\partial \sigma_{xx}}{\partial x}+\frac{\partial \sigma_{xy}}{\partial y}+\frac{\partial \sigma_{zx}}{\partial z}\\
\rho\frac{\partial^2 u_y}{\partial t^2} &= \frac{\partial \sigma_{xy}}{\partial x}+\frac{\partial \sigma_{yy}}{\partial y}+\frac{\partial \sigma_{yz}}{\partial z}\\
\rho\frac{\partial^2 u_z}{\partial t^2} &= \frac{\partial \sigma_{zx}}{\partial x}+\frac{\partial \sigma_{yz}}{\partial y}+\frac{\partial \sigma_{zz}}{\partial z}
\end{aligned} \tag{5.28}$$

The velocity–stress equations in a viscoelastic medium with parallel fractures can be derived by combining the constitutive equations, geometric equations, and equilibrium equations of motion.

Using the relation between particle velocities and particle displacement, the velocity equations can be written as

$$\begin{aligned}
\rho\frac{\upsilon_x}{\partial t} &= \frac{\partial \sigma_{xx}}{\partial x}+\frac{\partial \sigma_{xy}}{\partial y}+\frac{\partial \sigma_{zx}}{\partial z}\\
\rho\frac{\upsilon_y}{\partial t} &= \frac{\partial \sigma_{xy}}{\partial x}+\frac{\partial \sigma_{yy}}{\partial y}+\frac{\partial \sigma_{yz}}{\partial z}\\
\rho\frac{\upsilon_z}{\partial t} &= \frac{\partial \sigma_{zx}}{\partial x}+\frac{\partial \sigma_{yz}}{\partial y}+\frac{\partial \sigma_{zz}}{\partial z}
\end{aligned} \tag{5.29}$$

By merging the same terms with respect to the two items with/without time partial-derivative in the constitutive equations 5.26, the stress equations can be written as

$$\begin{aligned}
\sigma_{xx} &= \sigma_{xx}^R+\sigma_{xx}^I\\
\sigma_{yy} &= \sigma_{yy}^R+\sigma_{yy}^I\\
\sigma_{zz} &= \sigma_{zz}^R+\sigma_{zz}^I\\
\sigma_{yz} &= \sigma_{yz}^R+\sigma_{yz}^I\\
\sigma_{zx} &= \sigma_{zx}^R+\sigma_{zx}^I\\
\sigma_{xy} &= \sigma_{xy}^R+\sigma_{xy}^I
\end{aligned} \tag{5.30}$$

where

$$\begin{aligned}
\sigma_{xx}^R &= C_{11}^R e_{xx}+C_{12}^R e_{yy}+C_{13}^R e_{zz}+C_{14}^R e_{yz}+C_{15}^R e_{zx}+C_{16}^R e_{xy}\\
\sigma_{yy}^R &= C_{12}^R e_{xx}+C_{22}^R e_{yy}+C_{23}^R e_{zz}+C_{24}^R e_{yz}+C_{25}^R e_{zx}+C_{26}^R e_{xy}\\
\sigma_{zz}^R &= C_{13}^R e_{xx}+C_{23}^R e_{yy}+C_{33}^R e_{zz}+C_{34}^R e_{yz}+C_{35}^R e_{zx}+C_{36}^R e_{xy}\\
\sigma_{yz}^R &= C_{14}^R e_{xx}+C_{24}^R e_{yy}+C_{34}^R e_{zz}+C_{44}^R e_{yz}+C_{45}^R e_{zx}+C_{46}^R e_{xy}\\
\sigma_{zx}^R &= C_{15}^R e_{xx}+C_{25}^R e_{yy}+C_{35}^R e_{zz}+C_{45}^R e_{yz}+C_{55}^R e_{zx}+C_{56}^R e_{xy}\\
\sigma_{xy}^R &= C_{16}^R e_{xx}+C_{26}^R e_{yy}+C_{36}^R e_{zz}+C_{46}^R e_{yz}+C_{56}^R e_{zx}+C_{66}^R e_{xy}
\end{aligned} \tag{5.31}$$

$$
\begin{aligned}
\sigma_{xx}^I &= c_{11}^I \frac{\partial v_x}{\partial x} + c_{12}^I \frac{\partial v_y}{\partial y} + c_{13}^I \frac{\partial v_z}{\partial z} + c_{14}^I \left(\frac{\partial v_y}{\partial z} + \frac{\partial v_z}{\partial y}\right) + c_{15}^I \left(\frac{\partial v_x}{\partial z} + \frac{\partial v_z}{\partial x}\right) + c_{16}^I \left(\frac{\partial v_x}{\partial y} + \frac{\partial v_y}{\partial x}\right) \\
\sigma_{yy}^I &= c_{12}^I \frac{\partial v_x}{\partial x} + c_{22}^I \frac{\partial v_y}{\partial y} + c_{23}^I \frac{\partial v_z}{\partial z} + c_{24}^I \left(\frac{\partial v_y}{\partial z} + \frac{\partial v_z}{\partial y}\right) + c_{25}^I \left(\frac{\partial v_x}{\partial z} + \frac{\partial v_z}{\partial x}\right) + c_{26}^I \left(\frac{\partial v_x}{\partial y} + \frac{\partial v_y}{\partial x}\right) \\
\sigma_{zz}^I &= c_{13}^I \frac{\partial v_x}{\partial x} + c_{23}^I \frac{\partial v_y}{\partial y} + c_{33}^I \frac{\partial v_z}{\partial z} + c_{34}^I \left(\frac{\partial v_y}{\partial z} + \frac{\partial v_z}{\partial y}\right) + c_{35}^I \left(\frac{\partial v_x}{\partial z} + \frac{\partial v_z}{\partial x}\right) + c_{36}^I \left(\frac{\partial v_x}{\partial y} + \frac{\partial v_y}{\partial x}\right) \\
\sigma_{yz}^I &= c_{14}^I \frac{\partial v_x}{\partial x} + c_{24}^I \frac{\partial v_y}{\partial y} + c_{34}^I \frac{\partial v_z}{\partial z} + c_{44}^I \left(\frac{\partial v_y}{\partial z} + \frac{\partial v_z}{\partial y}\right) + c_{45}^I \left(\frac{\partial v_x}{\partial z} + \frac{\partial v_z}{\partial x}\right) + c_{46}^I \left(\frac{\partial v_x}{\partial y} + \frac{\partial v_y}{\partial x}\right) \\
\sigma_{zx}^I &= c_{15}^I \frac{\partial v_x}{\partial x} + c_{25}^I \frac{\partial v_y}{\partial y} + c_{35}^I \frac{\partial v_z}{\partial z} + c_{45}^I \left(\frac{\partial v_y}{\partial z} + \frac{\partial v_z}{\partial y}\right) + c_{55}^I \left(\frac{\partial v_x}{\partial z} + \frac{\partial v_z}{\partial x}\right) + c_{56}^I \left(\frac{\partial v_x}{\partial y} + \frac{\partial v_y}{\partial x}\right) \\
\sigma_{xy}^I &= c_{16}^I \frac{\partial v_x}{\partial x} + c_{26}^I \frac{\partial v_y}{\partial y} + c_{36}^I \frac{\partial v_z}{\partial z} + c_{46}^I \left(\frac{\partial v_y}{\partial z} + \frac{\partial v_z}{\partial y}\right) + c_{56}^I \left(\frac{\partial v_x}{\partial z} + \frac{\partial v_z}{\partial x}\right) + c_{66}^I \left(\frac{\partial v_x}{\partial y} + \frac{\partial v_y}{\partial x}\right)
\end{aligned}
\tag{5.32}
$$

Furthermore, the time partial derivatives of the real parts of the six stress components are

$$
\begin{aligned}
\frac{\sigma_{xx}^R}{\partial t} &= C_{11}^R \frac{\partial v_x}{\partial x} + C_{12}^R \frac{\partial v_y}{\partial y} + C_{13}^R \frac{\partial v_z}{\partial z} + C_{14}^R \left(\frac{\partial v_y}{\partial z} + \frac{\partial v_z}{\partial y}\right) + C_{15}^R \left(\frac{\partial v_x}{\partial z} + \frac{\partial v_z}{\partial x}\right) + C_{16}^R \left(\frac{\partial v_x}{\partial y} + \frac{\partial v_y}{\partial x}\right) \\
\frac{\sigma_{yy}^R}{\partial t} &= C_{12}^R \frac{\partial v_x}{\partial x} + C_{22}^R \frac{\partial v_y}{\partial y} + C_{23}^R \frac{\partial v_z}{\partial z} + C_{24}^R \left(\frac{\partial v_y}{\partial z} + \frac{\partial v_z}{\partial y}\right) + C_{25}^R \left(\frac{\partial v_x}{\partial z} + \frac{\partial v_z}{\partial x}\right) + C_{26}^R \left(\frac{\partial v_x}{\partial y} + \frac{\partial v_y}{\partial x}\right) \\
\frac{\sigma_{zz}^R}{\partial t} &= C_{13}^R \frac{\partial v_x}{\partial x} + C_{23}^R \frac{\partial v_y}{\partial y} + C_{33}^R \frac{\partial v_z}{\partial z} + C_{34}^R \left(\frac{\partial v_y}{\partial z} + \frac{\partial v_z}{\partial y}\right) + C_{35}^R \left(\frac{\partial v_x}{\partial z} + \frac{\partial v_z}{\partial x}\right) + C_{36}^R \left(\frac{\partial v_x}{\partial y} + \frac{\partial v_y}{\partial x}\right) \\
\frac{\sigma_{yz}^R}{\partial t} &= C_{14}^R \frac{\partial v_x}{\partial x} + C_{24}^R \frac{\partial v_y}{\partial y} + C_{34}^R \frac{\partial v_z}{\partial z} + C_{44}^R \left(\frac{\partial v_y}{\partial z} + \frac{\partial v_z}{\partial y}\right) + C_{45}^R \left(\frac{\partial v_x}{\partial z} + \frac{\partial v_z}{\partial x}\right) + C_{46}^R \left(\frac{\partial v_x}{\partial y} + \frac{\partial v_y}{\partial x}\right) \\
\frac{\sigma_{zx}^R}{\partial t} &= C_{15}^R \frac{\partial v_x}{\partial x} + C_{25}^R \frac{\partial v_y}{\partial y} + C_{35}^R \frac{\partial v_z}{\partial z} + C_{45}^R \left(\frac{\partial v_y}{\partial z} + \frac{\partial v_z}{\partial y}\right) + C_{55}^R \left(\frac{\partial v_x}{\partial z} + \frac{\partial v_z}{\partial x}\right) + C_{56}^R \left(\frac{\partial v_x}{\partial y} + \frac{\partial v_y}{\partial x}\right) \\
\frac{\sigma_{xy}^R}{\partial t} &= C_{16}^R \frac{\partial v_x}{\partial x} + C_{26}^R \frac{\partial v_y}{\partial y} + C_{36}^R \frac{\partial v_z}{\partial z} + C_{46}^R \left(\frac{\partial v_y}{\partial z} + \frac{\partial v_z}{\partial y}\right) + C_{56}^R \left(\frac{\partial v_x}{\partial z} + \frac{\partial v_z}{\partial x}\right) + C_{66}^R \left(\frac{\partial v_x}{\partial y} + \frac{\partial v_y}{\partial x}\right)
\end{aligned}
\tag{5.33}
$$

Equations (5.29)–(5.33) are the velocity–stress equations of a 3D TTI viscoelastic anisotropic fractured medium.

In the above derivations, the most general viscoelastic velocity–stress equations with arbitrary anisotropy are considered. For elastic medium, the stress equations can be simplified by setting imaginary terms to zero. For general HTI medium, VTI medium, and homogeneous isotropic medium, the equations can be simplified according to the corresponding characteristics of the stiffness matrixes. To derive the velocity–stress equations in a 2D medium (in *XOZ* plane), set all the "*y*"-related partial derivatives in the velocity–stress equations to zero.

5.3.2 3D RSG Finite-Difference Algorithm

In RSG finite-difference algorithm, variables with the same physical meanings are defined on the same grid point. Compared with the staggered-grid (SG) finite-difference scheme, the elasticity parameters need not to be interpolated when these parameters are defined at the same grid point. Therefore, it can be used for simulating wave propagation process in any heterogeneous and anisotropic medium. The method is suitable for the numerical simulation of wave propagation in a viscoelastic anisotropic fractured medium discussed in this chapter.

5.3.2.1 Finite-Difference Grid Setup

In RSG finite difference algorithm, the coordinate system need not be changed. The rotation of difference direction is performed. Based on the linear sum of the rotated differential terms, the spatial-derivatives of variables are computed under the Cartesian coordinates (Chu et al., 2009). The RSG difference grid is shown in Figure 5.3.

As shown above, in the rotated staggered grid, velocity and density are given at the grid points, and stress and stiffness coefficients are given at the intermediate points.

As shown in Figure 5.1, x, y, and z are three axes of the Cartesian coordinates. d_1, d_2, d_3, and d_4 denote the four diagonals and can be expressed as a combination of the three coordinate vectors:

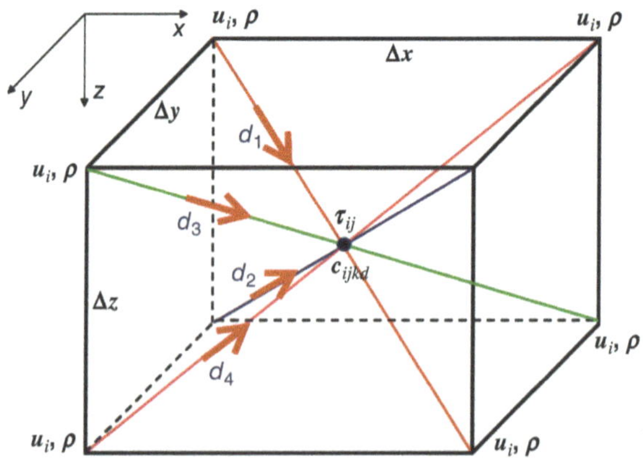

FIGURE 5.3 Schematic diagram of the rotated staggered grid *(Chu et al., 2009)*.

$$\begin{aligned}
\boldsymbol{d}_1 &= \frac{\Delta x}{\Delta l}\mathbf{x} + \frac{\Delta z}{\Delta l}\mathbf{z} + \frac{\Delta y}{\Delta l}\mathbf{y} \\
\boldsymbol{d}_2 &= \frac{\Delta x}{\Delta l}\mathbf{x} - \frac{\Delta z}{\Delta l}\mathbf{z} + \frac{\Delta y}{\Delta l}\mathbf{y} \\
\boldsymbol{d}_3 &= \frac{\Delta x}{\Delta l}\mathbf{x} + \frac{\Delta z}{\Delta l}\mathbf{z} - \frac{\Delta y}{\Delta l}\mathbf{y} \\
\boldsymbol{d}_4 &= \frac{\Delta x}{\Delta l}\mathbf{x} - \frac{\Delta z}{\Delta l}\mathbf{z} - \frac{\Delta y}{\Delta l}\mathbf{y} \\
\Delta l &= \sqrt{\Delta x^2 + \Delta z^2 + \Delta y^2}
\end{aligned} \tag{5.34}$$

The velocity–stress equations contain the partial derivatives of velocity and stress components with respect to x, y, and z in Cartesian coordinates. In order to solve the space partial derivative in the Cartesian coordinates during numerical simulation, RSG algorithm first calculates the space partial derivatives of the variables along the diagonals d_1, d_2, d_3, and d_4 in 3D space, then use these space partial derivatives to compute the space partial derivatives along the three axes x, y, and z in Cartesian coordinates, as

$$\begin{aligned}
\frac{\partial}{\partial x} &= \frac{\Delta l}{4\Delta x}\left(\frac{\partial}{\partial d_1} + \frac{\partial}{\partial d_2} + \frac{\partial}{\partial d_3} + \frac{\partial}{\partial d_4}\right) \\
\frac{\partial}{\partial y} &= \frac{\Delta l}{4\Delta y}\left(\frac{\partial}{\partial d_1} + \frac{\partial}{\partial d_2} - \frac{\partial}{\partial d_3} - \frac{\partial}{\partial d_4}\right) \\
\frac{\partial}{\partial z} &= \frac{\Delta l}{4\Delta z}\left(\frac{\partial}{\partial d_1} - \frac{\partial}{\partial d_2} + \frac{\partial}{\partial d_3} - \frac{\partial}{\partial d_4}\right)
\end{aligned} \tag{5.35}$$

For the finite-difference algorithm, the finite difference operators need to be substituted into the equations to replace the partial-derivative terms. Let D_x, D_y, and D_z denote the finite-difference operators along axes x, y, and z, respectively, corresponding to $\partial/\partial x$, $\partial/\partial y$, and $\partial/\partial z$. D_{d1}, D_{d2}, D_{d3}, and D_{d4} correspond to the partial derivatives $\partial/\partial d_1$, $\partial/\partial d_2$, $\partial/\partial d_3$, and $\partial/\partial d_4$. Thus, the finite-difference form of equation (5.35) can be expressed as

$$\begin{aligned}
D_x &= \frac{\Delta l}{4\Delta x}\left(D_{d1} + D_{d2} + D_{d3} + D_{d4}\right) \\
D_y &= \frac{\Delta l}{4\Delta y}\left(D_{d1} + D_{d2} - D_{d3} - D_{d4}\right) \\
D_z &= \frac{\Delta l}{4\Delta x}\left(D_{d1} - D_{d2} + D_{d3} - D_{d4}\right)
\end{aligned} \tag{5.36}$$

5.3.2.2 Stability Analysis on Finite-Difference Equations

The standard Neumann analytical method can be used to derive the stability criteria for the finite-difference algorithm. For RSG finite-difference algorithm, Saenger et al. (2000) derived the stability conditions for homogeneous medium using Neumann stability analysis. In 2D/3D, the RSG stability condition can be expressed as

$$\frac{\Delta t \upsilon_p}{\Delta h} \leq 1 / \sum_{n=1}^{N} \left| C_m^M \right| \tag{5.37}$$

where Δh is the spatial interval in the grid, υ_p is the P-wave velocity, and C_m^M are the finite difference coefficients.

In 3D wave numerical simulations, the stability conditions for second-order finite difference in time direction in the SSG in homogeneous medium is

$$\frac{\Delta t \upsilon_p}{\Delta h} \leq 1 / \sqrt{3} \sum_{n=1}^{N} \left| C_m^M \right| \tag{5.38}$$

Comparing (5.37) and (5.38), we can see, for the same finite-difference coefficients, the stability conditions of RSG is better than that of SSG. With the same finite-difference coefficients, RSG gives large time steps than that given by standard SG grid finite-difference and therefore computation time is reduced.

5.3.2.3 Finite Difference Operator

In the implementation of the finite difference algorithm, the derivatives of the corresponding variables at the intermediate points on grid are obtained by an approximate finite difference. Here, the corresponding finite difference coefficients are calculated by a finite difference discretization on the equations using the basic Taylor expansion.

Assume that the function $f(x)$ has continuous derivatives up to $2L+1$th order, and $x = x_0 \pm [(2m-1)/2]\Delta x$, then the $f(x)$'s $2L+1$-order Taylor expansion at x_0 is

$$f[x_0 \pm (2m-1)\Delta x / 2] = f(x_0) + \sum_{i=1}^{2L+1} \frac{\left[\pm \frac{(2m-1)\Delta x}{2}\right]^i}{i!} f^{(i)}(x_0) + O\left(\Delta x^{2L+2}\right) \tag{5.39}$$

From equation (5.39), the approximate $2L$-order staggered-grid finite difference expression for a first-order derivative can be derived as

$$\frac{\partial f}{\partial x}\Big|_{x=x_0} = \frac{1}{\Delta x} \sum_{m=1}^{L} C_m^{(L)} \{f[x_0 + \frac{(2m-1)\Delta x}{2}] - f[x_0 - \frac{(2m-1)\Delta x}{2}]\} + O\left(\Delta x^{2L}\right) \tag{5.40}$$

where $C_m^{(L)}$ is the finite difference coefficient, which can be determined by

$$\begin{bmatrix} 1 & 3 & \cdots & 2L-1 \\ 1 & 3^3 & \cdots & (2L-1)^3 \\ \vdots & \vdots & \cdots & \vdots \\ 1 & 3^{2L-1} & \cdots & (2L-1)^{2L-1} \end{bmatrix} \begin{bmatrix} C_1^{(L)} \\ C_2^{(L)} \\ \vdots \\ C_L^{(L)} \end{bmatrix} = \begin{bmatrix} 1 \\ 0 \\ \vdots \\ 0 \end{bmatrix} \tag{5.41}$$

Solving the equations, we found that

$$C_m^{(L)} = \frac{(-1)^{m+1} \prod_{i=1,i\neq m}^{L} (2i-1)^2}{(2m-1)\prod_{i=1,i\neq m}^{L-1} \left[(2m-1)^2 - (2i-1)^2\right]} \tag{5.42}$$

The finite difference coefficients for the several cases with different orders are given below.

For $L=1$, $C_1^{(1)} = 1$.

For $L=2$, $C_1^{(2)} = 1.125; C_2^{(2)} = -0.041667$.

For $L=3$, $C_1^{(3)} = 1.171875; C_2^{(3)} = -0.065104; C_3^{(3)} = 0.004688$.

For $L=4$, $C_1^{(4)} = 1.19629; C_2^{(4)} = -0.079753; C_3^{(4)} = 0.0095703;$

$C_4^{(4)} = -0.000697545$.

For $L=5$, $C_1^{(5)} = 1.121124; C_2^{(5)} = -0.0897217; C_3^{(5)} = 0.013843;$

$C_4^{(5)} = -0.00176566$; $C_5^{(5)} = 0.000118679$.

5.3.2.4 Discretized Velocity–Stress Finite Difference Equation

In RSG finite-difference algorithm, the 3D viscoelastic anisotropic velocity–stress equations needs to be discretized on the grids.

After difference discretization, the velocity equations are given below:

$$\begin{aligned} v_x\left(t+\frac{1}{2}\Delta t\right) &= v_x\left(t-\frac{1}{2}\Delta t\right) + \frac{\Delta t}{\rho}\left\{D_x\left[\sigma_{xx}(t)\right] + D_y\left[\sigma_{xy}(t)\right] + D_z\left[\sigma_{xz}(t)\right]\right\} \\ v_y\left(t+\frac{1}{2}\Delta t\right) &= v_y\left(t-\frac{1}{2}\Delta t\right) + \frac{\Delta t}{\rho}\left\{D_x\left[\sigma_{yx}(t)\right] + D_y\left[\sigma_{yy}(t)\right] + D_z\left[\sigma_{yz}(t)\right]\right\} \\ v_z\left(t+\frac{1}{2}\Delta t\right) &= v_z\left(t-\frac{1}{2}\Delta t\right) + \frac{\Delta t}{\rho}\left\{D_x\left[\sigma_{zx}(t)\right] + D_y\left[\sigma_{zy}(t)\right] + D_z\left[\sigma_{zz}(t)\right]\right\} \end{aligned} \tag{5.43}$$

The stress equations are discretized as

$$\begin{aligned}
\sigma_{xx}(t+\Delta t) &= \sigma_{xx}^{R}(t+\Delta t)+\sigma_{xx}^{I}(t+\Delta t)\\
\sigma_{yy}(t+\Delta t) &= \sigma_{yy}^{R}(t+\Delta t)+\sigma_{yy}^{I}(t+\Delta t)\\
\sigma_{zz}(t+\Delta t) &= \sigma_{zz}^{R}(t+\Delta t)+\sigma_{zz}^{I}(t+\Delta t)\\
\sigma_{yz}(t+\Delta t) &= \sigma_{yz}^{R}(t+\Delta t)+\sigma_{yz}^{I}(t+\Delta t)\\
\sigma_{zx}(t+\Delta t) &= \sigma_{zx}^{R}(t+\Delta t)+\sigma_{zx}^{I}(t+\Delta t)\\
\sigma_{xy}(t+\Delta t) &= \sigma_{xy}^{R}(t+\Delta t)+\sigma_{xy}^{I}(t+\Delta t)
\end{aligned} \tag{5.44}$$

where the real part of the stress $\sigma_{ij}^{R}(t+\Delta t)$, $(i, j = x, y, z)$ is expressed as

$$\begin{aligned}
\sigma_{xx}^{R}(t+\Delta t) = \sigma_{xx}^{R}(t)+\Delta t\Bigg\{ & c_{11}^{R}D_x\left[\upsilon_x\left(t+\frac{1}{2}\Delta t\right)\right]+c_{12}^{R}D_y\left[\upsilon_y\left(t+\frac{1}{2}\Delta t\right)\right]+c_{13}^{R}D_z\left[\upsilon_z\left(t+\frac{1}{2}\Delta t\right)\right]\\
&+c_{14}^{R}\left(D_z\left[\upsilon_y\left(t+\frac{1}{2}\Delta t\right)\right]+D_y\left[\upsilon_z\left(t+\frac{1}{2}\Delta t\right)\right]\right)+c_{15}^{R}\left(D_x\left[\upsilon_z\left(t+\frac{1}{2}\Delta t\right)\right]+D_z\left[\upsilon_x\left(t+\frac{1}{2}\Delta t\right)\right]\right)\\
&+c_{16}^{R}\left(D_y\left[\upsilon_x\left(t+\frac{1}{2}\Delta t\right)\right]+D_x\left[\upsilon_y\left(t+\frac{1}{2}\Delta t\right)\right]\right)\Bigg\}
\end{aligned}$$

$$\begin{aligned}
\sigma_{yy}^{R}(t+\Delta t) = \sigma_{yy}^{R}(t)+\Delta t\Bigg\{ & c_{12}^{R}D_x\left[\upsilon_x\left(t+\frac{1}{2}\Delta t\right)\right]+c_{22}^{R}D_y\left[\upsilon_y\left(t+\frac{1}{2}\Delta t\right)\right]+c_{23}^{R}D_z\left[\upsilon_z\left(t+\frac{1}{2}\Delta t\right)\right]\\
&+c_{24}^{R}\left(D_z\left[\upsilon_y\left(t+\frac{1}{2}\Delta t\right)\right]+D_y\left[\upsilon_z\left(t+\frac{1}{2}\Delta t\right)\right]\right)+c_{25}^{R}\left(D_x\left[\upsilon_z\left(t+\frac{1}{2}\Delta t\right)\right]+D_z\left[\upsilon_x\left(t+\frac{1}{2}\Delta t\right)\right]\right)\\
&+c_{26}^{R}\left(D_y\left[\upsilon_x\left(t+\frac{1}{2}\Delta t\right)\right]+D_x\left[\upsilon_y\left(t+\frac{1}{2}\Delta t\right)\right]\right)\Bigg\}
\end{aligned}$$

$$\begin{aligned}
\sigma_{zz}^{R}(t+\Delta t) = \sigma_{zz}^{R}(t)+\Delta t\Bigg\{ & c_{13}^{R}D_x\left[\upsilon_x\left(t+\frac{1}{2}\Delta t\right)\right]+c_{23}^{R}D_y\left[\upsilon_y\left(t+\frac{1}{2}\Delta t\right)\right]+c_{33}^{R}D_z\left[\upsilon_z\left(t+\frac{1}{2}\Delta t\right)\right]\\
&+c_{34}^{R}\left(D_z\left[\upsilon_y\left(t+\frac{1}{2}\Delta t\right)\right]+D_y\left[\upsilon_z\left(t+\frac{1}{2}\Delta t\right)\right]\right)+c_{35}^{R}\left(D_x\left[\upsilon_z\left(t+\frac{1}{2}\Delta t\right)\right]+D_z\left[\upsilon_x\left(t+\frac{1}{2}\Delta t\right)\right]\right)\\
&+c_{36}^{R}\left(D_y\left[\upsilon_x\left(t+\frac{1}{2}\Delta t\right)\right]+D_x\left[\upsilon_y\left(t+\frac{1}{2}\Delta t\right)\right]\right)\Bigg\}
\end{aligned}$$

$$\begin{aligned}
\sigma_{yz}^{R}(t+\Delta t) = \sigma_{yz}^{R}(t)+\Delta t\Bigg\{ & c_{14}^{R}D_x\left[\upsilon_x\left(t+\frac{1}{2}\Delta t\right)\right]+c_{24}^{R}D_y\left[\upsilon_y\left(t+\frac{1}{2}\Delta t\right)\right]+c_{34}^{R}D_z\left[\upsilon_z\left(t+\frac{1}{2}\Delta t\right)\right]\\
&+c_{44}^{R}\left(D_z\left[\upsilon_y\left(t+\frac{1}{2}\Delta t\right)\right]+D_y\left[\upsilon_z\left(t+\frac{1}{2}\Delta t\right)\right]\right)+c_{45}^{R}\left(D_x\left[\upsilon_z\left(t+\frac{1}{2}\Delta t\right)\right]+D_z\left[\upsilon_x\left(t+\frac{1}{2}\Delta t\right)\right]\right)\\
&+c_{46}^{R}\left(D_y\left[\upsilon_x\left(t+\frac{1}{2}\Delta t\right)\right]+D_x\left[\upsilon_y\left(t+\frac{1}{2}\Delta t\right)\right]\right)\Bigg\}
\end{aligned} \tag{5.45}$$

$$\begin{aligned}\sigma_{zx}^{R}(t+\Delta t)=\sigma_{zx}^{R}(t)+\Delta t\Big\{&c_{15}^{R}D_x\left[\upsilon_x\left(t+\frac{1}{2}\Delta t\right)\right]+c_{25}^{R}D_y\left[\upsilon_y\left(t+\frac{1}{2}\Delta t\right)\right]+c_{35}^{R}D_z\left[\upsilon_z\left(t+\frac{1}{2}\Delta t\right)\right]\\
&+c_{45}^{R}\left(D_z\left[\upsilon_y\left(t+\frac{1}{2}\Delta t\right)\right]+D_y\left[\upsilon_z\left(t+\frac{1}{2}\Delta t\right)\right]\right)+c_{55}^{R}\left(D_x\left[\upsilon_z\left(t+\frac{1}{2}\Delta t\right)\right]+D_z\left[\upsilon_x\left(t+\frac{1}{2}\Delta t\right)\right]\right)\\
&+c_{56}^{R}\left(D_y\left[\upsilon_x\left(t+\frac{1}{2}\Delta t\right)\right]+D_x\left[\upsilon_y\left(t+\frac{1}{2}\Delta t\right)\right]\right)\Big\}\end{aligned}$$

$$\begin{aligned}\sigma_{xy}^{R}(t+\Delta t)=\sigma_{xy}^{R}(t)+\Delta t\Big\{&c_{16}^{R}D_x\left[\upsilon_x\left(t+\frac{1}{2}\Delta t\right)\right]+c_{26}^{R}D_y\left[\upsilon_y\left(t+\frac{1}{2}\Delta t\right)\right]+c_{36}^{R}D_z\left[\upsilon_z\left(t+\frac{1}{2}\Delta t\right)\right]\\
&+c_{46}^{R}\left(D_z\left[\upsilon_y\left(t+\frac{1}{2}\Delta t\right)\right]+D_y\left[\upsilon_z\left(t+\frac{1}{2}\Delta t\right)\right]\right)+c_{56}^{R}\left(D_x\left[\upsilon_z\left(t+\frac{1}{2}\Delta t\right)\right]+D_z\left[\upsilon_x\left(t+\frac{1}{2}\Delta t\right)\right]\right)\\
&+c_{66}^{R}\left(D_y\left[\upsilon_x\left(t+\frac{1}{2}\Delta t\right)\right]+D_x\left[\upsilon_y\left(t+\frac{1}{2}\Delta t\right)\right]\right)\Big\}\end{aligned}$$

Since the velocity components calculated in the difference algorithm are only available at the intermediate points at each time interval, the stress components are at the grid point, as an approximation, $\upsilon_i\,(t+\Delta t/2)$ is used to replace $\upsilon_i\,(t+\Delta t)$ in calculating the imaginary part of the stress $\sigma_{ij}^{I}\,(t+\Delta t)$, $(i, j=x, y, z)$.

$$\begin{aligned}\sigma_{xx}^{I}(t+\Delta t)=\;&c_{11}^{I}D_x\left[\upsilon_x\left(t+\frac{1}{2}\Delta t\right)\right]+c_{12}^{I}D_y\left[\upsilon_y\left(t+\frac{1}{2}\Delta t\right)\right]+c_{13}^{I}D_z\left[\upsilon_z\left(t+\frac{1}{2}\Delta t\right)\right]\\
&+c_{14}^{I}\left(D_z\left[\upsilon_y\left(t+\frac{1}{2}\Delta t\right)\right]+D_y\left[\upsilon_z\left(t+\frac{1}{2}\Delta t\right)\right]\right)+c_{15}^{I}\left(D_x\left[\upsilon_z\left(t+\frac{1}{2}\Delta t\right)\right]+D_z\left[\upsilon_x\left(t+\frac{1}{2}\Delta t\right)\right]\right)\\
&+c_{16}^{I}\left(D_y\left[\upsilon_x\left(t+\frac{1}{2}\Delta t\right)\right]+D_x\left[\upsilon_y\left(t+\frac{1}{2}\Delta t\right)\right]\right)\end{aligned}$$

$$\begin{aligned}\sigma_{yy}^{I}(t+\Delta t)=\;&c_{12}^{I}D_x\left[\upsilon_x\left(t+\frac{1}{2}\Delta t\right)\right]+c_{22}^{I}D_y\left[\upsilon_y\left(t+\frac{1}{2}\Delta t\right)\right]+c_{23}^{I}D_z\left[\upsilon_z\left(t+\frac{1}{2}\Delta t\right)\right]\\
&+c_{24}^{I}\left(D_z\left[\upsilon_y\left(t+\frac{1}{2}\Delta t\right)\right]+D_y\left[\upsilon_z\left(t+\frac{1}{2}\Delta t\right)\right]\right)+c_{25}^{I}\left(D_x\left[\upsilon_z\left(t+\frac{1}{2}\Delta t\right)\right]+D_z\left[\upsilon_x\left(t+\frac{1}{2}\Delta t\right)\right]\right)\\
&+c_{26}^{I}\left(D_y\left[\upsilon_x\left(t+\frac{1}{2}\Delta t\right)\right]+D_x\left[\upsilon_y\left(t+\frac{1}{2}\Delta t\right)\right]\right)\end{aligned}$$

$$\begin{aligned}\sigma_{zz}^{I}(t+\Delta t)=\;&c_{13}^{I}D_x\left[\upsilon_x\left(t+\frac{1}{2}\Delta t\right)\right]+c_{23}^{I}D_y\left[\upsilon_y\left(t+\frac{1}{2}\Delta t\right)\right]+c_{33}^{I}D_z\left[\upsilon_z\left(t+\frac{1}{2}\Delta t\right)\right]\\
&+c_{34}^{I}\left(D_z\left[\upsilon_y\left(t+\frac{1}{2}\Delta t\right)\right]+D_y\left[\upsilon_z\left(t+\frac{1}{2}\Delta t\right)\right]\right)+c_{35}^{I}\left(D_x\left[\upsilon_z\left(t+\frac{1}{2}\Delta t\right)\right]+D_z\left[\upsilon_x\left(t+\frac{1}{2}\Delta t\right)\right]\right)\\
&+c_{36}^{I}\left(D_y\left[\upsilon_x\left(t+\frac{1}{2}\Delta t\right)\right]+D_x\left[\upsilon_y\left(t+\frac{1}{2}\Delta t\right)\right]\right)\end{aligned}$$

$$
\begin{aligned}
\sigma_{yz}^{I}(t+\Delta t) &= c_{14}^{I} D_x\left[\upsilon_x\left(t+\frac{1}{2}\Delta t\right)\right] + c_{24}^{I} D_y\left[\upsilon_y\left(t+\frac{1}{2}\Delta t\right)\right] + c_{34}^{I} D_z\left[\upsilon_z\left(t+\frac{1}{2}\Delta t\right)\right] \\
&+ c_{44}^{I}\left(D_z\left[\upsilon_y\left(t+\frac{1}{2}\Delta t\right)\right] + D_y\left[\upsilon_z\left(t+\frac{1}{2}\Delta t\right)\right]\right) + c_{45}^{I}\left(D_x\left[\upsilon_z\left(t+\frac{1}{2}\Delta t\right)\right] + D_z\left[\upsilon_x\left(t+\frac{1}{2}\Delta t\right)\right]\right) \\
&+ c_{46}^{I}\left(D_y\left[\upsilon_x\left(t+\frac{1}{2}\Delta t\right)\right] + D_x\left[\upsilon_y\left(t+\frac{1}{2}\Delta t\right)\right]\right)
\end{aligned} \tag{5.46}
$$

$$
\begin{aligned}
\sigma_{zx}^{I}(t+\Delta t) &= c_{15}^{I} D_x\left[\upsilon_x\left(t+\frac{1}{2}\Delta t\right)\right] + c_{25}^{I} D_y\left[\upsilon_y\left(t+\frac{1}{2}\Delta t\right)\right] + c_{35}^{I} D_z\left[\upsilon_z\left(t+\frac{1}{2}\Delta t\right)\right] \\
&+ c_{45}^{I}\left(D_z\left[\upsilon_y\left(t+\frac{1}{2}\Delta t\right)\right] + D_y\left[\upsilon_z\left(t+\frac{1}{2}\Delta t\right)\right]\right) + c_{55}^{I}\left(D_x\left[\upsilon_z\left(t+\frac{1}{2}\Delta t\right)\right] + D_z\left[\upsilon_x\left(t+\frac{1}{2}\Delta t\right)\right]\right) \\
&+ c_{56}^{I}\left(D_y\left[\upsilon_x\left(t+\frac{1}{2}\Delta t\right)\right] + D_x\left[\upsilon_y\left(t+\frac{1}{2}\Delta t\right)\right]\right)
\end{aligned}
$$

$$
\begin{aligned}
\sigma_{xy}^{I}(t+\Delta t) &= c_{16}^{I} D_x\left[\upsilon_x\left(t+\frac{1}{2}\Delta t\right)\right] + c_{26}^{I} D_y\left[\upsilon_y\left(t+\frac{1}{2}\Delta t\right)\right] + c_{36}^{I} D_z\left[\upsilon_z\left(t+\frac{1}{2}\Delta t\right)\right] \\
&+ c_{46}^{I}\left(D_z\left[\upsilon_y\left(t+\frac{1}{2}\Delta t\right)\right] + D_y\left[\upsilon_z\left(t+\frac{1}{2}\Delta t\right)\right]\right) + c_{56}^{I}\left(D_x\left[\upsilon_z\left(t+\frac{1}{2}\Delta t\right)\right] + D_z\left[\upsilon_x\left(t+\frac{1}{2}\Delta t\right)\right]\right) \\
&+ c_{66}^{I}\left(D_y\left[\upsilon_x\left(t+\frac{1}{2}\Delta t\right)\right] + D_x\left[\upsilon_y\left(t+\frac{1}{2}\Delta t\right)\right]\right)
\end{aligned}
$$

5.3.2.5 Boundary Condition for Wave Modeling in Viscoelastic Anisotropic Medium

According to Drossaert and Giannopoulos (2007) that proposed the complex frequency-shifted recursive-integration perfectly matched layer absorbing boundary condition (CFSRIPML) for an elastic isotropic medium, the complex frequency-shifted nonsplitting perfectly matched layer (PML) absorbing boundary condition is used for a viscoelastic, anisotropic fracture medium.

5.3.2.5.1 Basic principles for the recursive-integration based nonsplitting complex frequency-shifted PML boundary conditions

The basic idea for the recursive integration based nonsplitting complex frequency-shifted PML boundary condition is the extension of spatial coordinates and frequency-related viscosity,

$$
\begin{aligned}
\frac{\partial}{\partial \tilde{x}} &\rightarrow \frac{1}{\varepsilon_x}\frac{\partial}{\partial x} \\
\frac{\partial}{\partial \tilde{y}} &\rightarrow \frac{1}{\varepsilon_y}\frac{\partial}{\partial y} \\
\frac{\partial}{\partial \tilde{z}} &\rightarrow \frac{1}{\varepsilon_z}\frac{\partial}{\partial z}
\end{aligned} \tag{5.47}
$$

where

$$
\begin{aligned}
\varepsilon_x &= \kappa_x + \frac{\sigma_x}{\alpha_x + i\omega} \\
\varepsilon_y &= \kappa_y + \frac{\sigma_y}{\alpha_y + i\omega} \\
\varepsilon_z &= \kappa_z + \frac{\sigma_z}{\alpha_z + i\omega}
\end{aligned}
\tag{5.48}
$$

σ_i, κ_l, and α_i ($i = x, y, z$) are the attenuation parameters of the PML boundary, where σ_i is the attenuation parameter of the conventional PML boundary, and κ_i and α_i are the two auxiliary attenuation parameters that are introduced in the complex frequency-shifted PML boundary condition. κ_i is used to attenuate the boundary reflections for wave incident to PML boundaries with a finite angle. α_i is used to attenuate the low-frequency waves (Drossaert and Giannopoulos, 2007). The definition of the three PML attenuation parameters on the boundary shall meet the following requirements (Roden and Gedney, 2000; Zhang and Shen, 2010): (1) σ_i is equal to zero in the computation domain and gradually increased to its maximum from the border between computation domain and PML domain to the outer boundary of PML domain; (2) κ_i is equal to 1 in the computation domain and gradually increased to its maximum from the border between computation domain and PML domain to the outer boundary of PML domain; and (3) α_i is equal to zero in the computation domain and equal to its maximum at the border between computation domain and PML domain, and gradually decreased to zero from the border to the outer boundary of PML domain. They can be expressed by equations as

$$
\begin{aligned}
\sigma_i(l) &= \sigma_i^{\max}(l/d)^{o\sigma} \\
\kappa_i(l) &= \left(\kappa_i^{\max} - 1\right)(l/d)^{o\kappa} + 1 \\
\alpha_i(l) &= \alpha_i^{\max}\left(1 - (l/d)^{o\alpha}\right)
\end{aligned}
\tag{5.49}
$$

where l denotes the distance between the PML boundary and the border between the calculation domain and the boundary domain. d denotes the total length of PML. o* (* denotes σ, κ, or α) denotes the attenuation coefficient. $\sigma_i^{\max}$, $\kappa_i^{\max}$, and $\alpha_i^{\max}$ ($i = x, y, z$) are respectively the three maximum PML boundary attenuation coefficients, which are determined by the relation between medium properties (wave velocity for instance) and the seismic frequency or wavefield features (Zhang and Shen, 2010; Collino and Tsogka, 2001).

$$
\begin{aligned}
\sigma^{\max} &= -\frac{(O_\sigma + 1)\upsilon_p}{2d} \ln R \\
\kappa^{\max} &= \frac{C}{0.5PPW_0 \Delta h f_c} \\
\alpha^{\max} &= \pi f_c
\end{aligned}
\tag{5.50}
$$

where υ_p is the P-wave velocity in the medium. f_c is the dominant seismic frequency; PPW_0 is the minimum number of grid points per wavelength. The parameter R is determined by the following formula as

$$
\log_{10}(R) = -\frac{\log_{10}(N) - 1}{\log_{10}(2)} - 3 \tag{5.51}
$$

where N is the number of grid points for a single PML boundary.

Using the complex frequency-shifted recursive-integration nonsplitting PML boundary conditions, the original space partial derivatives are changed into

$$
\begin{aligned}
\tilde{S}_{xx} &= \frac{1}{\varepsilon_x}\frac{\partial \tilde{\sigma}_{xx}}{\partial x}, \tilde{S}_{yx} = \frac{1}{\varepsilon_x}\frac{\partial \tilde{\sigma}_{yx}}{\partial x}, \tilde{S}_{zr} = \frac{1}{\varepsilon_x}\frac{\partial \tilde{\sigma}_{zx}}{\partial x} \\
\tilde{S}_{xy} &= \frac{1}{\varepsilon_y}\frac{\partial \tilde{\sigma}_{xy}}{\partial y}, \tilde{S}_{yx} = \frac{1}{\varepsilon_y}\frac{\partial \tilde{\sigma}_{yy}}{\partial y}, \tilde{S}_{zx} = \frac{1}{\varepsilon_y}\frac{\partial \tilde{\sigma}_{zy}}{\partial y} \\
\tilde{S}_{xz} &= \frac{1}{\varepsilon_z}\frac{\partial \tilde{\sigma}_{xz}}{\partial z}, \tilde{S}_{yz} = \frac{1}{\varepsilon_z}\frac{\partial \tilde{\sigma}_{yz}}{\partial z}, \tilde{S}_{zz} = \frac{1}{\varepsilon_z}\frac{\partial \tilde{\sigma}_{zz}}{\partial z}
\end{aligned}
\tag{5.52}
$$

$$
\begin{aligned}
\tilde{E}_{xx} &= \frac{1}{\varepsilon_x}\frac{\partial \upsilon_x}{\partial x}, \tilde{E}_{yx} = \frac{1}{\varepsilon_x}\frac{\partial \upsilon_y}{\partial x}, \tilde{E}_{zx} = \frac{1}{\varepsilon_x}\frac{\partial \upsilon_z}{\partial x} \\
\tilde{E}_{xy} &= \frac{1}{\varepsilon_y}\frac{\partial \upsilon_x}{\partial y}, \tilde{E}_{yy} = \frac{1}{\varepsilon_y}\frac{\partial \upsilon_y}{\partial y}, \tilde{E}_{zx} = \frac{1}{\varepsilon_y}\frac{\partial \upsilon_z}{\partial y} \\
\tilde{E}_{xz} &= \frac{1}{\varepsilon_z}\frac{\partial \upsilon_x}{\partial z}, \tilde{E}_{yz} = \frac{1}{\varepsilon_z}\frac{\partial \upsilon_y}{\partial z}, \tilde{E}_{zz} = \frac{1}{\varepsilon_z}\frac{\partial \upsilon_z}{\partial z}
\end{aligned}
\tag{5.53}
$$

$\tilde{S}_{ij}$, $\tilde{E}_{ij}$ $(i, j = x, y, z)$ denote the stress and strain-related variables, respectively. $\tilde{S}_{ij}$, $\tilde{E}_{ij}$ can be used to derive the corresponding spatial partial derivatives.

5.3.2.5.2 3D discretized velocity–stress equations with the introduced boundary conditions

By introducing the boundary conditions into the discretized velocity–stress equations, the output equations are can be derived.

Velocity equations with complex frequency-shifted PML boundaries can be written as

$$\begin{aligned}
\upsilon_x\left(t+\frac{1}{2}\Delta t\right) &= \upsilon_x\left(t-\frac{1}{2}\Delta t\right)+\frac{\Delta t}{\rho}\left\{S_{xx}(t)+S_{xy}(t)+S_{xz}(t)\right\} \\
\upsilon_y\left(t+\frac{1}{2}\Delta t\right) &= \upsilon_y\left(t-\frac{1}{2}\Delta t\right)+\frac{\Delta t}{\rho}\left\{S_{yx}(t)+S_{yy}(t)+S_{yz}(t)\right\} \\
\upsilon_z\left(t+\frac{1}{2}\Delta t\right) &= \upsilon_z\left(t-\frac{1}{2}\Delta t\right)+\frac{\Delta t}{\rho}\left\{S_{zx}(t)+S_{zy}(t)+S_{zz}(t)\right\}
\end{aligned} \tag{5.54}$$

Stress equations with complex frequency-shifted PML boundaries can be written as

$$\begin{aligned}
\sigma_{xx}(t+\Delta t) &= \sigma_{xx}^R(t+\Delta t)+\sigma_{xx}^I(t+\Delta t) \\
\sigma_{yy}(t+\Delta t) &= \sigma_{yy}^R(t+\Delta t)+\sigma_{yy}^I(t+\Delta t) \\
\sigma_{zz}(t+\Delta t) &= \sigma_{zz}^R(t+\Delta t)+\sigma_{zz}^I(t+\Delta t) \\
\sigma_{yz}(t+\Delta t) &= \sigma_{yz}^R(t+\Delta t)+\sigma_{yz}^I(t+\Delta t) \\
\sigma_{zx}(t+\Delta t) &= \sigma_{zx}^R(t+\Delta t)+\sigma_{zx}^I(t+\Delta t) \\
\sigma_{xy}(t+\Delta t) &= \sigma_{xy}^R(t+\Delta t)+\sigma_{xy}^I(t+\Delta t)
\end{aligned} \tag{5.55}$$

where

$$\begin{aligned}
\sigma_{xx}^R(t+\Delta t) &= \sigma_{xx}^R(t)+\Delta t\left\{c_{11}^R E_{xx}\left(t+\frac{1}{2}\Delta t\right)+c_{12}^R E_{yy}\left(t+\frac{1}{2}\Delta t\right)+c_{13}^R E_{zz}\left(t+\frac{1}{2}\Delta t\right)\right. \\
&+c_{14}^R\left(E_{yz}\left(t+\frac{1}{2}\Delta t\right)+E_{zy}\left(t+\frac{1}{2}\Delta t\right)\right)+c_{15}^R\left(E_{zx}\left(t+\frac{1}{2}\Delta t\right)+E_{xz}\left(t+\frac{1}{2}\Delta t\right)\right) \\
&\left.+c_{16}^R\left(E_{xy}\left(t+\frac{1}{2}\Delta t\right)+E_{yx}\left(t+\frac{1}{2}\Delta t\right)\right)\right\}
\end{aligned}$$

$$\begin{aligned}
\sigma_{yy}^R(t+\Delta t) &= \sigma_{yy}^R(t)+\Delta t\left\{c_{12}^R E_{xx}\left(t+\frac{1}{2}\Delta t\right)+c_{22}^R E_{yy}\left(t+\frac{1}{2}\Delta t\right)+c_{23}^R E_{zz}\left(t+\frac{1}{2}\Delta t\right)\right. \\
&+c_{24}^R\left(E_{yz}\left(t+\frac{1}{2}\Delta t\right)+E_{zy}\left(t+\frac{1}{2}\Delta t\right)\right)+c_{25}^R\left(E_{zx}\left(t+\frac{1}{2}\Delta t\right)+E_{xz}\left(t+\frac{1}{2}\Delta t\right)\right) \\
&\left.+c_{26}^R\left(E_{xy}\left(t+\frac{1}{2}\Delta t\right)+E_{yx}\left(t+\frac{1}{2}\Delta t\right)\right)\right\}
\end{aligned}$$

$$\begin{aligned}
\sigma_{zz}^R(t+\Delta t) &= \sigma_{zz}^R(t)+\Delta t\left\{c_{13}^R E_{xx}\left(t+\frac{1}{2}\Delta t\right)+c_{23}^R E_{yy}\left(t+\frac{1}{2}\Delta t\right)+c_{33}^R E_{zz}\left(t+\frac{1}{2}\Delta t\right)\right. \\
&+c_{34}^R\left(E_{yz}\left(t+\frac{1}{2}\Delta t\right)+E_{zy}\left(t+\frac{1}{2}\Delta t\right)\right)+c_{35}^R\left(E_{zx}\left(t+\frac{1}{2}\Delta t\right)+E_{xz}\left(t+\frac{1}{2}\Delta t\right)\right) \\
&\left.+c_{36}^R\left(E_{xy}\left(t+\frac{1}{2}\Delta t\right)+E_{yx}\left(t+\frac{1}{2}\Delta t\right)\right)\right\}
\end{aligned}$$

$$\sigma_{yz}^{R}(t+\Delta t)=\sigma_{yz}^{R}(t)+\Delta t\left\{c_{14}^{R}E_{xx}\left(t+\frac{1}{2}\Delta t\right)+c_{24}^{R}E_{yy}\left(t+\frac{1}{2}\Delta t\right)+c_{34}^{R}E_{zz}\left(t+\frac{1}{2}\Delta t\right)\right.$$
$$+c_{44}^{R}\left(E_{yz}\left(t+\frac{1}{2}\Delta t\right)+E_{zy}\left(t+\frac{1}{2}\Delta t\right)\right)+c_{45}^{R}\left(E_{zx}\left(t+\frac{1}{2}\Delta t\right)+E_{xz}\left(t+\frac{1}{2}\Delta t\right)\right) \quad (5.56)$$
$$\left.+c_{46}^{R}\left(E_{xy}\left(t+\frac{1}{2}\Delta t\right)+E_{yx}\left(t+\frac{1}{2}\Delta t\right)\right)\right\}$$

$$\sigma_{zx}^{R}(t+\Delta t)=\sigma_{zx}^{R}(t)+\Delta t\left\{c_{15}^{R}E_{xx}\left(t+\frac{1}{2}\Delta t\right)+c_{25}^{R}E_{yy}\left(t+\frac{1}{2}\Delta t\right)+c_{35}^{R}E_{zz}\left(t+\frac{1}{2}\Delta t\right)\right.$$
$$+c_{45}^{R}\left(E_{yz}\left(t+\frac{1}{2}\Delta t\right)+E_{zy}\left(t+\frac{1}{2}\Delta t\right)\right)+c_{55}^{R}\left(E_{zx}\left(t+\frac{1}{2}\Delta t\right)+E_{xz}\left(t+\frac{1}{2}\Delta t\right)\right)$$
$$\left.+c_{56}^{R}\left(E_{xy}\left(t+\frac{1}{2}\Delta t\right)+E_{yx}\left(t+\frac{1}{2}\Delta t\right)\right)\right\}$$

$$\sigma_{xy}^{R}(t+\Delta t)=\sigma_{xy}^{R}(t)+\Delta t\left\{c_{16}^{R}E_{xx}\left(t+\frac{1}{2}\Delta t\right)+c_{26}^{R}E_{yy}\left(t+\frac{1}{2}\Delta t\right)+c_{36}^{R}E_{zz}\left(t+\frac{1}{2}\Delta t\right)\right.$$
$$+c_{46}^{R}\left(E_{yz}\left(t+\frac{1}{2}\Delta t\right)+E_{zy}\left(t+\frac{1}{2}\Delta t\right)\right)+c_{56}^{R}\left(E_{zx}\left(t+\frac{1}{2}\Delta t\right)+E_{xz}\left(t+\frac{1}{2}\Delta t\right)\right)$$
$$\left.+c_{66}^{R}\left(E_{xy}\left(t+\frac{1}{2}\Delta t\right)+E_{yx}\left(t+\frac{1}{2}\Delta t\right)\right)\right\}$$

$$\sigma_{xx}^{I}(t+\Delta t)=c_{11}^{I}E_{xx}\left(t+\frac{1}{2}\Delta t\right)+c_{12}^{I}E_{yy}\left(t+\frac{1}{2}\Delta t\right)+c_{13}^{I}E_{zz}\left(t+\frac{1}{2}\Delta t\right)$$
$$+c_{14}^{I}\left(E_{yz}\left(t+\frac{1}{2}\Delta t\right)+E_{zy}\left(t+\frac{1}{2}\Delta t\right)\right)+c_{15}^{I}\left(E_{zx}\left(t+\frac{1}{2}\Delta t\right)+E_{xz}\left(t+\frac{1}{2}\Delta t\right)\right) \quad (5.57)$$
$$+c_{16}^{I}\left(E_{xy}\left(t+\frac{1}{2}\Delta t\right)+E_{yx}\left(t+\frac{1}{2}\Delta t\right)\right)$$

$$\sigma_{yy}^{I}(t+\Delta t)=c_{12}^{I}E_{xx}\left(t+\frac{1}{2}\Delta t\right)+c_{22}^{I}E_{yy}\left(t+\frac{1}{2}\Delta t\right)+c_{23}^{I}E_{zz}\left(t+\frac{1}{2}\Delta t\right)$$
$$+c_{24}^{I}\left(E_{yz}\left(t+\frac{1}{2}\Delta t\right)+E_{zy}\left(t+\frac{1}{2}\Delta t\right)\right)+c_{25}^{I}\left(E_{zx}\left(t+\frac{1}{2}\Delta t\right)+E_{xz}\left(t+\frac{1}{2}\Delta t\right)\right)$$
$$+c_{26}^{I}\left(E_{xy}\left(t+\frac{1}{2}\Delta t\right)+E_{yx}\left(t+\frac{1}{2}\Delta t\right)\right)$$

$$\sigma_{zz}^I(t+\Delta t)=c_{13}^I E_{xx}\left(t+\frac{1}{2}\Delta t\right)+c_{23}^I E_{yy}\left(t+\frac{1}{2}\Delta t\right)+c_{33}^I E_{zz}\left(t+\frac{1}{2}\Delta t\right)$$
$$+c_{34}^I\left(E_{yz}\left(t+\frac{1}{2}\Delta t\right)+E_{zy}\left(t+\frac{1}{2}\Delta t\right)\right)+c_{35}^I\left(E_{zx}\left(t+\frac{1}{2}\Delta t\right)+E_{xz}\left(t+\frac{1}{2}\Delta t\right)\right)$$
$$+c_{36}^I\left(E_{xy}\left(t+\frac{1}{2}\Delta t\right)+E_{yx}\left(t+\frac{1}{2}\Delta t\right)\right)$$

$$\sigma_{yz}^I(t+\Delta t)=c_{14}^I E_{xx}\left(t+\frac{1}{2}\Delta t\right)+c_{24}^I E_{yy}\left(t+\frac{1}{2}\Delta t\right)+c_{34}^I E_{zz}\left(t+\frac{1}{2}\Delta t\right)$$
$$+c_{44}^I\left(E_{yz}\left(t+\frac{1}{2}\Delta t\right)+E_{zy}\left(t+\frac{1}{2}\Delta t\right)\right)+c_{45}^I\left(E_{zx}\left(t+\frac{1}{2}\Delta t\right)+E_{xz}\left(t+\frac{1}{2}\Delta t\right)\right)$$
$$+c_{46}^I\left(E_{xy}\left(t+\frac{1}{2}\Delta t\right)+E_{yx}\left(t+\frac{1}{2}\Delta t\right)\right)$$

$$\sigma_{zx}^I(t+\Delta t)=c_{15}^I E_{xx}\left(t+\frac{1}{2}\Delta t\right)+c_{25}^I E_{yy}\left(t+\frac{1}{2}\Delta t\right)+c_{35}^I E_{zz}\left(t+\frac{1}{2}\Delta t\right)$$
$$+c_{45}^I\left(E_{yz}\left(t+\frac{1}{2}\Delta t\right)+E_{zy}\left(t+\frac{1}{2}\Delta t\right)\right)+c_{55}^I\left(E_{zx}\left(t+\frac{1}{2}\Delta t\right)+E_{xz}\left(t+\frac{1}{2}\Delta t\right)\right)$$
$$+c_{56}^I\left(E_{xy}\left(t+\frac{1}{2}\Delta t\right)+E_{yx}\left(t+\frac{1}{2}\Delta t\right)\right)$$

$$\sigma_{xy}^I(t+\Delta t)=c_{16}^I E_{xx}\left(t+\frac{1}{2}\Delta t\right)+c_{26}^I E_{yy}\left(t+\frac{1}{2}\Delta t\right)+c_{36}^I E_{zz}\left(t+\frac{1}{2}\Delta t\right)$$
$$+c_{46}^I\left(E_{yz}\left(t+\frac{1}{2}\Delta t\right)+E_{zy}\left(t+\frac{1}{2}\Delta t\right)\right)+c_{56}^I\left(E_{zx}\left(t+\frac{1}{2}\Delta t\right)+E_{xz}\left(t+\frac{1}{2}\Delta t\right)\right)$$
$$+c_{66}^I\left(E_{xy}\left(t+\frac{1}{2}\Delta t\right)+E_{yx}\left(t+\frac{1}{2}\Delta t\right)\right)$$

where $S_{ij},(i,j=x,y,z)$ in equation (5.54) are

$$\begin{aligned}
S_{xx}(t)&=\xi_x D_x\left[\sigma_{xx}(t)\right]-\phi_x\,\Omega_{xx}(t-\Delta t)\\
S_{xy}(t)&=\xi_y D_y\left[\sigma_{xy}(t)\right]-\phi_y\,\Omega_{xy}(t-\Delta t)\\
S_{xz}(t)&=\xi_z D_z\left[\sigma_{xz}(t)\right]-\phi_z\,\Omega_{xz}(t-\Delta t)\\
S_{yx}(t)&=\xi_x D_x\left[\sigma_{yx}(t)\right]-\phi_x\,\Omega_{yx}(t-\Delta t)\\
S_{yy}(t)&=\xi_y D_y\left[\sigma_{yy}(t)\right]-\phi_y\,\Omega_{yy}(t-\Delta t)\\
S_{yz}(t)&=\xi_z D_z\left[\sigma_{yz}(t)\right]-\phi_z\,\Omega_{yz}(t-\Delta t)\\
S_{zx}(t)&=\xi_x D_x\left[\sigma_{zx}(t)\right]-\phi_x\,\Omega_{zx}(t-\Delta t)\\
S_{zy}(t)&=\xi_y D_y\left[\sigma_{zy}(t)\right]-\phi_y\,\Omega_{zy}(t-\Delta t)\\
S_{zz}(t)&=\xi_z D_z\left[\sigma_{zz}(t)\right]-\phi_z\,\Omega_{zz}(t-\Delta t)
\end{aligned}\tag{5.58}$$

where

$$
\begin{aligned}
\Omega_{xx}(t) &= \Omega_{xx}(t-\Delta t) + \Lambda_x S_{xx}(t) - \alpha_x D_x\left[\sigma_{xx}(t)\right] \\
\Omega_{xy}(t) &= \Omega_{xy}(t-\Delta t) + \Lambda_y S_{xy}(t) - \alpha_y D_y\left[\sigma_{xy}(t)\right] \\
\Omega_{xz}(t) &= \Omega_{xz}(t-\Delta t) + \Lambda_z S_{xz}(t) - \alpha_z D_z\left[\sigma_{xz}(t)\right] \\
\Omega_{yx}(t) &= \Omega_{yx}(t-\Delta t) + \Lambda_x S_{yx}(t) - \alpha_x D_x\left[\sigma_{yx}(t)\right] \\
\Omega_{yy}(t) &= \Omega_{yy}(t-\Delta t) + \Lambda_y S_{yy}(t) - \alpha_y D_y\left[\sigma_{yy}(t)\right] \\
\Omega_{yz}(t) &= \Omega_{yz}(t-\Delta t) + \Lambda_z S_{yz}(t) - \alpha_z D_z\left[\sigma_{yz}(t)\right] \\
\Omega_{zx}(t) &= \Omega_{zx}(t-\Delta t) + \Lambda_x S_{zx}(t) - \alpha_x D_x\left[\sigma_{zx}(t)\right] \\
\Omega_{zy}(t) &= \Omega_{zy}(t-\Delta t) + \Lambda_y S_{zy}(t) - \alpha_y D_y\left[\sigma_{zy}(t)\right] \\
\Omega_{zz}(t) &= \Omega_{zz}(t-\Delta t) + \Lambda_z S_{zz}(t) - \alpha_z D_z\left[\sigma_{zz}(t)\right]
\end{aligned}
\tag{5.59}
$$

And $E_{ij},(i,j=x,y,z)$ in equations (5.58) and (5.59) can be written as

$$
\begin{aligned}
E_{xx}\left(t+\frac{1}{2}\Delta t\right) &= \xi_x D_x\left[\upsilon_x\left(t+\frac{1}{2}\Delta t\right)\right] - \phi_x \psi_{xx}\left(t-\frac{1}{2}\Delta t\right) \\
E_{xy}\left(t+\frac{1}{2}\Delta t\right) &= \xi_y D_y\left[\upsilon_x\left(t+\frac{1}{2}\Delta t\right)\right] - \phi_y \psi_{xy}\left(t-\frac{1}{2}\Delta t\right) \\
E_{xz}\left(t+\frac{1}{2}\Delta t\right) &= \xi_z D_z\left[\upsilon_x\left(t+\frac{1}{2}\Delta t\right)\right] - \phi_z \psi_{xz}\left(t-\frac{1}{2}\Delta t\right) \\
E_{yx}\left(t+\frac{1}{2}\Delta t\right) &= \xi_x D_x\left[\upsilon_y\left(t+\frac{1}{2}\Delta t\right)\right] - \phi_x \psi_{yx}\left(t-\frac{1}{2}\Delta t\right) \\
E_{yy}\left(t+\frac{1}{2}\Delta t\right) &= \xi_y D_y\left[\upsilon_y\left(t+\frac{1}{2}\Delta t\right)\right] - \phi_y \psi_{yy}\left(t-\frac{1}{2}\Delta t\right) \\
E_{yz}\left(t+\frac{1}{2}\Delta t\right) &= \xi_z D_z\left[\upsilon_y\left(t+\frac{1}{2}\Delta t\right)\right] - \phi_z \psi_{yz}\left(t-\frac{1}{2}\Delta t\right) \\
E_{zx}\left(t+\frac{1}{2}\Delta t\right) &= \xi_x D_x\left[\upsilon_z\left(t+\frac{1}{2}\Delta t\right)\right] - \phi_x \psi_{zx}\left(t-\frac{1}{2}\Delta t\right) \\
E_{zy}\left(t+\frac{1}{2}\Delta t\right) &= \xi_y D_y\left[\upsilon_z\left(t+\frac{1}{2}\Delta t\right)\right] - \phi_y \psi_{zy}\left(t-\frac{1}{2}\Delta t\right) \\
E_{zz}\left(t+\frac{1}{2}\Delta t\right) &= \xi_z D_z\left[\upsilon_z\left(t+\frac{1}{2}\Delta t\right)\right] - \phi_z \psi_{zz}\left(t-\frac{1}{2}\Delta t\right)
\end{aligned}
\tag{5.60}
$$

$$\begin{aligned}
\psi_{xx}\left(t+\frac{1}{2}\Delta t\right) &= \psi_{xx}\left(t-\frac{1}{2}\Delta t\right)+\Lambda_x E_{xx}\left(t+\frac{1}{2}\Delta t\right)-\alpha_x D_x\left[v_x\left(t+\frac{1}{2}\Delta t\right)\right]\\
\psi_{xy}\left(t+\frac{1}{2}\Delta t\right) &= \psi_{xy}\left(t-\frac{1}{2}\Delta t\right)+\Lambda_y E_{xy}\left(t+\frac{1}{2}\Delta t\right)-\alpha_y D_y\left[v_x\left(t+\frac{1}{2}\Delta t\right)\right]\\
\psi_{xz}\left(t+\frac{1}{2}\Delta t\right) &= \psi_{xz}\left(t-\frac{1}{2}\Delta t\right)+\Lambda_z E_{xz}\left(t+\frac{1}{2}\Delta t\right)-\alpha_z D_z\left[v_x\left(t+\frac{1}{2}\Delta t\right)\right]\\
\psi_{yx}\left(t+\frac{1}{2}\Delta t\right) &= \psi_{yx}\left(t-\frac{1}{2}\Delta t\right)+\Lambda_x E_{yx}\left(t+\frac{1}{2}\Delta t\right)-\alpha_x D_x\left[v_y\left(t+\frac{1}{2}\Delta t\right)\right]\\
\psi_{yy}\left(t+\frac{1}{2}\Delta t\right) &= \psi_{yy}\left(t-\frac{1}{2}\Delta t\right)+\Lambda_y E_{yy}\left(t+\frac{1}{2}\Delta t\right)-\alpha_y D_y\left[v_y\left(t+\frac{1}{2}\Delta t\right)\right]\\
\psi_{yz}\left(t+\frac{1}{2}\Delta t\right) &= \psi_{yz}\left(t-\frac{1}{2}\Delta t\right)+\Lambda_z E_{yz}\left(t+\frac{1}{2}\Delta t\right)-\alpha_z D_z\left[v_y\left(t+\frac{1}{2}\Delta t\right)\right]\\
\psi_{zx}\left(t+\frac{1}{2}\Delta t\right) &= \psi_{zx}\left(t-\frac{1}{2}\Delta t\right)+\Lambda_x E_{zx}\left(t+\frac{1}{2}\Delta t\right)-\alpha_x D_x\left[v_z\left(t+\frac{1}{2}\Delta t\right)\right]\\
\psi_{zy}\left(t+\frac{1}{2}\Delta t\right) &= \psi_{zy}\left(t-\frac{1}{2}\Delta t\right)+\Lambda_y E_{zy}\left(t+\frac{1}{2}\Delta t\right)-\alpha_y D_y\left[v_z\left(t+\frac{1}{2}\Delta t\right)\right]\\
\psi_{zz}\left(t+\frac{1}{2}\Delta t\right) &= \psi_{zz}\left(t-\frac{1}{2}\Delta t\right)+\Lambda_z E_{zz}\left(t+\frac{1}{2}\Delta t\right)-\alpha_z D_z\left[v_z\left(t+\frac{1}{2}\Delta t\right)\right]
\end{aligned} \tag{5.61}$$

When compute of S_{ij} and E_{ij}, the multiplication factors Λ_i, ζ_i, and ϕ_i ($i = x, y, z$) are, respectively

$$\begin{aligned}
\Lambda_x &= \sigma_x + \alpha_x \kappa_x\\
\Lambda_y &= \sigma_y + \alpha_y \kappa_y\\
\Lambda_z &= \sigma_z + \alpha_z \kappa_z
\end{aligned} \tag{5.62}$$

$$\begin{aligned}
\xi_x &= \frac{1+\frac{1}{2}\Delta t\alpha_x}{\frac{1}{2}\Delta t(\sigma_x+\alpha_x\kappa_x)+\kappa_x}\\
\xi_y &= \frac{1+\frac{1}{2}\Delta t\alpha_y}{\frac{1}{2}\Delta t(\sigma_y+\alpha_y\kappa_y)+\kappa_y}\\
\xi_z &= \frac{1+\frac{1}{2}\Delta t\alpha_z}{\frac{1}{2}\Delta t(\sigma_z+\alpha_z\kappa_z)+\kappa_z}
\end{aligned} \tag{5.63}$$

$$\begin{aligned}
\phi_x &= \frac{\Delta t}{\frac{1}{2}\Delta t\left(\sigma_x + \alpha_x \kappa_x\right) + \kappa_x} \\
\phi_y &= \frac{\Delta t}{\frac{1}{2}\Delta t\left(\sigma_y + \alpha_y \kappa_y\right) + \kappa_y} \\
\phi_z &= \frac{\Delta t}{\frac{1}{2}\Delta t\left(\sigma_z + \alpha_z \kappa_z\right) + \kappa_z}
\end{aligned} \tag{5.64}$$

5.4 ANALYSIS OF WAVE PROPAGATION CHARACTERISTICS IN VISCOELASTIC ANISOTROPIC FRACTURED MEDIUM

In order to investigate the effects of solid viscoelasticity and fracture parameters on seismic wave propagation characteristics in the fractured medium, we model the anisotropic viscoelastic characteristics of fractured reservoirs based on the constructed effective medium model.

Two conceptual models are designed as shown in Figure 5.4, forward modeling is performed and seismic wave propagation features are analyzed. The first model is designed as a homogeneous model, and the solid velocity is the same as that of the third-layer medium in Table 5.1. The source is excited in the center of model 1. The second model (refer to Table 5.1 for properties) is designed as a three-layer horizontally layered model, where the thicknesses of the first layer, second layer, and third layer is 2000, 500, and 500 m, respectively. In model 2, fractures are developed in the second layer and the source is excited at the left end of earth surface. The 16-order RSG finite-difference algorithm is used in forward modeling and an explosive source with dominant frequency of 30 Hz is used for excitation. Properties of medium and fracture fluid are varied. Seismic wave propagation characteristics are analyzed from both seismic wavefield snapshots and seismic profiles.

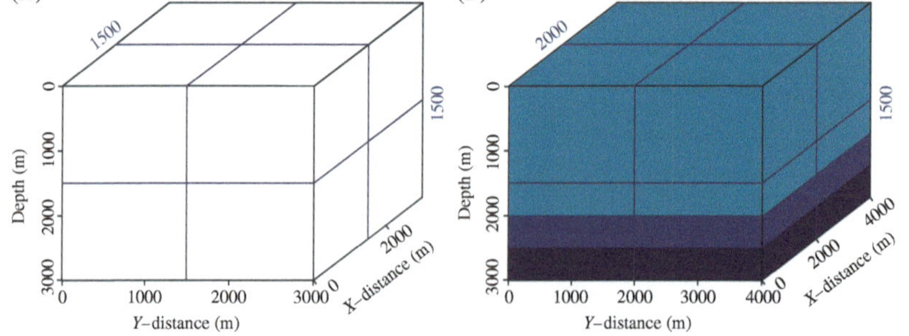

FIGURE 5.4 Schematics diagrams of the conceptual models. (A) A homogeneous model; (B) a three-layered model, where the second layer contains viscoelastic fractures.

TABLE 5.1 Model Parameters

Rock matrix velocity			
Layer (thickness/m)	**P-wave velocity V_P (m/s)**	**S-wave velocity V_S (m/s)**	**Density ρ (kg/m^3)**
Layer 1 (2000)	4482	2587	2500
Layer 2 (500)	5300	2975	2750
Layer 3 (500)	6220	3386	2790

Fracture fluid parameters (Pointer et al., 2000)				
Fluid type	**P-wave velocity V_P (m/s)**	**Density ρ (kg/m^3)**	**Bulk modulus (GPa)**	**Viscosity (Pa·s)**
Oil	1299	800	1.35	0.003
Gas	620	65	0.025	0.00002
Water	1500	1000	2.25	0.001

The basic parameters of the models are shown in Table 5.1. The various properties in the study are given correspondingly.

5.4.1 Effects of *Q*-Value Variations of Solid

The solid viscoelasticity mainly affects seismic wave attenuation in the medium and it is characterized by P-wave and S-wave quality factors. In this section, the effects of various *Q*-values on seismic wave propagation are investigated using seismic simulation. In the simulations, P-wave *Q*-values are 25, 50, 100, and 200, respectively. *Q*-values of S-wave are 10, 20, 40, and 80, respectively.

5.4.1.1 Homogeneous Model

For different *Q* values, the wavefield snapshots at 225 ms after source excitation in a homogeneous medium are shown in Figure 5.5.

As can be seen from the wavefield snapshots, the amplitude increases with increasing Q_p.

To clearly analyze the variations of seismic energy for different solid *Q*-values, the middle traces in wavefield snapshots are plotted in Figure 5.6.

In Figure 5.6, the dashed dotted curve corresponds to $Q_p = 200$, indicating the lowest solid anelasticity, and the solid curve corresponds to $Q_p = 25$, indicating the strongest solid anelasticity. It is obvious that wavefront amplitude declines with the increase of solid anelasticity (Q_p decrease). The positions of peaks and troughs of the four wavefront curves basically coincide, which indicates that the changes in solid viscoelasticity have almost no effect on group velocities. Wavelength increases slightly with the decrease of Q_p. The principle

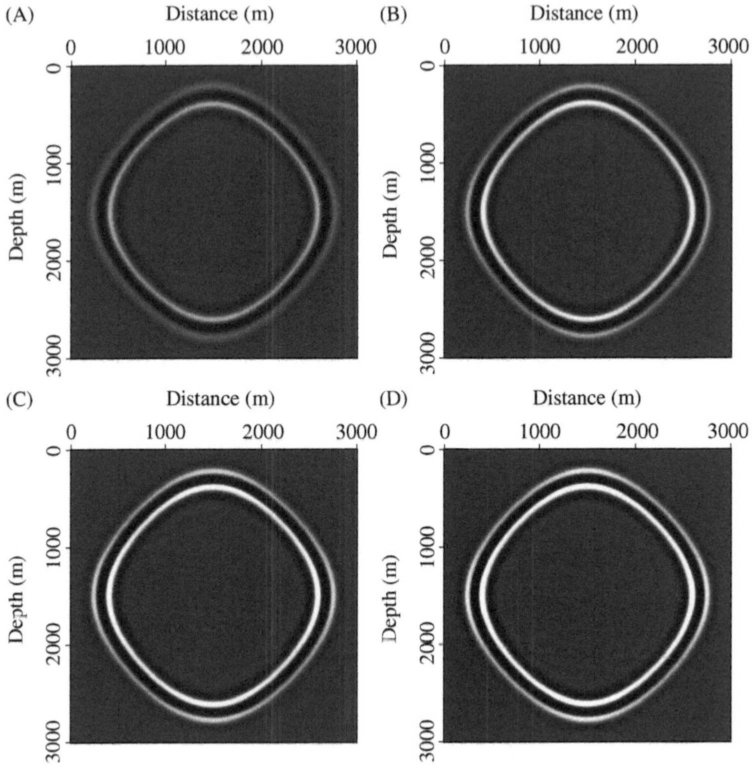

FIGURE 5.5 Wavefield snapshots in a homogeneous model with different Q-values. (A) $Q_p = 25$, (B) $Q_p = 50$, (C) $Q_p = 100$, (D) $Q_p = 200$.

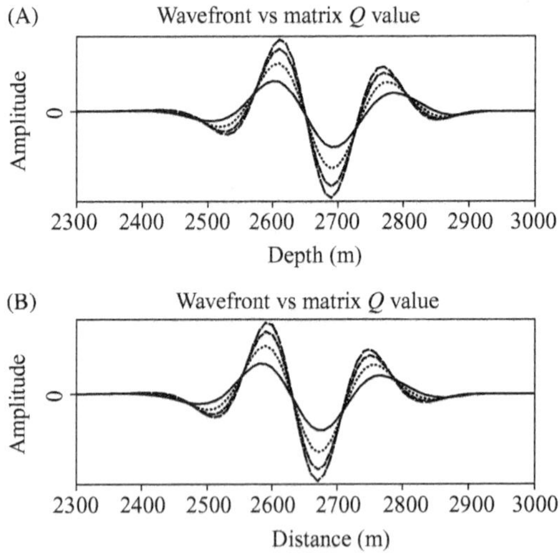

FIGURE 5.6 Wavefront curves of the middle traces in wavefield snapshots in the homogeneous model with different Q-values. (A) Longitudinal middle traces. (B) Transverse middle traces. Solid curves for $Q_p = 25$, dotted curves for $Q_p = 50$, dashed curves for $Q_p = 100$, and dashed dotted curves for $Q_p = 200$.

frequency of seismic waves decreases accordingly with the increase of solid anelasticity.

5.4.1.2 Layered Model

Reflection seismic profiles of the target layer in a layered model with different solid Q-values are shown in Figure 5.7.

As can be seen from Figure 5.7, with the increase of solid Q-value, the reflection amplitude from the bottom interface of the viscoelastic fractured layer is strengthened.

Two traces of near offset and far offset are extracted from the seismic profile to analyze the effect of solid viscoelasticity on reflected waves at different offsets.

Based on the single trace of simulated reflected waves shown in Figure 5.8, the reflection from the bottom interface of viscoelastic fractured layer can be analyzed. With the decrease of solid Q-value, the viscoelastic fractured layer causes more attenuation to seismic waves, and the received reflection energies

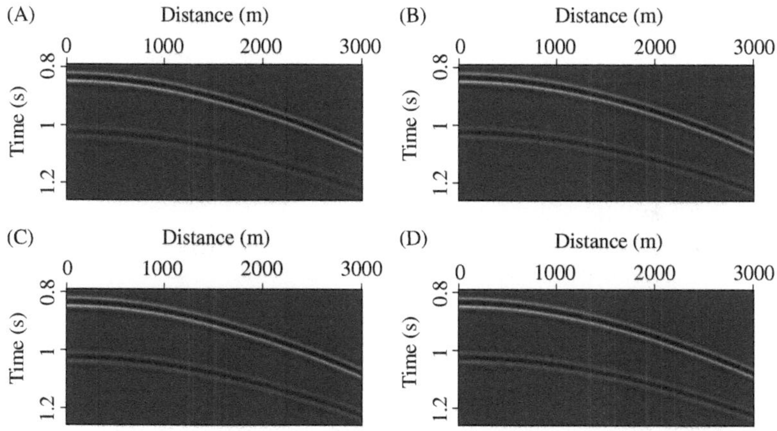

FIGURE 5.7 Seismic profiles of a layered model with different solid Q-values. (A) $Q_p = 25$, (B) $Q_p = 50$, (C) $Q_p = 100$, (D) $Q_p = 200$.

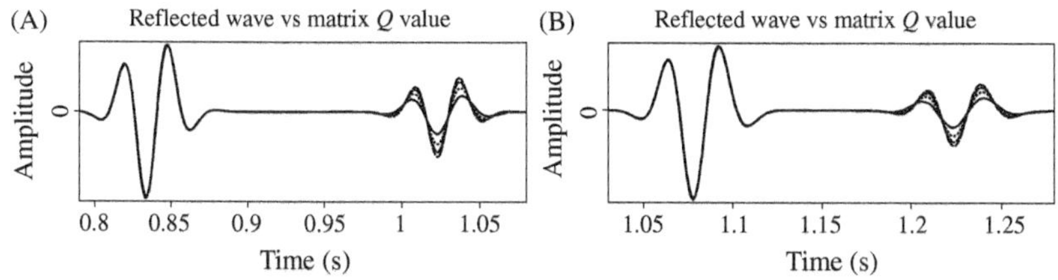

FIGURE 5.8 (A) Near-offset and (B) far-offset reflection curves in seismic profiles of the layered model with different solid Q-values. Solid curves denote $Q_p = 25$ and dotted curves denote $Q_p = 50$. Dashed curves denote $Q_p = 100$ and dashed dotted curves denote $Q_p = 200$.

at surface get weaker. The wavelength of reflections gradually increases, which means the principal wave frequency decreases with anelasticity. Besides, the variation of reflection seismic waves versus solid Q-value at near offset traces is consistent with those of far offset traces.

5.4.2 Effects of Reservoir Fluid Changes

Different states of fluid saturation in reservoir also affect seismic wave propagation. In this section, three cases are considered, reservoir is fully saturated with oil, gas, or water, respectively. Relevant fluid properties are shown in Table 5.1.

5.4.2.1 Homogeneous Model

With different types of reservoir fluid, the wavefield snapshots in a homogeneous model at 225 ms after source excitation are shown in Figure 5.9.

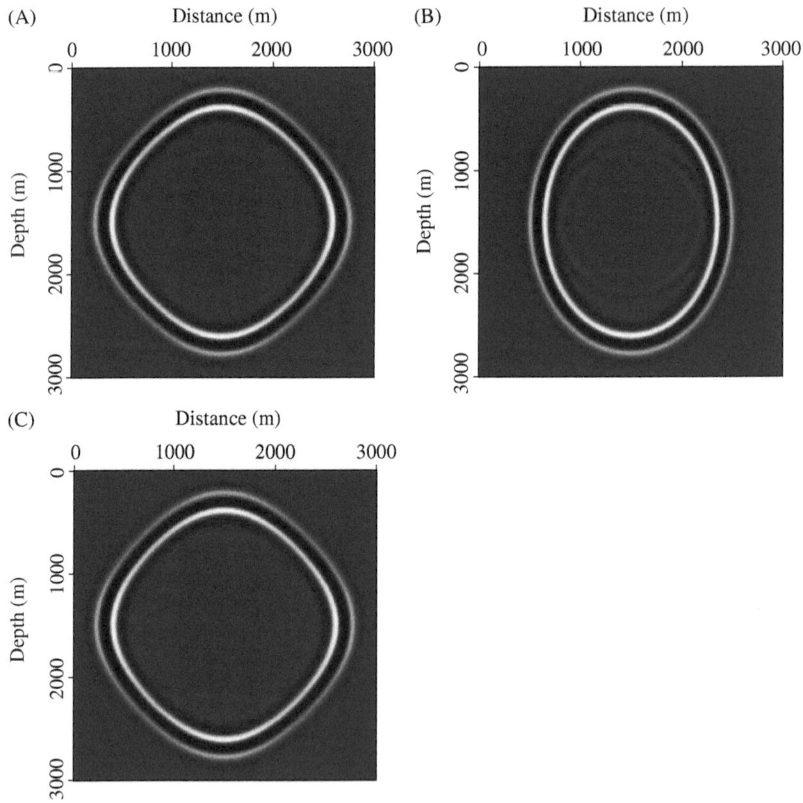

FIGURE 5.9 Wavefield snapshots in a homogeneous model with different reservoir fluids. (A) Oil. (B) Gas. (C) Water.

As can be seen from Figure 5.9, the shape and energy of seismic wavefront are almost the same when the fractures are fully saturated with oil or water, respectively. If the fractures are saturated with gas, seismic wavefront is obviously different from the other two cases. The velocity of waves propagating parallel with the fracture plane is significantly higher than that propagating vertical to the fracture plane. The velocity of seismic waves propagating vertically to the fracture plane decreases significantly when the fractures are filled with gas.

The middle traces from the wavefield snapshots are extracted, as is shown in Figure 5.10.

The results of the longitudinal single-trace curves show that, when fractures are saturated with different types of fluid, seismic wavefront locations and shapes are basically the same in the direction parallel with fracture plane. In the transverse single-trace curves, the wavefront is located closer to the source when fractures are saturated with gas, since there is lower wave velocity. The modeling results indicate that fluid types in fractures almost have no effect on the direct longitudinal waves propagating parallel to fracture plane, however they have significant effect on the waves propagating in a direction vertically to fracture plane. The wave velocity is the highest when fractures are saturated with water and decrease significantly when fractures contain gas.

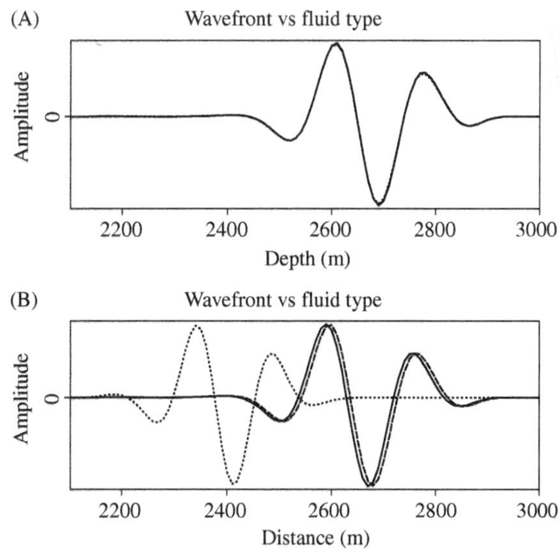

FIGURE 5.10 Wavefront curves of the middle traces in wavefield snapshots in a homogeneous model with different fluids. (A) Longitudinal middle traces. (B) Transverse middle traces. Solid curves denote the model saturated with oil. Dotted curves denote the model with gas while dashed curves denote the model with water.

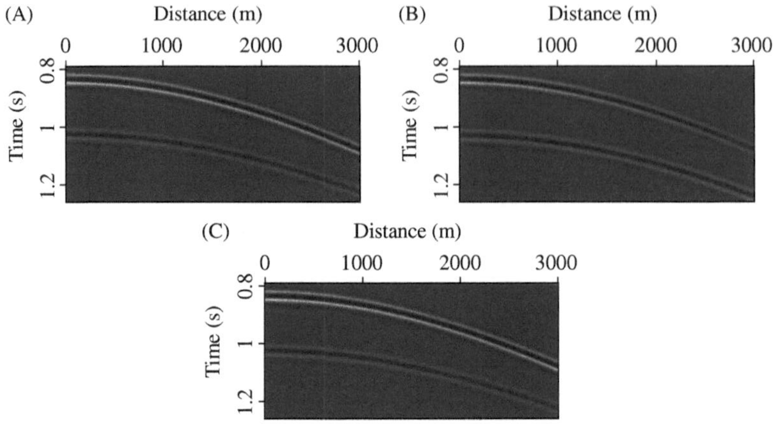

FIGURE 5.11 Seismic profiles of a layered model with different reservoir fluids. (A) Oil. (B) Gas. (C) Water.

5.4.2.2 Layered Model

Reflection seismic profiles of the target layer in a layered model with different reservoir fluid are shown in Figure 5.11.

The two traces of near offset and far offset are extracted, respectively, as shown in Figure 5.12 from seismic profiles to analyze the effect of saturation fluid types on reflection waves at different offsets.

As can be seen from the seismic profiles and single-trace synthesized records, the seismic profiles are generally the same for the two cases when fractures are fully saturated with oil or fully saturated with water, and the reflected wave have got significant changes when fractures are saturated with gas. For the gas case, it shows that the reflection amplitude of the top interface of viscoelastic fractured layer is weaker while that of the bottom interface of the layer is stronger, with a longer travel time. The changes are more obvious at the far offset traces than those at near offset. When fractures are saturated with gas, the velocity of seismic waves in the viscoelastic fractured layer gradually declines with the increase of wave incidence angle (i.e., the increase of offset), therefore travel time increases, furthermore, the decrease of velocity results in the decrease of wave

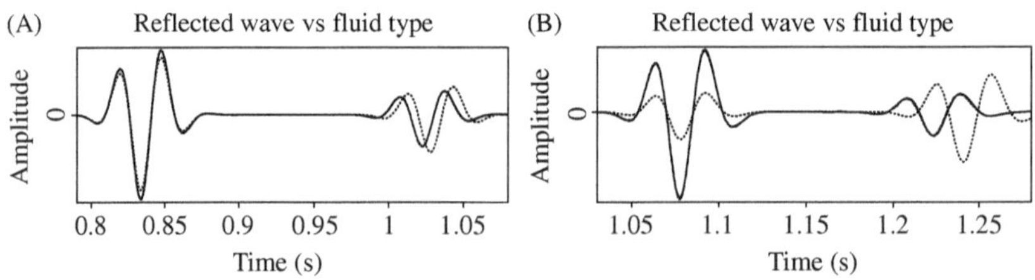

FIGURE 5.12 (A) Near-offset and (B) far-offset reflection curves in seismic profiles of the layered model with different reservoir fluids. Solid curves denote the model saturated with oil and dotted curves denote the model with gas. Dashed curves denote the model with water.

impedance of the fractured reservoir layer, which, in turn, lead to the decrease of wave-impedance difference between the reservoir and the upper layer, so that the reflection amplitude from the top interface are weakened, while on the contrary, the wave-impedance difference between the reservoir and the lower layer increases, leading to the strengthened reflection from the reservoir bottom and these responses will become even more obvious at larger offsets.

5.4.3 Effects of Variation in Fracture Bulk Density

The bulk density and aspect ratio of fractures are two important parameters for describing fracture development system. Changes in fracture properties will inevitably affect seismic wave propagation in the medium. In this chapter, we first consider the effect of the changes in fracture bulk density.

5.4.3.1 Homogeneous Model

With different fracture bulk densities, the wavefield snapshots in a homogeneous model at 225 ms after source excitation are shown in Figure 5.13.

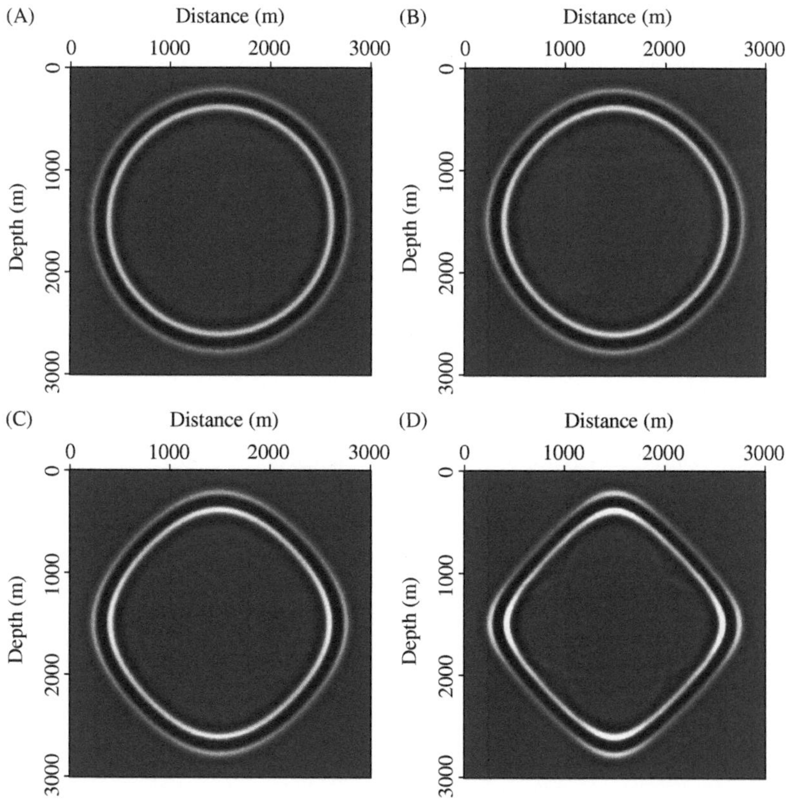

FIGURE 5.13 Wavefield snapshots in a homogeneous model with different fracture bulk densities. (A) $e = 0.01$, (B) $e = 0.05$, (C) $e = 0.1$, and (D) $e = 0.2$.

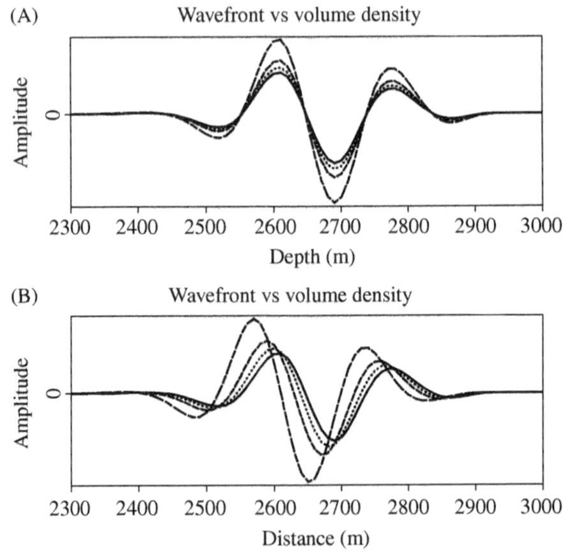

FIGURE 5.14 Wavefront curves of the middle traces in wavefield snapshots in a homogeneous model with different fracture bulk densities. (A) Longitudinal middle traces. (B) Transverse middle traces. Solid curves denote $e = 0.01$, dotted curves denote $e = 0.05$, dashed curves denote $e = 0.1$, and dashed dotted curves denote $e = 0.2$.

In the wavefield snapshots with different fracture densities, it can be seen that the anisotropy of viscoelastic fractured medium gradually becomes stronger with the increase of fracture bulk density.

The middle traces are extracted from the wavefield snapshots, as is shown in Figure 5.14.

As is shown in Figure 5.14, changes in fracture bulk density almost have no effect on the P-wave velocity in the direction parallel with the fracture plane, while in the direction vertical to the fracture plane, P-wave velocity gradually decreases with the increase of fracture bulk density.

5.4.3.2 *Layered Model*

Reflection seismic profiles of the target layer in a layered model with different fracture densities are shown in Figure 5.15.

Figure 5.15 shows that the travel time of reflected waves from the bottom interface of viscoelastic fractured reservoir increases with the fracture bulk density at far offsets. To some extent, the velocity of seismic waves propagating in a viscoelastic fractured medium decreases with the increase of incidence angle (within a range of around 0°–45°).

The two traces of near offset and far offset are extracted respectively as shown in Figure 5.16 from seismic profiles to analyze the effect of variation in fracture bulk density on reflection waves at different offsets.

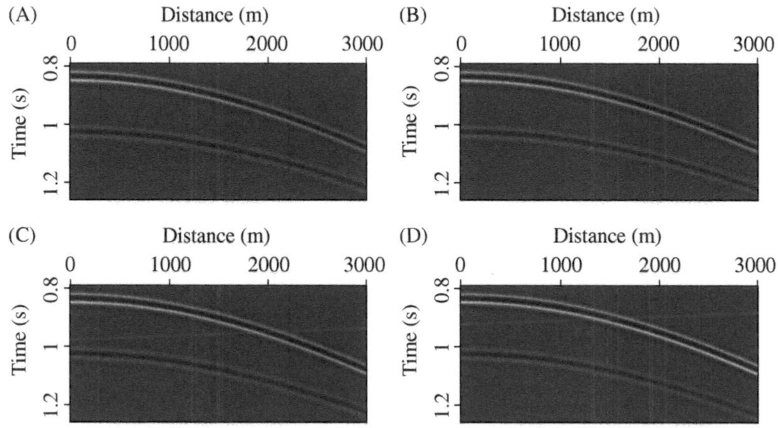

FIGURE 5.15 Seismic profiles of a layered model with different fracture bulk densities. (A) $e = 0.01$, (B) $e = 0.05$, (C) $e = 0.1$, and (D) $e = 0.2$.

In the single-trace curves, it can be seen that fracture bulk density has very slight effect on the reflected wave propagation at near offset, while at far offset, the amplitude of the reflected waves from the top interface of viscoelastic fractured layer increases with fracture bulk density, and the travel time of reflected waves from the bottom interface gradually increases with fracture bulk density. The main reason lies in that at far offsets (for large incidence angles), the velocity of seismic wave propagation in viscoelastic fractured medium decreases with an increasing fracture density.

5.4.4 Effects of Variation of Fracture Dip

Fractures developed in subsurface usually have dominant geological dips and azimuths. The changes of dip and azimuth mainly affect the anisotropy of fractured medium. In the wave simulation of this section, the fracture azimuth is given as 90° and all relevant parameters are fixed except for the fracture dip.

5.4.4.1 Homogeneous Model

The wavefield snapshots with different developed fracture dips are shown in Figure 5.16. The angles between the fracture plane and the plane *XOY* are shown by a red line in the figure.

In the wavefield snapshots in Figure 5.17, it is shown that: (1) when the dip is equal to 0°, that is, the parallel fractures are distributed horizontally, the simulated direct wavefield that are parallel to the fracture plane (plane *XOY*) has the wavefront in a circular shape, while the wavefronts in the other two planes are nearly rhombic; (2) the wavefronts in the three planes are anisotropic when the dip is equal to 30°, because all the three Cartesian coordinate planes are in certain angles with the fracture plane; the angle between the axis *X* and

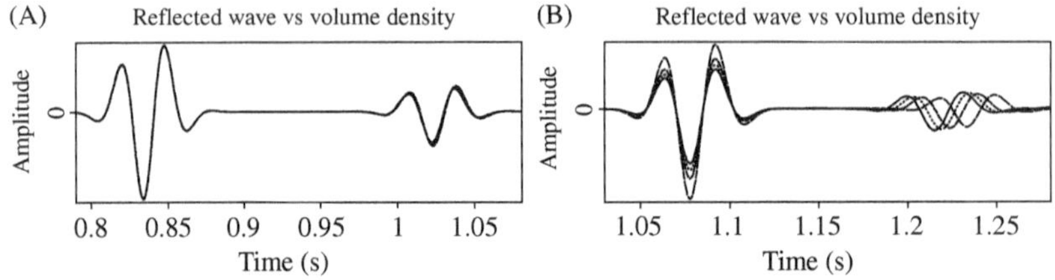

FIGURE 5.16 (A) Near-offset and (B) far-offset reflection curves in seismic profiles of the layered model with different fracture densities. Solid curves denote $e = 0.01$, dotted curves denote $e = 0.05$, dashed curves denote $e = 0.1$, and dashed dotted curves denote $e = 0.2$.

the symmetry axis of the anisotropic rhombic wavefront in plane *XOZ* is equal to 30°, that is, the fracture dip angle; the angle between the plane *XOY* and fracture plane is smaller (30°) and the wavefront is ellipse in shape; the angle between the plane *YOZ* and fracture plane is larger (60°) and the wavefront is

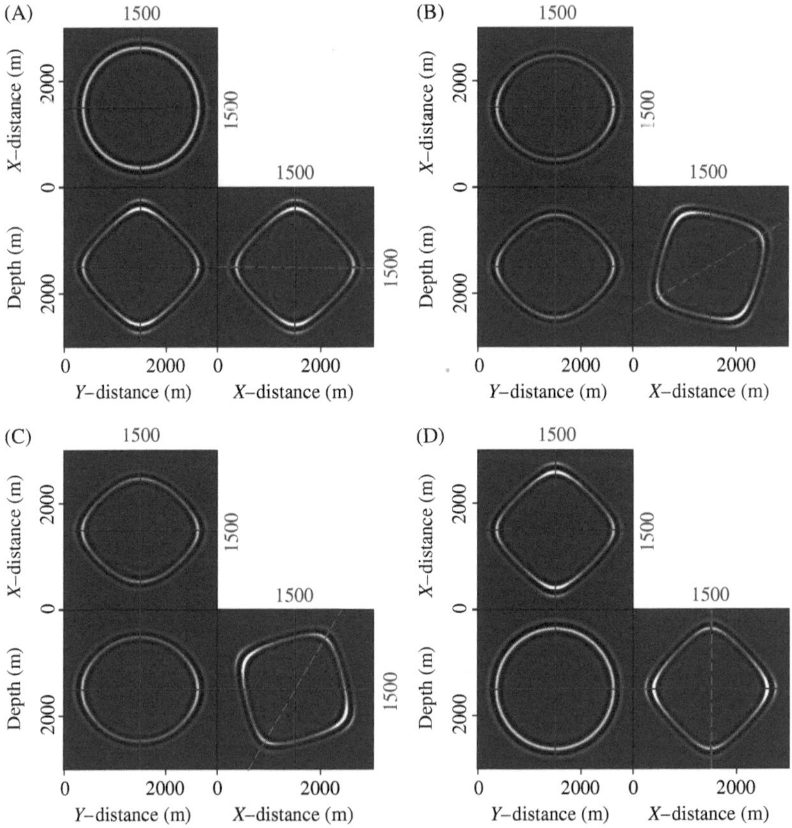

FIGURE 5.17 Wavefield snapshots in a homogeneous model with different fracture dips. (A) Dip = 0°, (B) dip = 30°, (C) dip = 60°, and (D) dip = 90°.

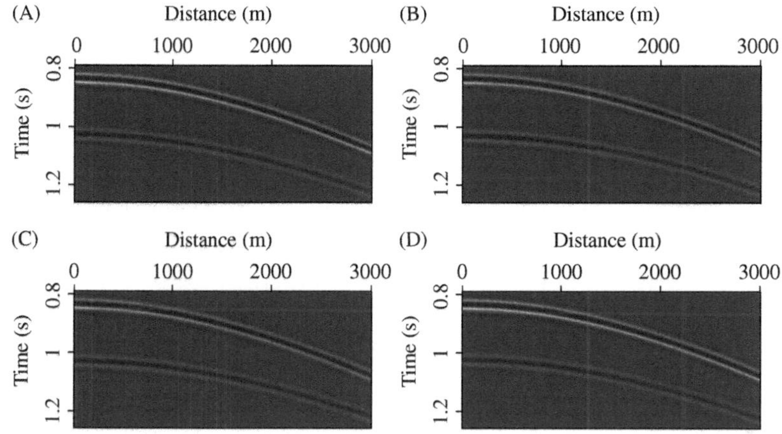

FIGURE 5.18 Seismic profiles of a layered model with different fracture dips. (A) Dip = 0°, (B) dip = 30°, (C) dip = 60°, and (D) dip = 90°.

nearly rhombic and significantly anisotropic; (3) when the dip is equal to 60°, the wavefronts on the three planes are anisotropic, similar to those at the dip of 30°; the angle between the axis X and the symmetry axis of the anisotropic rhombic wavefront in plane XOZ is equal to 60°; the anisotropy increases with the angle between the other two planes and the fracture plane; and (4) when the dip is equal to 90°, the wavefronts are similar to those at the dip of 0°; the shape of wavefront in the plane YOZ that is parallel to the fracture plane is circular, showing no anisotropy; the other two planes show similar anisotropy features and the shapes are nearly rhombic.

5.4.4.2 Layered Model

Reflection seismic profiles of the target layer in a layered model with different fracture dips are shown in Figure 5.18.

The two traces of near offset and far offset are extracted respectively from seismic profiles to analyze the effects of variation in fracture dip on reflection waves at different offsets (Figure 5.19).

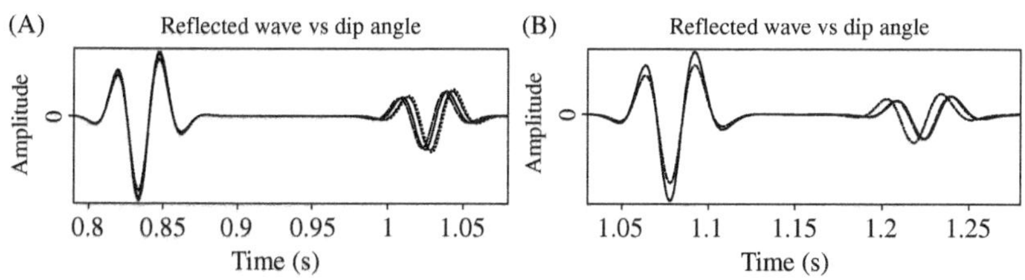

FIGURE 5.19 (A) Near-offset and (B) far-offset reflection curves in seismic profiles of the layered model with different fracture dips. Solid curves denote dip = 0°, dotted curves denote dip = 30°, dashed curves denote dip = 60°, and dashed dotted curves denote dip = 90°.

From the seismic profiles and single-trace curves, no obvious feature for the effect of fracture dip on the reflection events is seen. For the two cases of fracture dip at 0° and 90°, seismic wave reflection characteristics are basically similar. For the cases of fracture dip at 30° and 60°, the characteristics are also similar. At near offset, when the fracture dip is equal to 0° or 90°, the seismic wave velocity in viscoelastic fractured layer is relatively higher, which results in stronger reflection amplitude on the top interface of viscoelastic fractured layer and weaker reflection amplitude on the bottom interface, as well as a shorter travel time. At far offset, when the fracture dip is 0° or 90°, the seismic wave velocity in viscoelastic fractured layer is lower, which results in weaker reflection amplitude on the top interface of the fractured reservoir and stronger reflection amplitude on the bottom interface, as well as a longer travel time. With the input fracture and fluid parameters, the anisotropy of medium is exhibited as the nearly rhombic wavefronts for seismic wave propagating along and vertical to the fracture development directions.

5.4.5 Effects of Variation in Fracture Azimuth

Fracture azimuth is a major factor that impacts the elasticity anisotropy in a fractured medium. In wave simulation of this section, the fracture dip is equal to 90° and all other parameters are fixed except the azimuth angle of fractures. We study the effects of fracture azimuth on wave propagation assuming different values for the azimuth.

5.4.5.1 *Homogeneous Model*

The wavefield snapshots with different fracture azimuths are shown in Figure 5.20. The angles between the fracture plane and the plane *XOZ* are shown by a red line in the figure.

In the wavefield snapshots with different fracture azimuths, it is shown that: (1) when the azimuth is equal to 0°, that is, fractures are parallel with the plane *XOZ*, the wavefront of direct wavefield in the plane is circular in shape and exhibits no anisotropy, and the wavefront shapes on the other two planes are rhombic; (2) the wavefronts in the three planes are anisotropic when the azimuth is 30°, since all the three Cartesian coordinate planes are in certain angles with the fracture plane; the angle between the axis *X* and the symmetry axis of anisotropic rhombic wavefront in plane *XOY* is equal to 30°, that is, the fracture azimuth; the angle between the plane *XOZ* and the fracture plane is smaller (30°) and the wavefront shape is ellipse; the angle between the plane *YOZ* and the fracture plane is larger (60°) and the wavefront shape is rhombic and significantly anisotropic; (3) when the azimuth is equal to 60°, the wavefronts on the three planes are anisotropic, similar to the one of azimuth at 30°; the angle between the axis *X* and the symmetry axis of anisotropic rhombic wavefront in plane *XOY* is 60°; the anisotropy increases with the increasing angle between the other two planes and the fracture plane; and (4) when the azimuth is equal

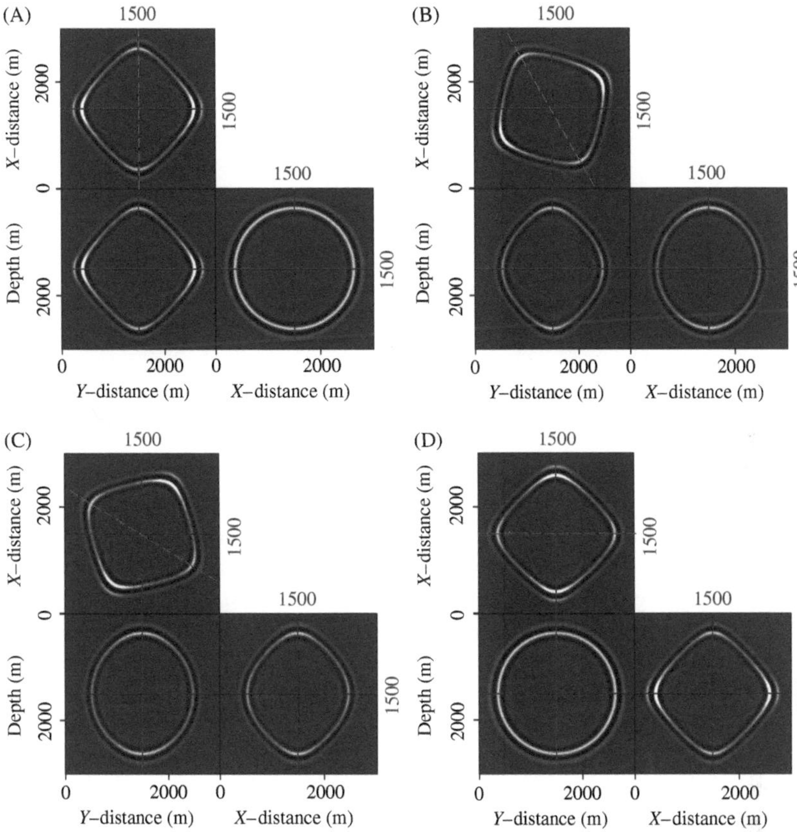

FIGURE 5.20 Wavefield snapshots in a homogeneous model with different fracture azimuths: (A) azi = 0°, (B) azi = 30°, (C) azi = 60°, and (D) azi = 90°.

to 90°, the wavefront is similar to the one of azimuth at 0°; wavefront in the plane *YOZ* that is parallel with the fracture plane is circular in shape, showing no anisotropy; the wavefront shapes are rhombic in the other two planes and show similar anisotropy.

5.4.5.2 Layered Model

By using a synthesized 3D acquirement system, the simulated seismic reflection waves are received. The effects of the relationship between locations of seismic test line and fracture development directions on seismic wave propagation are analyzed.

The source is located at the zero coordinates in the layered model. Seismic profiles of the reflected waves are obtained as is shown in Figure 5.21.

It is shown in the seismic profiles of the reflections that: (1) the reflected waves at different azimuths from the top interface of target layer basically show

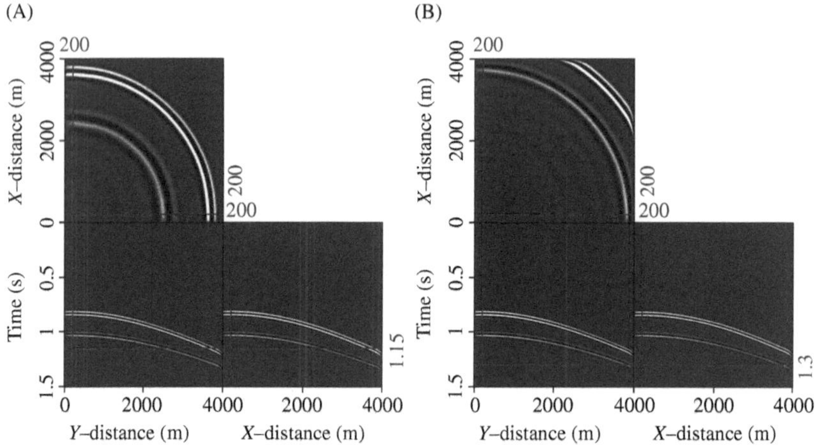

FIGURE 5.21 Synthesized seismograms of a layered model. (A) 1.15 s. (B) 1.3 s.

no anisotropy; circular wavefront is observed in the transverse section; on the two longitudinal seismic profiles which are parallel with and vertical to the fracture plane (Figure 5.21A), respectively, there are the same locations of reflection events that are labeled by the 1.15 s time lines, since the reflection waves from the fractured-reservoir top that are received at surface have not transmitted into the viscoelastic fractured layer; and (2) the reflected waves at different azimuths from the bottom interface of target layer show a certain degree of azimuthal anisotropy, that is, the velocity of seismic waves propagating parallel with the fracture plane is higher than that propagating vertical to the plane. An ellipse wavefront of reflected waves is observed on the transverse section; seismic wave velocity in *Y* direction is higher than that in *X* direction, and the wavefront in *Y* direction is closer to the model boundary, which is more obvious in the seismograms (1.3 s) at far offset; in the longitudinal profile (Figure 5.21B), the travel time of bottom-interface reflected waves which propagate vertical to the fracture plane (plane *XOZ*) and are labeled by the 1.3 s time lines is larger, with a relatively lower velocity.

In the synthesized seismograms, single-trace curves are extracted from the 1.15 s of test line at the azimuths of 0° and 90°, and from the far offset of test line at the azimuths of 0° and 90°, seismic wave propagation characteristics are compared and analyzed between different azimuths. In the case that the azimuth is equal to 0°, the test line is vertical to the fracture plane, while for the azimuth of 90°, the test line is parallel with the fracture plane.

In the comparison results of single-trace curves, the waveforms from 3500 to 4000 m of Figure 5.22A indicate the reflected waves from the top interface of fractured layer. As the waves have not transmitted through the fractured reservoir, the positions where the reflected waves are received are the same when azimuths are different. The waveforms at around 2500 m indicate the reflected waves from the bottom of fractured layer, which are affected by the anisotropy

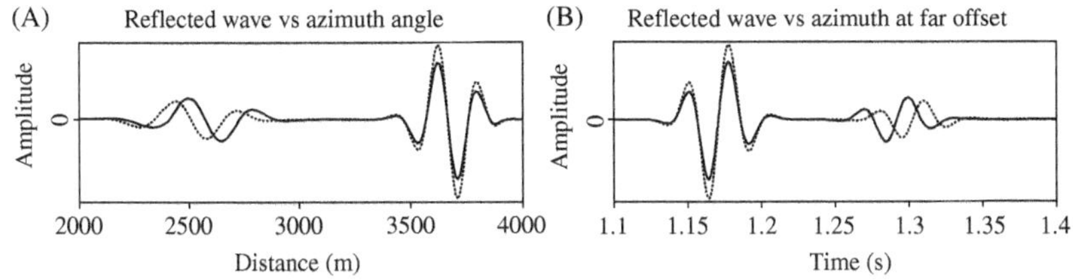

FIGURE 5.22 Contrast of single-trace seismogram curves in the test linea between azimuths 0° and 90°. (A) At 1.15 s. (B) At far offset. Solid curve denotes the azimuth of 90° in test line and dotted curve denotes the azimuth of 0° in test line.

of fracture system. The wave velocity is higher for the test line azimuth at 90° and the travel distance of reflected waves that are received at surface is longer. In Figure 5.22B, the waveforms at around 1.15 s indicate the reflected waves from the top of fractured layer. Since the waves have not transmitted into the fractured layer, the travel times are the same when the azimuths are different. At around 1.3 s, the travel time at different test line azimuths is different and it is smaller when the azimuth is equal to 90° (parallel with the fracture plane), since seismic waves propagate faster in the direction parallel with the fracture plane. In summary, when seismic test lines are set up in different azimuths, based on the received data, the fracture development properties can be estimated in accordance with the offset characteristics at a fixed wave travel time or the travel time characteristics at a fixed offset.

5.5 CONCLUSIONS

In this chapter, on the basis of previous studies, we presented an effective medium model that is suitable for describing the wave behaviors in the viscoelastic anisotropic fractured reservoirs. Three-dimensional seismic forwarding modeling is implemented by applying a rotated staggered grid finite-difference scheme. The complex frequency-shifted recursive integration nonsplit PML boundary condition is applied to absorb the boundary reflection. Through numerical results, we analyzed the effects of the types of rocks, the presence of fluids, and the fracture development patterns on wave behaviors in viscoelastic anisotropic fractured medium, and we got some meaningful conclusions.

1. Numerical simulation of seismic wavefield can be used to visually demonstrate the propagation of seismic waves in Earth's medium. For the viscoelastic anisotropic fractured medium considered in this chapter, space derivatives are implemented using finite difference in the diagonal directions in RSG algorithm, with no need of interpolation for elasticity parameters of the medium. Therefore, it has higher precision and better stability and is an ideal finite difference algorithm for numerical simulation of viscoelasticity and anisotropy. In addition, by applying complex frequency-shifted recursive-integration nonsplitting PML boundary conditions, boundary reflection can

be effectively attenuated and this makes wavefield simulation results more reliable and accurate.

2. Wavefield numerical simulation results show that seismic wave velocity in viscoelastic fractured reservoirs show anisotropy features. The seismic wavefronts in the propagation plane are nearly ellipse or rhombus, rather than circular, since the velocity of seismic waves propagating parallel with the fracture plane is higher, while that of waves propagating in the direction at a certain angle with fracture plane is lower. Besides, attenuation of seismic wave propagation in viscoelastic fractured reservoirs causes decease of wave amplitude, increase of wavelength and loss of principle frequency.
3. Wavefield numerical simulation results show that the characteristics of seismic wave propagation in viscoelastic fractured reservoirs are also affected by the factors such as mineral grain (solid), reservoir fluid and fracture developing properties. The seismic wave attenuation increases with the decrease of solid Q-values, namely, the increase of medium anelasticity. With the increase of fracture bulk density or fracture aspect ratio, or when the fracture is saturated with gas, the anisotropy in seismic wave propagation in fractured medium is more significant. Differences in fracture dips and fracture azimuths may result in differences of azimuths and strengths of anisotropy in fractured medium. In addition, the relevant properties will not only cause seismic anisotropy changes, but also cause differences in reflection strength and travel time of reflected seismic waves. Moreover, due to the different incidence angles for seismic waves transmitting into the fractured reservoirs and then being received at different offsets on surface, there is corresponding changes in wave propagation velocity and attenuation characteristics, making reflection seismic waves that are received at different offsets show different response characteristics.
4. The factors that induce seismic attenuation include the viscoelasticity in solid and the gas saturation in fractures. The factors that induce anisotropy include the increase in fracture bulk density and aspect ratio and gas saturation in fractures. Based on the above studies, further investigations need to combine the studies on viscoelastic parameter estimation and Amplitude Versvs Azimuth (AVAZ) inversion. Areas with high attenuation or strong anisotropy features should be identified in seismic exploration to define the development zone of fractured oil/gas reservoirs.

REFERENCES

Aki, K., Richards, P.G., 1980. Quantitative Seismology. W. H. Freeman and Company, San Francisco, California.

Carcione, J.M., 1993. Seismic modeling in viscoelastic media. Geophysics 58, 110–1110.

Carcione, J.M., Kosloff, D., Kosloff, R., 1988. Wave propagation simulation in a linear viscoelastic medium. Geophys. J. 95 (3), 597–611.

Carcione, J.M., Poletto, F., Gei, D., 2004. 3-D wave simulation in anelastic media using the Kelvin–Voigt constitutive equation. J. Computat. Phys. 196 (1), 282–297.

Casula, G., Carcione, J., 1992. Generalized mechanical model analogies of linear viscoelastic behaviour. Bollettino Geofisica Teorica Applicata.

Chu, C., Stoffa, P.L., Seif, R., 2009. High-order rotated staggered finite difference modeling of 3d elastic wave propagation in general anisotropic media. In 2009, SEG., Annual Meeting, 291–295.

Collino, F., Tsogka, C., 2001. Application of the perfectly matched absorbing layer model to the linear elastodynamic problem in anisotropic heterogeneous media. Geophysics 66 (1), 294–307.

Dong, Z., McMechan, G.A., 1995. 3-D viscoelastic anisotropic modeling of data from a multicomponent: multiazimuth seismic experiment in Northeast Texas. Geophysics 60 (4), 1128–1138.

Drossaert, F.H., Giannopoulos, A., 2007. A nonsplit complex frequency-shifted PML based on recursive integration for Fdtd modeling of elastic waves. Geophysics 72 (2), T9–T17.

Du Qi-zhen, Sun Rui-yan, Zhang Qiang, 2011. Numerical simulation of three-component elastic wavefield in 2D TTI media in the condition of the combined boundary. Oil Geophys. Prospect. 46 (2), 187–195.

Fan, Z.-X., Teng, Y.-C., 1988. The numerical modeling of seismic wave in the transversely isotropic and viscoelastic media. Chin. J. Geophys. 31 (2), 184–210.

Hudson, J., 1986. A higher order approximation to the wave propagation constants for a cracked solid. Geophys. J. Royal Astronom. Soc. 87 (1), 265–274.

Hudson, J., 1980. Overall properties of a cracked solid. Mathematical Proceedings of the Cambridge Philosophical Society. Cambridge University Press, 371–384.

Hudson, J., 1981. Wave speeds and attenuation of elastic waves in material containing cracks. Geophys. J. Royal Astronom. Soc. 64 (1), 133–150.

Jia, L.-K., 2011. Numerical Modeling of Seismic Wave in Viscoelastic Media[D]. China University of Petroleum.

Nie, J., Yang, D., 2008. Viscoelastic Bisq model for low-permeability sandstone with clay. Chin. Phys. Lett. 25 (8), 3079.

Nie, J., Yang, D., Yang, H., 2008. A generalized viscoelastic Biot/Squirt model for clay-bearing sandstones in a wide range of permeabilities. Appl. Geophy. 5 (4), 249–260.

Nie, J.X., Ba, J., Yang, D.H., Yan, X.F., Yuan, Z.Y., Qiao, H.P., 2012. BISQ model based on a Kelvin-Voigt viscoelastic frame in a partially saturated porous medium. Appl. Geophys. 9 (2), 213–222.

Pointer, T., Liu, E., Hudson, J.A., 2000. Seismic Wave Propagation in Cracked Porous Media[J]. Geophys. J. Int. 142 (1), 199–231.

Robertsson, J., Blanch, J., Symes, W., 1994. Viscoelastic finite difference modeling. Geophysics 59 (9), 1444–1456.

Roden, J.A., Gedney, S.D., 2000. Convolutional PML (CPML): an efficient FDTD implementation of the CFS-PML for arbitrary media. Microwave Optic. Technol. Lett. 27 (5), 334–338.

Saenger, E.H., Bohlen, T., 2002. 3d finite-difference modeling of viscoelastic and anisotropic wave propagation using the rotated staggered grid. SEG Tech. Prog. Expanded Abstracts 21 (1), 2019–2022.

Saenger, E.H., Bohlen, T., 2004. Finite-difference modeling of viscoelastic and anisotropic wave propagation using the rotated staggered grid. Geophysics 69 (2), 583–591.

Saenger, E.H., Gold, N., Shapiro, S.A., 2000. Modeling the propagation of elastic waves using a modified finite-difference grid. Wave Motion 31 (1), 77–92.

Schoenberg, M., 1980. Elastic wave behavior across linear slip interfaces. J. Acous. Soc. Am. 68 (5), 1516–1521.

Schoenberg, M., Douma, J., 1988. Elastic wave propagation in media with parallel fractures and aligned cracks. Geophys. Prosp. 36 (6), 571–590.

Schoenberg, M., Sayers, C.M., 1995. Seismic anisotropy of fractured rock. Geophysics 60 (1), 204–211.

Shan, Qi-tong, 2007. Simulation and inversion of wave equation in viscoelastic media[D]. China University of Petroleum.

Stekl, I., Pratt, R.G., 1998. Accurate viscoelastic modeling by frequency-domain finite differences using rotated operators. Geophysics 63 (5), 1779–1794.

Thomsen, L., 1995. Elastic anisotropy due to aligned cracks in porous Rock1. Geophys. Prosp. 43 (6), 805–829.

Thomsen, L., 1986. Weak elastic anisotropy. Geophysics 51 (10), 1954–1966.

Virieux, J., 1986. P-Sv wave propagation in heterogeneous media: velocity-stress finite-difference method. Geophysics 51 (4), 889–901.

Virieux, J., 1984. Sh-wave propagation in heterogeneous media: velocity-stress finite-difference method. Geophysics 49 (11), 1933–1942.

Yan, H., Liu, Y., 2012. Rotated staggered grid high-order finite-difference numerical modeling for wave propagation in viscoelastic TTI media. Chin. J. Geophys. 55 (4), 1354–1365.

Zhang, W., Shen, Y., 2010. Unsplit complex frequency-shifted PML implementation using auxiliary differential equations for seismic wave modeling. Geophysics 75 (4), T141–T154.

Chapter 6

Reverse-Time Migration: Principles, Practical Issues, and Recent Developments

Houzhu (James) Zhang
Geophysical Technology Department, ConocoPhillips, Houston, Texas, US

Chapter Outline

Seismic Exploration of Hydrocarbons in Heterogeneous Reservoirs. http://dx.doi.org/10.1016/B978-0-12-420151-4.00007-6

6.1 INTRODUCTION TO SEISMIC IMAGING AND REVERSE-TIME MIGRATION

6.1.1 Seismic Imaging

Seismic imaging is a process to obtain the information of the underground structures using sound waves recorded on the ground or sea surface. The Earth is an elastic body and waves can be propagated from one place to another once initiated at certain locations. In seismic exploration, sources are fired on the surface of the Earth, mechanical waves travel into the deep subsurface. Similar to all types of waves, mechanical waves can be reflected, diffracted, or refracted once they reach some locations with contrasts in seismic properties, such as velocity or density. These reflected, diffracted, or refracted waves, or generally scattered waves, may come back to the surface. Therefore, the scattered waves carry the information of the media they pass through. If recorded, they can be used to estimate the subsurface information, such as geological structures. One of the major tools to obtain subsurface structure is seismic imaging or seismic migration.

Seismic imaging methods can be classified into two major categories: ray-based methods and wave equation-based methods (Leveille et al., 2011). The former includes Kirchhoff depth migration (KMIG) and beam migration (BMIG). The latter mainly includes one-way wave equation migration (WEM) and two-way WEM or reverse-time migration (RTM). It is now a widely recognized practice that BMIG, because of its high efficiency, is mainly used for velocity model building such as postmigration tomography. KMIG is used for imaging in areas with moderate velocity contrast, whereas RTM is best used in geological environments where strong velocity contrasts are present.

We will focus on RTM in this chapter. This overview is divided into the following sections. Section 6.1 gives a general introduction to RTM and lists its major procedures. Section 6.2 outlines the general theory and wave equations used in wave extrapolation in isotropic media and the imaging condition for stacked image. Because of the critical importance of anisotropy in subsurface imaging, Section 6.3 covers RTM in VTI and TTI media and most recently developed RTM in orthorhombic media. Imaging condition plays a central role in RTM. Section 6.4 discusses different imaging conditions. Although RTM is considered as the best imaging algorithm for complicated area, it is not free of problems. Section 6.5 discusses commonly observed issues such as the low-frequency noise, illumination issues in subsalt imaging, and physical noises that usually happen in anisotropic RTM. Acquisition has significant impact on seismic imaging. Section 6.6 summarizes the impact of recent developments of data acquisitions on imaging and model building. The final Section 6.7 gives summary and conclusions.

6.1.2 Overview of Reverse-Time Migration

RTM was designed simultaneously by several authors (McMechan, 1983; Whitmore, 1983; Baysal et al., 1983). Unlike Kirchhoff migration, in which trace events are explicitly put at the locations where the total travel time from source to image point and from image point to the receiver equals the time of the events. RTM backward extrapolates the data records at the surface based on wave equations. This is similar to WEM (Claerbout, 1985). RTM can be implemented for both poststack migration and prestack migration. During prestack migration, in additional to the extrapolation of receiver data, we also need to forward extrapolate the source wavelet, usually in the same given model. The image is formed at the time where the total time of source wavefield and the receiver wavefield equals to the time of the data. This is the location where the reflection happens (Claerbout, 1985).

Since RTM uses two-way wave equation for wavefield extrapolations, it can properly handle wave propagation in a complicated area such as subsalt, where multiple arrivals and shadow zone happen. It can also image arbitrary dips where WEM usually fails because of the use of one-way propagator. With the emerging of new types of acquisitions such as broadband seismic data and long offset data, RTM now can handle very low frequency, as well as large offset data. RTM is the primary imaging engine for Gulf of Mexico and offshore West Africa where strong velocity contrasts are present.

6.2 THEORY AND GENERAL PROCEDURES OF REVERSE-TIME MIGRATION

Prestack RTM contains three major steps: forward extrapolation of source wavefield, backward extrapolation of receiver wavefield, and application of imaging conditions. This three-step strategy is also the underlying imaging principle for KMIG, BMIG, and WEM although they appear very different in terms of actual numerical implementations. The two extrapolation steps are nothing but forward modeling in different types of media, with possibly different boundary conditions and possible filters applied to the input traces and sources.

The application of imaging condition for pre-stack migration contains several pieces of critical information: (1) When and where the source wave should meet the receiver side wave to make the most coherent contribution; (2) How to represent the image; (3) What information contained in the image can help us to improve velocity model building. In addition to the theoretical aspects, how the imaging condition can be implemented numerically is equally important.

In the subsections that follow, we first describe the basic wave equations for RTM in isotropic media and imaging condition for stacked image. This is the most fundamental imaging condition used in RTM. More complicated forms of wave equations for anisotropic media will be given in Section 6.3. Various other forms of imaging conditions will be given in Section 6.4.

6.2.1 Basic Wave Equations for Reverse-Time Migration in Isotropic Media

For common shot RTM, the source wavefield in isotropic media with constant density can be forward extrapolated by

$$\left\{\frac{1}{c^2(\mathbf{x})}\frac{\partial^2}{\partial t^2}-\nabla^2\right\}p_s(\mathbf{x},t)=S(t)\delta(\mathbf{x}-\mathbf{x}_s), \tag{6.1}$$

where $c(\mathbf{x})$ is the velocity field, $p_s(\mathbf{x},t)$ is the acoustic pressure field for the source side, $S(t)$ is the source wavelet, and $\mathbf{x}_s$ is the source position. Since during RTM, we need to preserve the spectrum of the data, $S(t)$ is usually a wavelet with broad spectrum. For different representations of data such as data decomposed into a set of plane waves, the right side of (6.1) must use the corresponding forms.

The receiver wavefield can be calculated by backward extrapolating the data recorded at the surface

$$\left\{\frac{1}{c_1^2(\mathbf{x})}\frac{\partial^2}{\partial t^2}-\nabla^2\right\}p_r(\mathbf{x},t)=\sum_{\mathbf{x}_r}D(t,\mathbf{x}_r), \tag{6.2}$$

where $p_r(\mathbf{x},t)$ is the acoustic pressure field for the receiver side and $D(t,\mathbf{x}_r)$ is the data trace at position $\mathbf{x}_r$. Here $c_1(\mathbf{x})$ can be different from $c(\mathbf{x})$ in (6.1).

Because of the reasons explained in Section 6.5, after applying the imaging condition, some postmigration steps have to be applied on the image, for example, the Laplacian operator (Zhang and Sun, 2008). In order for the final image to have the same spectrum as that of the input data, a spectrum filter and a phase rotation have to be applied on either the source wavelet or the receiver traces before input to RTM.

6.2.2 Fundamental Imaging Condition

After both the source and receiver wavefield have been extrapolated, the final image can be computed. The most widely used and most important type of image is the correlation-based stacked image which is computed by

$$I(\mathbf{x})=I(\mathbf{x},\tau)_{\tau=0}=\sum_{\mathbf{x}_s}p_s(\mathbf{x},t;\mathbf{x}_s)\otimes p_r(\mathbf{x},t;\mathbf{x}_s), \tag{6.3}$$

where $\otimes$ symbol represents time-domain cross-correlation. τ is the variable of cross-correlation. In (6.3), $p_s(\mathbf{x},t;\mathbf{x}_s)$ is the source wavefield with the shot location at $\mathbf{x}_s$ and $p_r(\mathbf{x},t;\mathbf{x}_s)$ is the receiver wavefield from the shot gather at the same shot location. Imaging condition (6.3) implies that the final image is the reflection where the source wavefield and receiver wavefield meet at the same spatial point and at the same time. This is the zero-time and zero-subsurface offset imaging condition (Claerbout, 1971).

Various imaging conditions for different image output and for different purposes can be defined. These forms of image will be given in Section 6.4.

Equations (6.1), (6.2), and (6.3) give the three basic steps of RTM. Equations (6.1) and (6.2) are only for RTM in isotropic media. Equations for anisotropic media can be found in Section 6.3. Equation (6.3) is valid for both isotropic and anisotropic media.

6.3 REVERSE-TIME MIGRATION IN ANISOTROPIC MEDIA

At depths that have economic interests to seismic exploration, the subsurface is elastic solid. During tectonics, different forces possibly in different direction have been applied on the rocks. Therefore, the seismic properties of the rock may be different in different directions. This direction-dependent property is called anisotropy. Another important mechanism of anisotropy is fine layering (Backus, 1962). When seismic wavelength is much larger than the scale of layers, the media behaves equivalently transversely isotropic even though each layer is isotropic.

Wave propagation in anisotropic media is fundamentally different from that in isotropic media. In isotropic media, phase propagation direction is always the same as the energy propagation (Cerveny, 2001). This fact does not hold in anisotropic media. In anisotropic media, for a given spatial location, waves propagating in different directions experience different resistance (Thomsen, 2002), thus they may travel at different speed. As a result, the phase and energy travel in different directions. This causes significant difficulties in understanding both theoretical aspect of wave phenomenon and numerical algorithm designs. For detail of anisotropy from theoretical point of view, see Cerveny (2001).

In this section, we will first discuss how to parameterize seismic properties in anisotropic media. Three important types of anisotropy will be discussed, namely, vertical transversely isotropy (VTI), tilted transversely isotropy (TTI),and orthorhombic anisotropy. The corresponding wave equations will be given. Modeling in anisotropic media is significantly more complicated and expensive than that in isotropic media. Examples will be presented to show the differences between isotropic, VTI, TTI, and orthorhombic images.

6.3.1 Seismic Properties and Parameterization in VTI and TTI Media

In anisotropic media, seismic wave travels in different speeds in different directions. For most general anisotropic media, we need 21 parameters to fully characterize its mechanics that can be represented by a 6×6 symmetric matrix. The simplest anisotropic media is polar anisotropy with one predefined direction called symmetry direction (Cerveny, 2001; Tsvankin, 2001; Helbig, 1994). In polar anisotropic media, wave propagation is isotropic in the plane perpendicular to the symmetry direction. In the planes containing the symmetry direction, wave propagation is angle dependent. According to the symmetry theory, five elastic constants are needed for characterizing polar anisotropy, namely, C_{11}, C_{13}, C_{33}, C_{55}, and C_{66}. Phase and group velocities and other properties can be expressed using these five constants.

Although the elastic constants are proper choices for mechanical analysis, they are not convenient for seismic analysis. In order to better and more conveniently characterize the wave propagation in polar anisotropic media, Thomsen (1986) proposed the following five parameters to replace the previous elastic constants:

$$V_{p0} = \sqrt{\frac{C_{33}}{\rho}}, \tag{6.4}$$

$$V_{s0} = \sqrt{\frac{C_{44}}{\rho}}, \tag{6.5}$$

$$\varepsilon = \frac{C_{11} - C_{33}}{2C_{33}}, \tag{6.6}$$

$$\delta = \frac{(C_{13} + C_{44})^2 - (C_{33} - C_{44})^2}{2C_{33}(C_{33} - C_{44})}, \tag{6.7}$$

$$\gamma = \frac{C_{66} - C_{44}}{2C_{44}}. \tag{6.8}$$

In (6.4)–(6.8), C_{ij} are elastic constants; V_{p0} is the P-wave velocity in the symmetry direction; V_{s0} is the shear-wave velocity in the symmetry direction; and ε, δ, and γ are called Thomsen parameters. Usually, ε, δ, and γ are small (weak anisotropy). Under the assumption of weak anisotropy, seismic properties such as phase velocity, group velocity, and phase and group angles can be greatly simplified. It is easy to show that isotropy is a special case of polar anisotropy with $\varepsilon = \delta = \gamma = 0$.

Figure 6.1 shows typical geological structures for VTI (A) and TTI media (B). In VTI media, the symmetry direction points downward. In TTI media, two angles are needed for the definition of the symmetry axis, the dip angle θ, and the azimuth angle ϕ at each location. Therefore, at each spatial location, seven parameters are needed to define TTI seismic properties; (C) and (D) are for orthorhombic media. They will be discussed in detail in Section 6.3.3.

Using Thomsen's parameters, the elastic constants can be written as:

$$C_{11} = (1 + 2\varepsilon)A, \tag{6.9}$$

$$C_{33} = A, \tag{6.10}$$

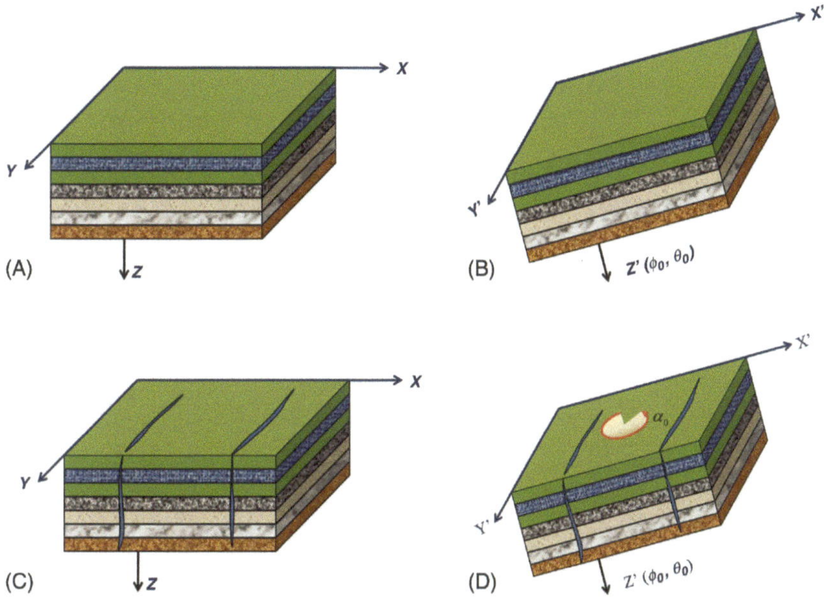

FIGURE 6.1 Geological models for typical types of anisotropy. (A) VTI media; (B) TTI media; (C) vertical orthorhombic media (VORT); and (D) tilted orthorhombic media (TORT).

$$C_{44} = f^2 A, \tag{6.11}$$

$$C_{66} = (1+2\gamma) f^2 A, \tag{6.12}$$

$$C_{13} = A\left\{\sqrt{\left(1-f^2\right)^2 + 2\delta(1-f^2)} - f^2\right\}, \tag{6.13}$$

where

$$f = V_{s0} / V_{p0}, \tag{6.14}$$

$$A = \rho V_{p0}^2. \tag{6.15}$$

In order to be a physically valid elastic model, both the five elastic constants and the Thomsen's parameters have to be constrained. The physical constraints will be discussed in details in Section 6.5.2.

For implementation and efficiency purposes, (pseudo) acoustic assumption is usually used. Under this assumption, V_{s0} is artificially set to zero. The 6 × 6 matrix can be replaced by the following 2 × 2 matrix for VTI and TTI media:

$$\rho V_{p0}^2 \begin{bmatrix} 1+2\varepsilon & \sqrt{1+2\delta} \\ \sqrt{1+2\delta} & 1 \end{bmatrix}. \tag{6.16}$$

Matrix (6.16) is the fundamental seismic property matrix used in the derivation of wave equations for RTM in acoustic VTI/TTI media.

Using Thomsen's parameters, the phase velocity can be written as

$$\frac{V^2(\theta)}{V_0^2} = 1+\varepsilon\sin^2\theta - \frac{f}{2} \pm \frac{f}{2}\sqrt{1+\frac{4\sin^2\theta}{f}(2\delta\cos^2\theta - \varepsilon\cos 2\theta)+\frac{4\varepsilon^2\sin^4\theta}{f^2}}, \tag{6.17}$$

where θ is the (phase) angle between symmetry axis and slowness direction. In (6.17), plus sign is for P-wave and minus sign is for SV-wave.

Group velocity of the P-wave and SV-wave is

$$V_G = V\sqrt{1+\left(\frac{1}{V}\frac{dV}{d\theta}\right)^2}. \tag{6.18}$$

Group velocity in (6.18) must be evaluated at the group angle that is computed by Tsvankin (2001) as

$$\tan(\psi) = \frac{\tan(\theta)+\frac{1}{V}\frac{dV}{d\theta}}{1-\tan(\theta)\frac{1}{V}\frac{dV}{d\theta}} = \tan(\theta+\Delta\theta), \tag{6.19}$$

where $\Delta\theta$ is the angle between energy-propagation direction and phase-propagation direction due to the existence of anisotropy.

Figure 6.2 shows the relation between group velocity (V_G), group angle (Ψ), phase velocity (V), and phase angle (θ) for homogeneous VTI media. The direction of the group velocity is the ray starting from the center, the phase velocity is normal to the wavefront. An important relation between group velocity and phase velocity is

$$V = V_G\cos(\Delta\theta). \tag{6.20}$$

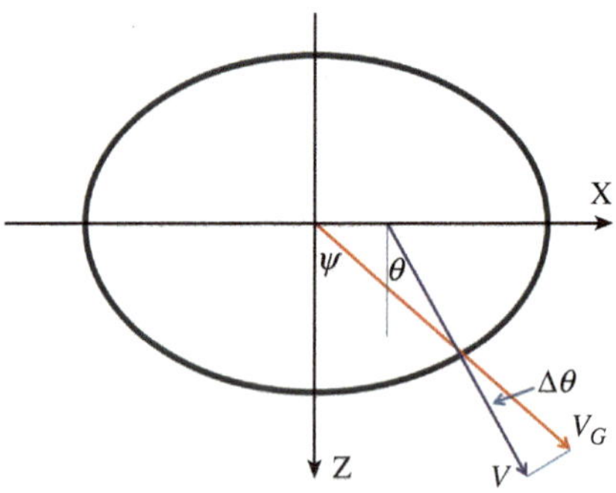

FIGURE 6.2 Group velocity (V_G), group angle (ψ), phase velocity (V), and phase angle (θ) for homogeneous VTI media. The black thick curve is wavefront.

6.3.2 Wave Propagation and Reverse-Time Migration in VTI and TTI Media

Acoustic wave equations in VTI/TTI media can be derived in various ways (Alkhalifah, 2000; Alkhalifah, 2003; Zhou et al., 2006; Zhang and Zhang, 2008; Duveneck et al., 2008). The temporal fourth-order wave equation by Alkhalifah (2000) is difficult for investigation and implementation. Zhou et al. (2006) split the fourth-order wave equation into two second-order wave equations on the basis of the dispersion relation (6.17). However, the system derived this way is not physically stable. The most straightforward way, as demonstrated in Duveneck et al. (2008), is to start with the original first-order wave equation (Aki and Richards, 2009)

$$\begin{cases} \rho\dfrac{\partial v_1}{\partial t} = \dfrac{\partial \sigma_{11}}{\partial x_1}+\dfrac{\partial \sigma_{12}}{\partial x_2}+\dfrac{\partial \sigma_{13}}{\partial x_3} \\ \rho\dfrac{\partial v_2}{\partial t} = \dfrac{\partial \sigma_{21}}{\partial x_1}+\dfrac{\partial \sigma_{22}}{\partial x_2}+\dfrac{\partial \sigma_{23}}{\partial x_3} \\ \rho\dfrac{\partial v_3}{\partial t} = \dfrac{\partial \sigma_{31}}{\partial x_1}+\dfrac{\partial \sigma_{32}}{\partial x_2}+\dfrac{\partial \sigma_{33}}{\partial x_3} \end{cases}, \tag{6.21}$$

where σ_{ij} is the stress tensor and υ_i are the particle velocity components.

The detailed derivation for wave equations in VTI media is available in Duveneck et al. (2008). The final wave equation for VTI media (assuming $V_{s0} = 0$) is

$$\begin{cases} \dfrac{1}{V_{p0}^2}\dfrac{\partial^2 r}{\partial t^2} = (1+2\varepsilon)\left(\partial_x^2+\partial_y^2\right)r+\sqrt{1+2\delta}\,\partial_z^2 p \\ \dfrac{1}{V_{p0}^2}\dfrac{\partial^2 p}{\partial t^2} = \sqrt{1+2\delta}\left(\partial_x^2+\partial_y^2\right)r+\partial_z^2 p \end{cases}, \tag{6.22}$$

where r is the horizontal principal stress component (in xy plane) and P is the vertical principal stress component. For marine data, either r or P can be used for imaging since they are equal in water layer. For imaging using VSP data, the average of r and P, $(2r+p)/3$, should be used for imaging.

Formulation (6.22) is very simple and its numerical implementation is also straightforward. The computing cost is approximately 1.5 times higher than for isotropic imaging (compare with (6.1)). Since (6.22) was derived based on the assumption of $V_{s0} = 0$, a special issue exists- the shear-wave noise. This particular issue will be discussed in detail in Section 6.5.3.

Equation (6.22) is kinematically consistent with the elastic wave equation (6.21), but they are different in terms of amplitude because of the absence of shear waves in (6.22). This can be seen from the following numerical examples.

Figure 6.3 shows the wavefront at time 1 s for different values of V_{s0} in a homogeneous VTI media. V_{p0} is fixed at 1500 m/s, $\varepsilon = 0.2$, and $\delta = -0.2$. Modeling is done using elastic wave equation (6.21). The travel time for both P-wave and SV-wave can also be computed by (6.17), (6.18), and (6.19). But they cannot give the amplitude information. The outer convex wavefront is for

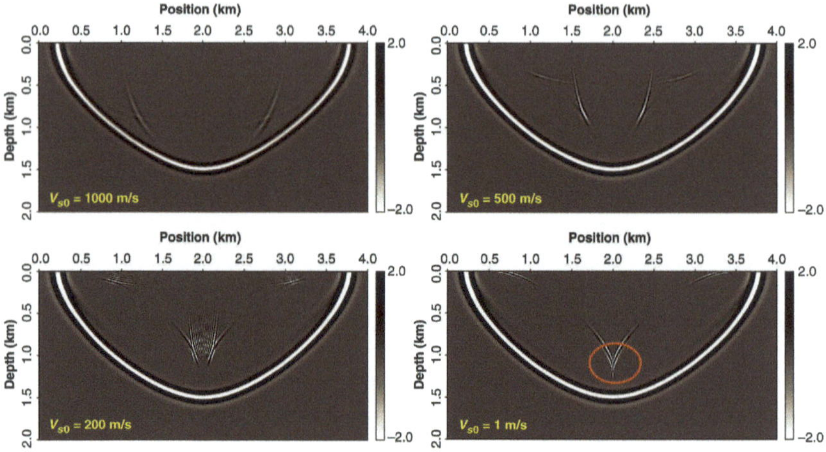

FIGURE 6.3 Wavefront of wave propagation at 1 s in VTI media. V_{p0} is fixed at 1500 m/s, $\varepsilon = 0.2$, and $\delta = -0.2$.

P-wave and the inner is for SV-wave. With the decrease of V_{s0}, the amplitude of P-wave wavefront changes slightly, but the kinematics remains almost the same. The SV wavefront behavior changes dramatically. When V_{s0} is close to zero, the SV-wave wavefront becomes a singular point with very large amplitude (in red circle). This is the case for modeling using system (6.22).

In marine environment, since the water layer is isotropic, the SV-wave vanishes. Although the SV-waves may be generated in the sediments due to the P-wave incidence (Fowler et al., 2010), they are usually very weak compared with P-waves. See Section 6.5.3 for additional information.

For wave propagation in TTI media, we need to introduce the angles of the symmetry axis as part of the wave equation. Assume the symmetry axis can be described by the dip angle (θ) and the azimuth angle (ϕ) of the bedding plane at each spatial location, the TTI wave equation (again assume $V_{s0} = 0$) can be written as:

$$\begin{cases} \dfrac{1}{V_{p0}^2}\dfrac{\partial^2 r}{\partial t^2} = (1+2\varepsilon)\left(\nabla^T R_1 R_1^T \nabla + \nabla^T R_2 R_2^T \nabla\right) r + \sqrt{1+2\delta}\,\nabla^T R_3 R_3^T \nabla p \\ \dfrac{1}{V_{p0}^2}\dfrac{\partial^2 p}{\partial t^2} = \sqrt{1+2\delta}\left(\nabla^T R_1 R_1^T \nabla + \nabla^T R_2 R_2^T \nabla\right) r + \nabla^T R_3 R_3^T \nabla p \end{cases}, \tag{6.23}$$

where R_1, R_2, and R_3 are the first, second, and third columns, respectively, of the following matrix (Zhang and Zhang, 2009, 2011)

$$R = \begin{pmatrix} \cos\phi\cos\theta & -\sin\phi & \cos\phi\sin\theta \\ \sin\phi\cos\theta & \cos\phi & \sin\theta\sin\phi \\ -\sin\theta & 0 & \cos\theta \end{pmatrix}. \tag{6.24}$$

Both (6.22) and (6.23) are physically stable systems and it is not necessary to include shear-wave terms to dampen the SV-waves (Fowler et al., 2010; Fletcher et al., 2009).

System (6.23) is considerably more complicated than its VTI counterpart (6.22) and the numerical implementation is also much less straightforward. Equation (6.22) contains second-order derivatives only, whereas (6.23) contains primarily first-order derivatives. Second-order derivative can be efficiently and accurately computed using either pseudospectral method (Zhang and Zhang, 2008) or high-order central finite-difference method. The design of high-order finite-difference for first-order derivatives is much harder and pseudo-spectral method may be a preferred choice (Zhang and Zhang, 2008; Canuto et al., 2006).

Figure 6.4 shows the comparison of the wavefronts in isotropic, VTI, and TTI media. The modeling was done using elastic equation (6.21).

Since VTI assumes vertical symmetry axis that rarely happens in reality, it cannot simulate the wave propagation correctly. TTI RTM is able to produce significantly better image than either isotropic image or VTI image. Figure 6.5 shows an example from Walker Ridge in the Gulf of Mexico (Huang et al., 2009). The TTI image is much more focused and coherent in the target below the salt body. In the deep mini basin, the dip angles can reach more than 50°. With such high dipping bedding, VTI is significantly inaccurate.

6.3.3 Seismic Properties, Wave Propagation, and Reverse-Time Migration in Orthorhombic Media

Media with orthorhombic symmetry is common in geological environments with fracture systems developed in different directions. For example, two sets of orthogonal vertical fractures or one set of vertical fractures in a VTI background produce an orthorhombic media with two vertical symmetry planes and one horizontal symmetry plane (Tsvankin, 2001). For a variety of different reasons, the bedding direction can be tilted (tilted orthorhombic media). It has been shown that in such a system, simple polar anisotropy such as VTI or TTI is not sufficient to describe the seismic-wave properties (Schoenberg and Helbig, 1997). More general anisotropic media with lower symmetry is needed.

In orthorhombic elastic media, there are nine independent elastic constants needed to fully define the constitutive relation:

$$\begin{bmatrix} C_{11} & C_{12} & C_{13} & 0 & 0 & 0 \\ C_{12} & C_{22} & C_{23} & 0 & 0 & 0 \\ C_{13} & C_{23} & C_{33} & 0 & 0 & 0 \\ 0 & 0 & 0 & C_{44} & 0 & 0 \\ 0 & 0 & 0 & 0 & C_{55} & 0 \\ 0 & 0 & 0 & 0 & 0 & C_{66} \end{bmatrix}. \tag{6.25}$$

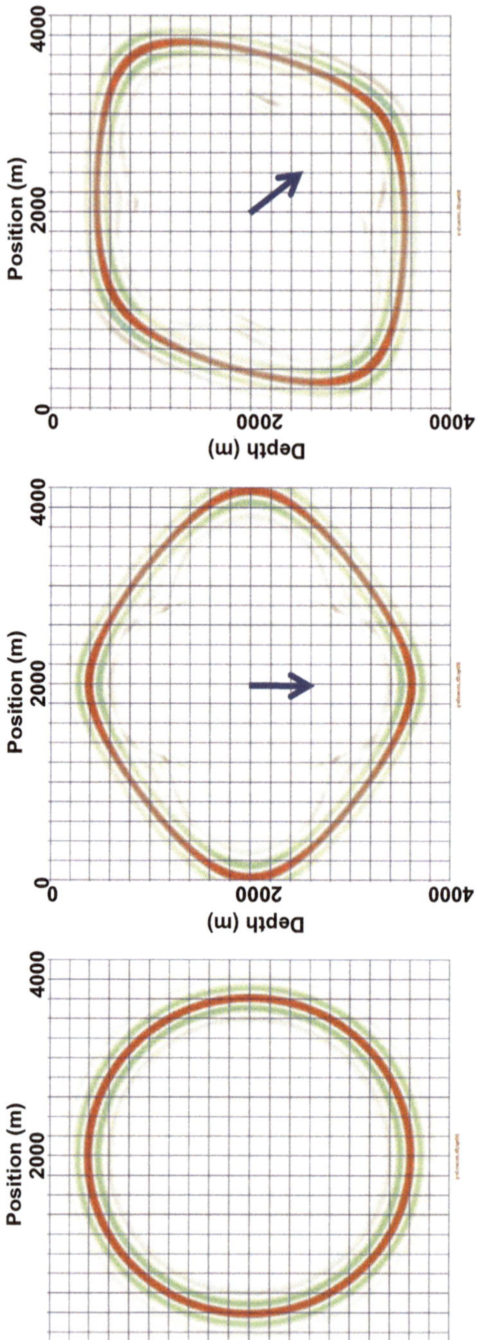

FIGURE 6.4 Wavefront of wave propagation at 1 s in isotropic (left), VTI (middle), and TTI (right) media. V_{p0} =2000 m/s, V_{s0} =1000 m/s, $\varepsilon = 0.25$, and $\delta = -0.25$. For TTI, $\varphi = 0^\circ$ and $\theta = 35^\circ$. The blue arrows point to the directions of symmetry axis.

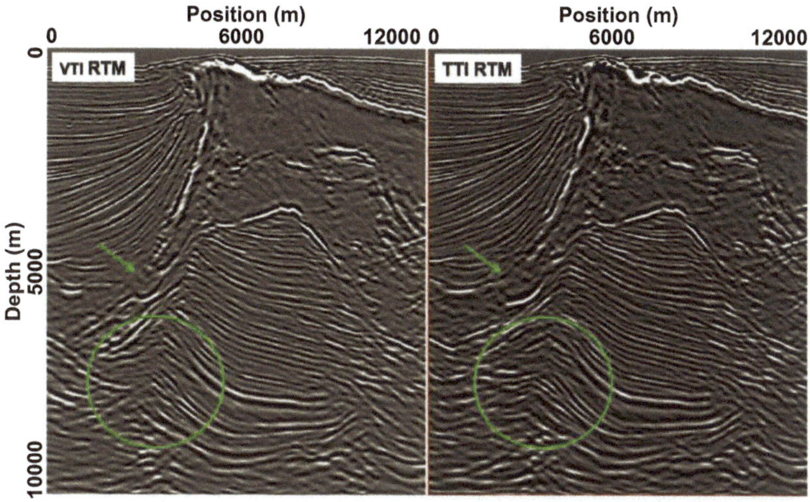

FIGURE 6.5 Comparison between VTI RTM (left) and TTI RTM (right) image for a WAZ dataset in Walker Ridge in the Gulf of Mexico *(Huang et al., 2009)*.

As in VTI/TTI media, the elastic constant matrix is not convenient for seismic analysis. We can design a set of Thomsen-style parameters to characterize the seismic properties in orthorhombic media (Tsvankin, 1997):

$$V_{p0} = \sqrt{\frac{C_{33}}{\rho}}, \tag{6.26}$$

$$V_{s0} = \sqrt{\frac{C_{55}}{\rho}}, \tag{6.27}$$

$$\varepsilon_2 = \frac{C_{11} - C_{33}}{2C_{33}}, \tag{6.28}$$

$$\delta_2 = \frac{(C_{13} + C_{55})^2 - (C_{33} - C_{55})^2}{2C_{33}(C_{33} - C_{55})}, \tag{6.29}$$

$$\gamma_2 = \frac{C_{66} - C_{44}}{2C_{44}}, \tag{6.30}$$

$$\varepsilon_1 = \frac{C_{22} - C_{33}}{2C_{33}}, \tag{6.31}$$

$$\delta_1 = \frac{(C_{23}+C_{44})^2-(C_{33}-C_{44})^2}{2C_{33}(C_{33}-C_{44})}, \tag{6.32}$$

$$\gamma_1 = \frac{C_{66}-C_{55}}{2C_{55}}, \tag{6.33}$$

$$\delta_3 = \frac{(C_{12}+C_{66})^2-(C_{11}-C_{66})^2}{2C_{11}(C_{11}-C_{66})}. \tag{6.34}$$

In this definition, V_{p0} is the vertical P-wave velocity, V_{s0} is vertical velocity of shear wave that polarizes in the x_1 direction. ε_2 is the VTI parameter ε in the symmetry plane $[x_1, x_3]$ normal to the x_2-axis, and δ_2 is the VTI parameter δ in the symmetry plane $[x_1, x_3]$. γ_2 is the VTI parameter γ in the symmetry plane $[x_1, x_3]$ normal to the x_2-axis. ε_1 is the VTI parameter ε in the symmetry plane $[x_2, x_3]$. δ_1 is the VTI parameter δ in the symmetry plane $[x_2, x_3]$. γ_1 is the VTI parameter γ in the symmetry plane $[x_2, x_3]$. δ_3 is the VTI parameter δ in the symmetry plane $[x_1, x_2]$.

Obviously, when $\delta_3 = 0$, $\varepsilon_1 = \varepsilon_2 = \varepsilon, \delta_1 = \delta_2 = \delta$, and $\gamma_1 = \gamma_2 = \gamma$, orthorhombic media reduces to VTI/TTI media.

With Thomsen's parameters, all the analysis done for VTI/TTI media, such as the approximation of phase and group velocity under weak anisotropy, can be equally applied to orthorhombic media (Tsvankin, 2001).

For pseudoacoustic orthorhombic media, (6.25) can be replaced by the following 3 × 3 matrix

$$\rho V_{p0}^2 \begin{bmatrix} 1+2\varepsilon_2 & (1+2\varepsilon_2)\sqrt{1+2\delta_3} & \sqrt{1+2\delta_2} \\ (1+2\varepsilon_2)\sqrt{1+2\delta_3} & 1+2\varepsilon_1 & \sqrt{1+2\delta_1} \\ \sqrt{1+2\delta_2} & \sqrt{1+2\delta_1} & 1 \end{bmatrix}. \tag{6.35}$$

Matrix (6.35) is the fundamental seismic property matrix used in derivation of wave equations for RTM in vertical and tilted acoustic orthorhombic media. When $\delta_3 = 0$, $\varepsilon_1 = \varepsilon_2 = \varepsilon, \delta_1 = \delta_2 = \delta$, and $\gamma_1 = \gamma_2 = \gamma$, the second row and second column in (6.35) are redundant and (6.35) reduces to (6.16).

Figure 6.1C and D shows examples of vertical and tilted orthorhombic media: one set of fractures embedded in VTI/TTI media. For tilted orthorhombic media, in addition to the nine seismic parameters defined in (6.26) to (6.34), three angles are needed to define the orientation of the location symmetry coordinate with respect to the global Cartesian coordinates (Zhang and Zhang, 2011). More specifically, similar to (6.24), the rotation matrix can be written as

$$R = \begin{pmatrix} \cos\phi & -\sin\phi & 0 \\ \sin\phi & \cos\phi & 0 \\ 0 & 0 & 1 \end{pmatrix} \begin{pmatrix} \cos\theta & 0 & \sin\theta \\ 0 & 1 & 0 \\ -\sin\theta & 0 & \cos\theta \end{pmatrix} \begin{pmatrix} \cos\alpha & -\sin\alpha & 0 \\ \sin\alpha & \cos\alpha & 0 \\ 0 & 0 & 1 \end{pmatrix}, \tag{6.36}$$

where (ϕ,θ) are used to define the vertical axis at each spatial point as we did for the symmetry axis in TTI. The third angle, α, is introduced to rotate the elastic tensor on the local x–y plane and to represent the orientation of the first crack for an orthorhombic media that is composed of two orthogonal crack systems.

Having defined all the seismic properties, we can now derive the governing equations for RTM in orthorhombic media. The most straightforward way to derive the RTM equations (under acoustic assumption and constant density) is to follow the procedure given by Duveneck et al. (2008) for TTI and Zhang and Zhang (2011).

In vertical orthorhombic media, the wave equations can be written as (ignore the variation of density):

$$\begin{cases} \dfrac{\partial^2 \sigma_{11}}{\partial t^2} = M_{11}\dfrac{\partial}{\partial x_1}\left(\dfrac{\partial \sigma_{11}}{\partial x_1}\right) + M_{12}\dfrac{\partial}{\partial x_2}\left(\dfrac{\partial \sigma_{22}}{\partial x_2}\right) + M_{13}\dfrac{\partial}{\partial x_3}\left(\dfrac{\partial \sigma_{33}}{\partial x_3}\right) \\ \dfrac{\partial^2 \sigma_{22}}{\partial t^2} = M_{12}\dfrac{\partial}{\partial x_1}\left(\dfrac{\partial \sigma_{11}}{\partial x_1}\right) + M_{22}\dfrac{\partial}{\partial x_2}\left(\dfrac{\partial \sigma_{22}}{\partial x_2}\right) + M_{23}\dfrac{\partial}{\partial x_3}\left(\dfrac{\partial \sigma_{33}}{\partial x_3}\right) \\ \dfrac{\partial^2 \sigma_{33}}{\partial t^2} = M_{13}\dfrac{\partial}{\partial x_1}\left(\dfrac{\partial \sigma_{11}}{\partial x_1}\right) + M_{23}\dfrac{\partial}{\partial x_2}\left(\dfrac{\partial \sigma_{22}}{\partial x_2}\right) + M_{33}\dfrac{\partial}{\partial x_3}\left(\dfrac{\partial \sigma_{33}}{\partial x_3}\right) \end{cases}, \quad (6.37)$$

where

$$M = V_{p0}^2 \begin{bmatrix} 1+2\varepsilon_2 & (1+2\varepsilon_2)\sqrt{1+2\delta_3} & \sqrt{1+2\delta_2} \\ (1+2\varepsilon_2)\sqrt{1+2\delta_3} & 1+2\varepsilon_1 & \sqrt{1+2\delta_1} \\ \sqrt{1+2\delta_2} & \sqrt{1+2\delta_1} & 1 \end{bmatrix} = V_{p0}^2 N. \quad (6.38)$$

Or in more compact form

$$\frac{1}{V_{p0}^2}\frac{\partial^2 \sigma}{\partial t^2} = ND^T D\sigma. \quad (6.39)$$

In (6.39), $\sigma = (\sigma_{11},\sigma_{22},\sigma_{33})^T$ is the principal stress vector, $D = \mathrm{diag}(\partial_1,\partial_2,\partial_3)$, and N is a matrix that contains anisotropic parameters only.

Equations (6.37), along with (6.38), deserves further analysis. Each equation of (6.37) can be written in the following form (no summation over i)

$$\frac{\partial^2 \sigma_{ii}}{\partial t^2} = M_{ii}\frac{\partial}{\partial x_i}\left(\frac{\partial \sigma_{ii}}{\partial x_i}\right) + f_i(t,\mathbf{x}). \quad (6.40)$$

It is obviously from (6.37) that M_{ii} controls the wave propagation and defines the propagation velocity mainly in the ith direction. ε_2 mainly impacts the propagation in x_1 direction and ε_1 mainly impacts the propagation in x_2

direction. So (6.40) intuitively defines a system that allows each of the principal stress components to propagate at different speeds in different directions. Similar formulation was used by Dong and McMechan (1991) but for one anisotropic scalar wave equation.

For wave propagation in tilted orthorhombic media, (6.39) remains the same. D is defined as $D = \mathrm{diag}\left(R_1^T \partial_1, R_2^T \partial_2, R_3^T \partial_3\right)$, where R_1, R_2, and R_3 are the first, second, and third columns, respectively, of the matrix defined in (6.36) (Zhang and Zhang, 2011).

A more rigorous formulation can be derived following the procedure (for TTI) in Duveneck and Bakker (2011). But the equations will be much more complicated in terms of both theoretical derivation and numerical implementation. The detailed formulation is given in Zhang and Zhang (2011).

A numerical solution with fourth-order temporal finite-difference is given as

$$\begin{aligned}\boldsymbol{\sigma}^{n+1} &= 2\boldsymbol{\sigma}^{n} - \boldsymbol{\sigma}^{n-1} + \left(\Delta t V_{p0}\right)^2 N\tilde{D}^T\tilde{D}\boldsymbol{\sigma}^{n} + \frac{1}{12}\left(\Delta t V_{p0}\right)^2 \\ &\quad N\tilde{D}^T\tilde{D}\left\{\left(\Delta t V_{p0}\right)^2 N\tilde{D}^T\tilde{D}\boldsymbol{\sigma}^{n}\right\},\end{aligned} \tag{6.41}$$

where $\tilde{D}$ is the spatially discretized operator of D. $\tilde{D}$ can be computed using either finite-difference or pseudo-spectrum method and Δt is the time step for extrapolation.

Figure 6.6 compares the wavefronts in VTI (A) and vertical orthorhombic (VORT) (B) media and Figure 6.7 compares the wavefronts in TTI (A) and tilted orthorhombic (TORT) (B) media. This set of parameters is taken from Schoenberg and Helbig (1997) and are listed in Table 6.1. They represent typical seismic properties of many shales such as Greenhorn shale (Thomsen, 1986). The values of the three angles are assigned arbitrarily. The wavefronts in orthorhombic media are significantly different from those in VTI and TTI media. Since the wavefront has the same kinematics as the RTM impulse responses, it is expected that the RTM image in orthorhombic media is significantly different from the images in VTI and TTI media. The inner, weak wavefronts are for shear waves.

Figure 6.8 shows a full-azimuth (FAZ) example for RTM in orthorhombic media in the Keathley Canyon in the central Gulf of Mexico (Wu et al., 2013). The input dataset contains seven azimuth sectors. TTI tomography failed to flatten the gathers in all azimuth directions. This suggests different anisotropy in different azimuth sectors. Tomography based on tilted orthorhombic media (Li et al., 2012) allows more freedom for model building. In Figure 6.8, the image of tilted orthorhombic media is significantly better in the subsalt than that of TTI media. The salt body is more clearly defined. The top of the salt body is more focused and the resolution of the small faults in the basin above the top of salt is much higher.

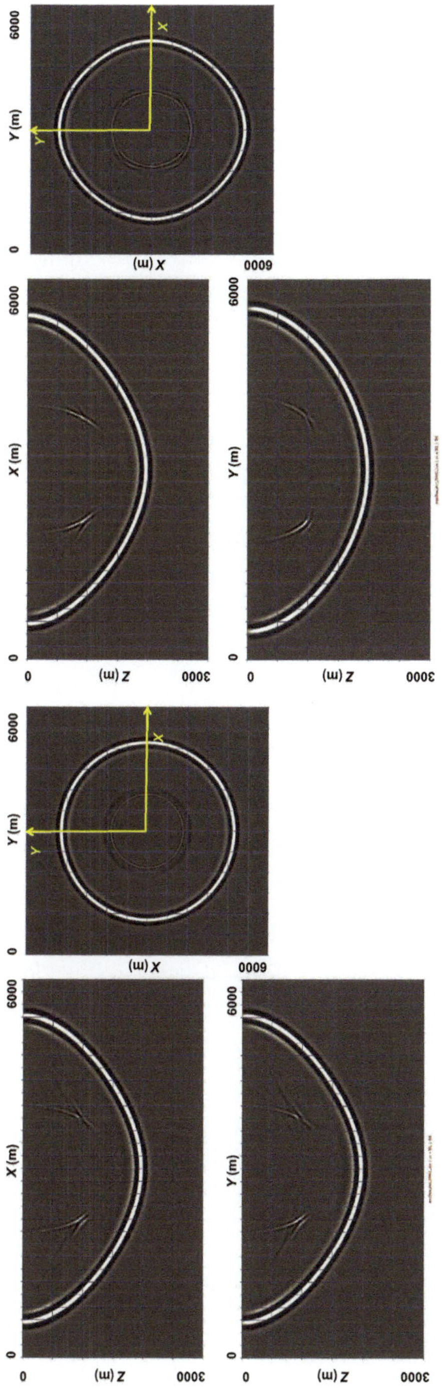

FIGURE 6.6 (A) Wavefronts in VTI media. Top left, view in inline direction; bottom left, view in crossline direction; right, depth slide. (B) Wavefronts in vertical orthorhombic media. Top left, view in inline direction; bottom left, view in crossline direction; right, depth slide.

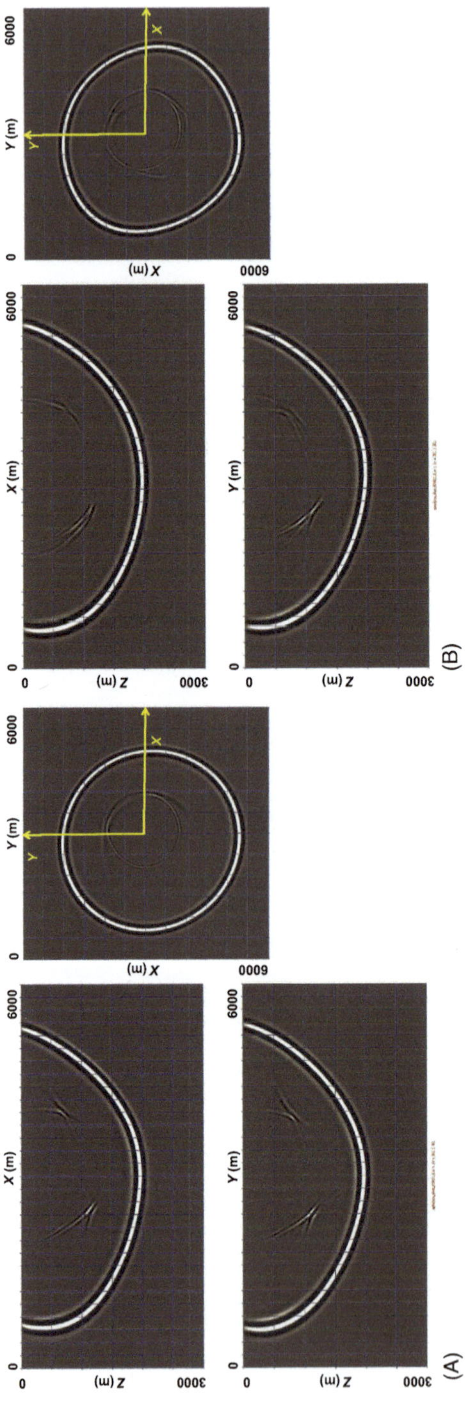

FIGURE 6.7 (A) Wavefronts in TTI media. Top left, view in inline direction; bottom left, view in crossline direction; right, depth slide. (B) Wavefronts in tilted orthorhombic media. Top left, view in inline direction; bottom left, view in crossline direction; right, depth slide.

TABLE 6.1 Seismic Properties of Shale Rock Used for the Computation of the Wavefronts in Figures 6.6 and 6.7

	VTI	TTI	VORT	TORT
V_{p0} (m/s)	2436	2436	2436	2436
V_{s0} (m/s)	1265	1265	1265	1265
ρ (kg/m^3)	1000	1000	1000	1000
ε_1			0.3286	0.3286
δ_1			0.0825	0.0825
γ_1			0.1819	0.1819
ε_2	0.2579	0.2579	0.2579	0.2579
δ_2	−0.0775	−0.0775	−0.0775	−0.0775
γ_2	0.0455	0.0455	0.0455	0.0455
δ_3			−0.1064	−0.1064
φ (°)		35		35
θ (°)		20		20
α (°)				30

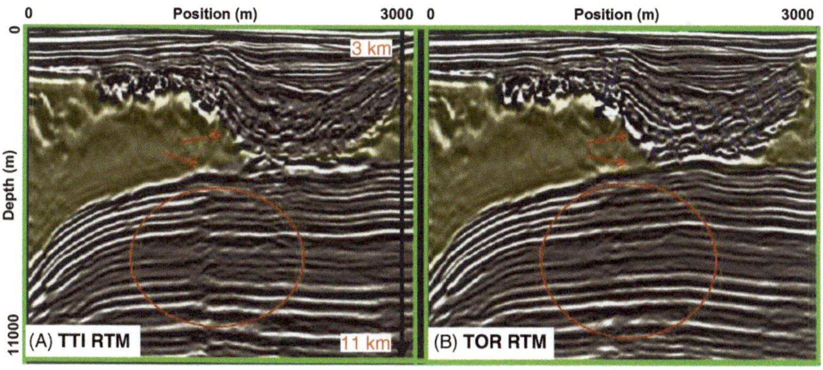

FIGURE 6.8 (A) Stacked image of TTI RTM. (B) Stacked image of tilted orthorhombic RTM *(Wu et al., 2013).*

6.4 GATHER REPRESENTATIONS OF IMAGES

The final stacked image given in eq. (6.3) is the most fundamental and most widely used representation of image from RTM. However, experience with real-data processing demonstrated that in order to obtain a good final image, having stacked image alone is not sufficient. The fundamental reason is that we do not have an accurate subsurface model for migration and the stacked image usually does not provide a quantitatively connection between the image mismatch and model errors (Leveille et al., 2011). When the migration velocity has errors, the locations of the best focused image may not be at zero time or zero subsurface offsets. Any model representation, any actual numerical models and wave equations including the most general anisotropic wave equation are always approximate to the real subsurface geology. There are always events in the seismic data that cannot be modeled by theory. These events will be contributors to the noise in the image. Stacked image is not sufficient for identifying and removing these migration noises.

The so-called extended imaging conditions describe how image changes when applying a shift in time or offset of the source and receiver wavefield. Unlike common image gathers from Kirchhoff migration or BMIG, image gathers from RTM lack intuitive understanding. Generating gathers from RTM is more involved in terms of both algorithm understandings and computation cost.

Based on the domain in which gathers are defined, image gathers can be divided into three types: reflection-angles gathers, 3D-angle gathers, and time/space-shift gathers. While time/space-shift gathers are defined in the natural physical space, angle gathers are defined in the converted space. Each of the gathers can be used for both velocity model update and image improvement.

6.4.1 Space, Time-Shift Gathers, and Angle Domain Conversion

Space and time-shift gathers can be defined as

$$I(\mathbf{x},\mathbf{h},\tau)=\sum_{\mathbf{x}_s,t} p_s\,(\mathbf{x}-\mathbf{h},t-\tau;\mathbf{x}_s)\otimes p_r\,(\mathbf{x}+\mathbf{h},t+\tau;\mathbf{x}_s), \tag{6.42}$$

where τ is the time shift (usually small) between the source wavefield and receiver wavefield and $\mathbf{h}$ is the spatial shift between the source wavefield and receiver wavefield at image location $\mathbf{x}$ (Sava and Fomel, 2006). The conventional stacked image is a slide of $I(\mathbf{x},\mathbf{h},\tau)$ when both $\mathbf{h}$ and τ vanish. It should be emphasized that, the variable $\mathbf{h}$ defined in (6.42) is the subsurface offset, not the offset defined on the acquisition surface as that defined in KMIG or BMIG common image gathers. Surface gather can be easily generated by KMIG and has clear geophysical explanation and therefore, they are widely used for post migration velocity update. For a given migration velocity, if the model has errors, the depths of the images from data at different surface offsets are different. The difference of depths between an arbitrary offset and near offset is the migration depth residual. Migration depth residual is a direct measurement of the mismatches in the migration velocity along different ray paths (Stork, 1992).

Figure 6.9 shows time-shift gathers for Sigbee2a model with incorrect velocity model (Sava and Fomel, 2006). Obviously, best-focused image does not happen at $\tau=0$ due to the velocity errors. The τ's for best-focused image are different at different depths and the slope of the time-shift gathers (with respect to τ) are dependent on the local propagation velocity. The former feature can be

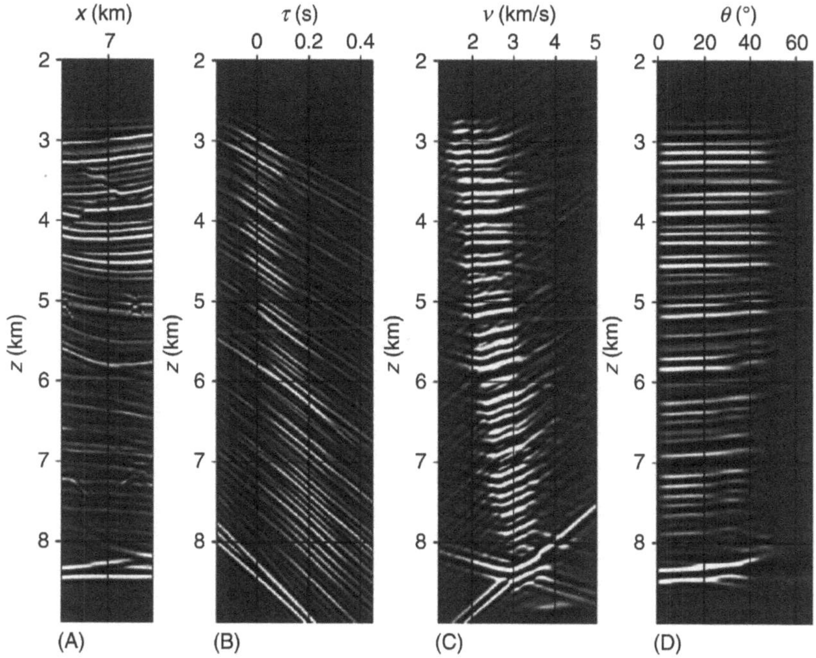

FIGURE 6.9 Time-shift imaging condition gathers at $x = 7$ km. The panels depict (A) the image, (B) the time-shift gather, (C) the slant-stacked gather, and (D) the angle gather.

used for migration velocity analysis, whereas the latter can be used for identifying and removing certain types of migration noise.

Figure 6.10 shows time-shift gathers for a typical salt geological model in the Gulf of Mexico. According to the definition in (6.42), the effective events have finite positive dips in the gathers. The vertical strong stripes that have infinite dips are due to the cross-correlation between reflected source wavefield and downward receiver wavefield. This is the low-frequency component frequently seen above top of salt or above water bottom in raw RTM-stacked image. The events with negative dips are diving waves or inter bedded multiples that travels to shallow depth with increasing time. Compared with the areas above top of salt, the gather in the subsalt is simple. Since different events have different apparent dips in time-shift gathers, noises can be easily removed.

Figure 6.11 shows the comparison of images without (top) and with (bottom) noise removal in time-shift gathers using a real-data example (Khalil et al., 2013). Obviously, the image after noise removal has more coherent and more continuous events. Some noise appearing in time-shift gathers can be removed by other approaches as well, such as Laplacian filter. However time-shift gather not only provides a simple systematic geophysical way to identify different events but also makes it possible to remove the coherent noises using very simple methods such as $\tau - p$ transform.

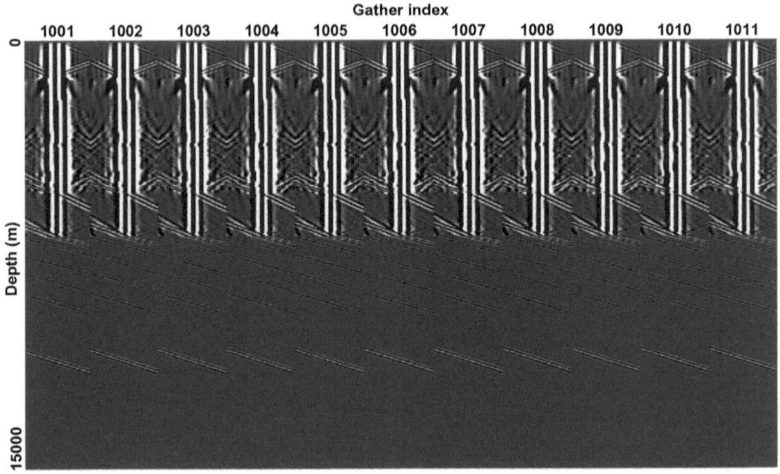

FIGURE 6.10 Time-shift gather of a typical example from the Gulf of Mexico. See text for details.

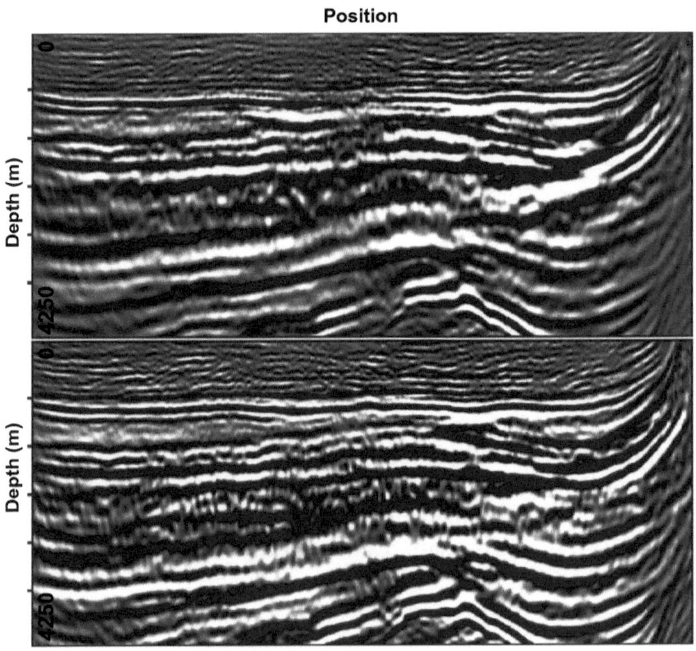

FIGURE 6.11 Comparison between images without (top) and with (bottom) noise removal using time-shift gathers *(Khalil et al., 2013).*

Both time-shift gathers and space-shift gathers can be converted into reflection-angle gathers that can be used for postmigration processing such as angle-domain tomography.

Time-shift gathers can be converted into reflection-angle gathers using (Sava and Fomel, 2006)

$$\cos^2\theta(\mathbf{x}) = V^2(\mathbf{x})\left\{\left(\frac{\partial\tau}{\partial x}\right)^2+\left(\frac{\partial\tau}{\partial y}\right)^2+\left(\frac{\partial\tau}{\partial z}\right)^2\right\}, \tag{6.43}$$

where $\frac{\partial\tau}{\partial x_i}$ is the slope in the ith direction in time-shift gathers. Figure 6.9D shows a reflection-angle gather that is computed using slant stacking of the time-shift gather.

If we defined $p_t = \sqrt{\left(\frac{\partial\tau}{\partial x}\right)^2+\left(\frac{\partial\tau}{\partial y}\right)^2+\left(\frac{\partial\tau}{\partial z}\right)^2}$ as the magnitude of the slope, then from (6.43)

$$p_t = \frac{V(\mathbf{x})}{\cos\theta(\mathbf{x})}, \tag{6.44}$$

which can be used to estimate the local apparent velocity.

Space shift gathers can be converted into reflection-angle gathers using

$$\tan\theta(\mathbf{x}) = \left|\frac{k_h}{k_m}\right|, \tag{6.45}$$

where k_h is the wavenumber in the offset direction and k_m is the wavenumber for the middle point coordinate.

It should be pointed out that reflection-angle gathers can also be computed from the source and receiver wavefield using local plane wave decomposition as demonstrated in Xie and Wu (2002).

6.4.2 Reverse-Time Migration 3D-Angle Gathers

Although time-shift gathers and space-shift gathers can be converted into reflection-angle gathers, RTM 3D-angle gathers can provide not only the reflection angle information but also the 3D orientation of the reflections, the dip, and the azimuth angles.

Figure 6.12 shows the definitions of angles used in RTM 3D-angle gathers from ray point of view. The angles should be computed from the wavefield at each local spatial point using, for example, local plane-wave decomposition.

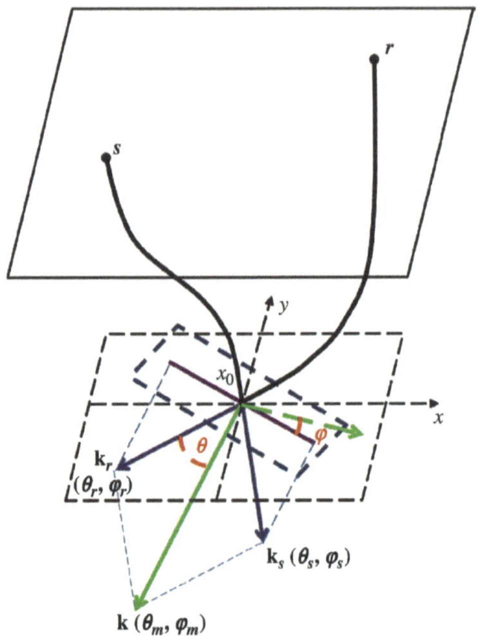

FIGURE 6.12 Definitions of angles for RTM 3D-angle gathers *(Xu et al., 2011)*. See text for details.

The basic procedure to compute the dip and azimuth angles is given by Xu et al. (2011) as

$$\cos\theta(\mathbf{x}) = \frac{\mathbf{k}\cdot\mathbf{k}_r}{|\mathbf{k}||\mathbf{k}_r|}, \tag{6.46}$$

and

$$\cos\phi(\mathbf{x}) = \frac{(\mathbf{k}_s\times\mathbf{k}_r)\cdot\{\mathbf{n}\times(\mathbf{k}_s+\mathbf{k}_r)\}}{|\mathbf{k}_s\times\mathbf{k}_r||\mathbf{n}\times(\mathbf{k}_s+\mathbf{k}_r)|}. \tag{6.47}$$

In (6.46) and (6.47), $\mathbf{k}_s$ is the wavenumber at the image point of the source wavefield, $\mathbf{k}_r$ is the wavenumber at the image point of the receiver wavefield. $\mathbf{k}$ is the wavenumber in the direction of the image vector, $\mathbf{k}=\mathbf{k}_s+\mathbf{k}_r$. $\theta(\mathbf{x})$ is the local dip angle, $\phi(\mathbf{x})$ is the local azimuth angle, and $\mathbf{n}$ is the unit vector in x coordinate, $\mathbf{n}=(1,0,0)^T$. Once the wavenumbers are computed for both source side and receiver side, (6.46) and (6.47) can be used to determine the dip and azimuth angles, respectively. So the basic step is to compute propagation directions. Several methods can be used for computing the propagation directions. Based on the computation domains, they can be approximately classified into vector variable-based methods, plane wave decomposition, and space/time-shift method (for details, see Hu et al., 2014).

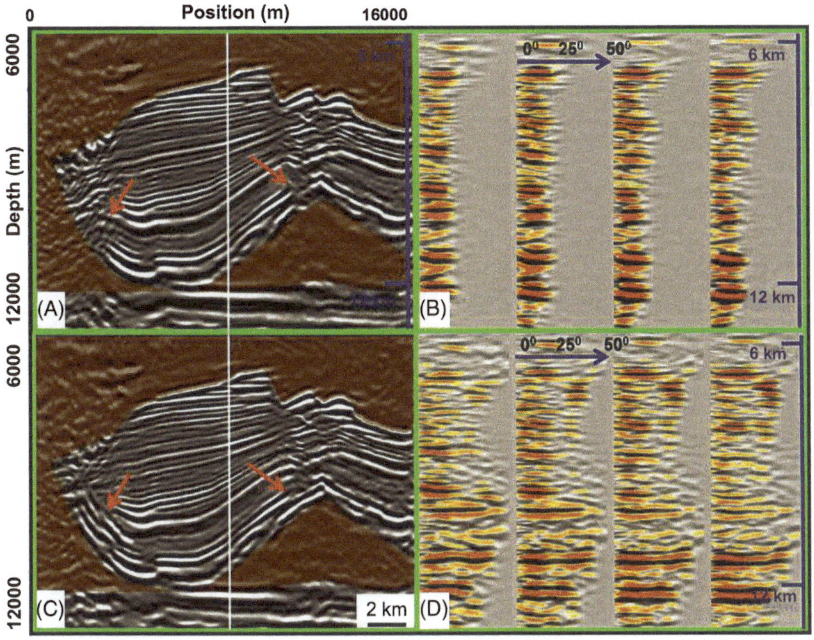

FIGURE 6.13 RTM 3D-angle gathers. (A) RTM image, (B) RTM 3D-angle gathers using 0–9 km offset data, (C) RTM image, and (D) RTM 3D-angle gathers using 0–18 km offset data. Red arrows in (A) and (C) denote steeply dipping subsalt events *(Wu and Li, 2014)*.

The results of RTM 3D-angle gathers are limited by the information contained in the acquired data. For typical wide azimuth data (WAZ), with largest offset 10 km, the maximum reflection angles available in subsalt areas may be only up to 20–30°. This range of reflection angles imposes great challenges for angle domain subsalt tomography using 3D-angle gathers. With the increasing offsets of modern acquisitions such as StagSeis and Coil Shooting, 3D-angle gathers will be able to provide more significant updates for velocity model in complicated such as subsalt areas. For examples using long offset and full azimuth for tomography, see Section 6.6 for details.

Figure 6.13 shows one real-data example for 3D-angle gathers in deep water of Gulf of Mexico. Figure 6.13A shows the RTM stacks and RTM-angle gathers migrated using 0–9 and 0–18 km offsets. The RTM stack images show that the ultra-long offsets help imaging the steeply dipping salt flanks (Figure 6.13C). The RTM 3D-angle gathers have much larger angle coverage with ultra-long offsets (from about 20° for 9 km maximum offsets to about 40° for 18 km maximum offsets, Figure 6.13B and D). The additional illumination is obtained from the extended offset ranges (9–18 km). This increased angle coverage provides better input for subsalt velocity updates.

6.5 PRACTICAL ISSUES IN REVERSE-TIME MIGRATION

Although RTM has many advantages compared with ray-based imaging methods or one-way wave equation imaging, it does have its own issues. In this section, we discuss some problems observed in real-data processing.

6.5.1 Low-Frequency Noise in Reverse-Time Migration Image and Impedance Sensitivity Kernel

When the conventional imaging condition (6.3) is used during migration, strong low-frequency noise is usually seen above water bottom or above top of salt in raw RTM image. This is due to the cross-correlation between upgoing wavefield (reflected at the sharp velocity boundary) and the downgoing wavefield.

Figure 6.14 shows one typical example in the Gulf of Mexico. Figure 6.14A is the direct output from RTM and Figure 6.14B is the image after postmigration

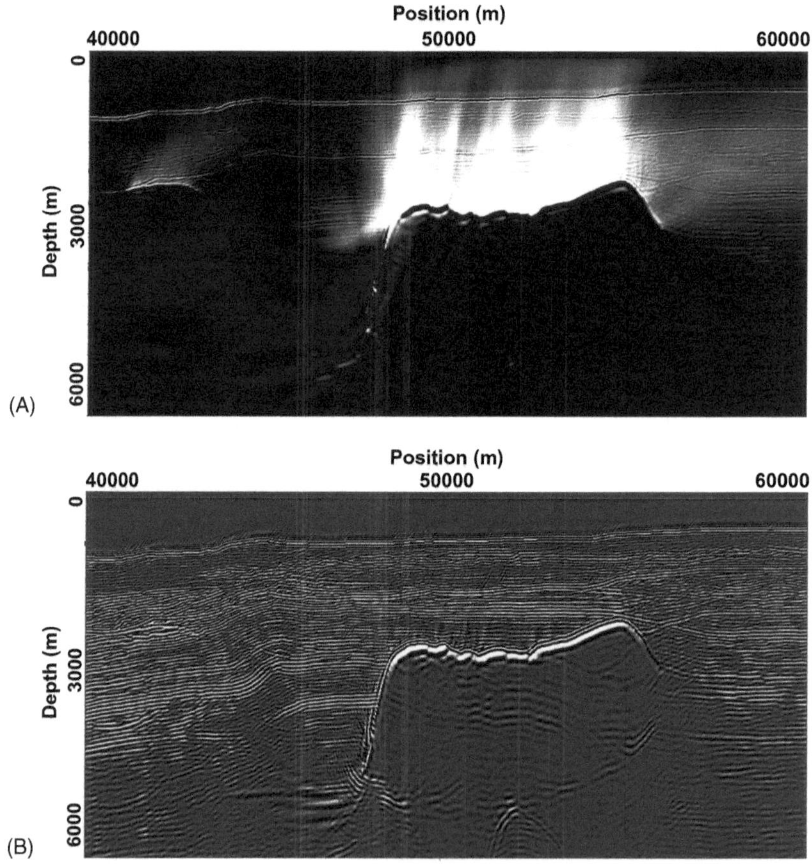

FIGURE 6.14 (A) Image output from RTM. (B) After applying Laplacian filter to (A). Both are parts of BP 2004 model *(Billette and Brandsberg-Dahl, 2005)*.

processing. The low-frequency noise is clearly seen above the top of salt. This is the low-frequency noise issue of RTM. The noise itself can be easily removed by applying a Laplacian filter on the raw RTM image. However, the understanding of the causes and its geophysical implication deserves further analysis.

During wavefield extrapolation in RTM, unlike one-way wave equation, the two-way wave equation used in RTM allows wave to propagate in all directions. When sharp velocity contrast exists, both source wavefield and receiver wavefield can be reflected from the sharp boundaries. Therefore, reflected source/receiver wavefield shares the same wavepaths with the downgoing receiver/source wavefield. The image produced at this "head-on collision" has strong amplitude. This happens at all the points along the entire wavepath and along all wavepaths.

There are various ways to solve the low-frequency noise problem. The most widely used method is to apply the Laplacian filter to boost the high-frequency component. In order to produce the right image; however, the low-frequency components in the source wavelet or input data have to be boosted before being used for RTM. The filter is defined as

$$F_{\mathrm{LB}}(\omega)=\begin{cases}\dfrac{1}{i\omega}, & \text{3D}\\[2mm] \dfrac{1}{\sqrt{i\omega}}, & \text{2D}\end{cases}, \tag{6.48}$$

where $F_{LB}(\omega)$ stands for low-boost filter, which is an integration operator.

Since the low-frequency noise is produced at the "head-on collision" along the common wavepaths between the source wavefield and the receiver wavefield, the reflection angles at these points are close to 90°. Therefore, in the common angle-image gathers, these events appear at the far end of the angle range. Therefore, the low-frequency noise can be removed by muting the large angle reflections in common angle-image gathers. This can be seen from the time-shift gather in Section 6.4.1 and Figure 6.10.

For seismic imaging, the low frequency is considered as noises and must be removed during or after migration. But further analysis showed that more significant geophysical information is contained in the low-frequency image. Recently, in-depth investigations of the relation between inversion and imaging has been carried out by various authors and new findings have been made (Zhu et al., 2009; Douma et al., 2010). Douma et al. (2010) theoretically proved the following relation between the image obtained by (6.3) and the impedance sensitivity kernel obtained by inversion for second-order wave equations (6.1) and (6.2)

$$K_Z(\mathbf{x})=\frac{1}{2}\left\{\begin{aligned}&\frac{V(\mathbf{x})}{Z(\mathbf{x})}\nabla^2 I(\mathbf{x})+\int f(\mathbf{x},t)\,p^{\perp}(\mathbf{x},T-t)\,dt\\&+\int f^{\perp}(\mathbf{x},T-t)\,p(\mathbf{x},t)\,dt\end{aligned}\right\}, \tag{6.49}$$

where $I(\mathbf{x})$ is defined in (6.3), $Z(\mathbf{x})$ is the impedance model, $f^{\perp}(\mathbf{x},t)$ is the adjoint source, and $p(\mathbf{x},t)$ is the adjoint receiver wavefield. Equation (6.3) clearly indicates that, for far wavefield, impedance sensitivity inversion kernel is a scaled version of image by (6.3) after Laplacian filtering. However, it does not suffer from the contamination of low-frequency noise. If the inversion process is parameterized by impedance and velocity, then impedance kernel can be used for imaging purpose and the velocity kernel can be used for model update. Similar conclusions have been drawn for elastic imaging (Zhu et al., 2009) from elastic inversion sensitivity kernels.

6.5.2 Physical Constraints for Anisotropic Parameters

Wave propagation in elastic media is controlled by the motion equation (Newton's second law)

$$\rho \frac{\partial \mathbf{v}}{\partial t} = \nabla \bullet \sigma + \mathbf{f}, \tag{6.50}$$

and the constitutive equation (Hooke's law)

$$\frac{\partial \sigma}{\partial t} = C \frac{\partial \varepsilon}{\partial t}. \tag{6.51}$$

Here ρ is the density, $\mathbf{v}$ is the particle velocity vector, C is the elastic matrix, σ is the stress tensor, ε is the strain tensor, and $\mathbf{f}$ is the body force.

For (6.51) to be a valid system, C cannot be arbitrarily chosen. It has to follow the energy constraints: the strain energy of any deformation of the elastic solid must be positive or

$$\sigma_{ij}\varepsilon_{ij} > 0. \tag{6.52}$$

In other words, the elastic tensor has to be positive definite. This constraint, expressed in the 6 × 6 matrix, can be written as (Carcione, 2001):

$$c_{I(I)} > 0, \det \begin{pmatrix} c_{I(I)} & c_{IJ} \\ c_{IJ} & c_{J(J)} \end{pmatrix}, \ldots, \det(c_{IJ}) > 0. \tag{6.53}$$

That is, the determinants of all principal submatrices must be positive. In practice, under certain assumptions such as acoustic VTI; however, some of the inequalities in (6.53) can be relaxed to be nonnegative.

We now derive explicit formulas for each of the commonly used types of media: VTI acoustic and elastic (including TTI) and orthorhombic acoustic and elastic (vertical or tilted) media. The constraints derived that follow are not only valid for the first-order wave equation (6.50), but also valid for the second-order wave equations (6.22), (6.23), and (6.37).

For VTI/TTI acoustic, the 6 × 6 matrix is reduced to the following 2 × 2 matrix (also see eq. (6.16))

$$\rho V_{p0}^2 \begin{bmatrix} 1+2\varepsilon & \sqrt{1+2\delta} \\ \sqrt{1+2\delta} & 1 \end{bmatrix}. \tag{6.54}$$

The conditions in (6.53) can be written as

$$V_{p0} > 0, \varepsilon \geq -\frac{1}{2}, \delta \geq -\frac{1}{2}, \varepsilon \geq \delta. \tag{6.55}$$

The last inequality in (6.55) is the most widely used condition in VTI/TTI model building and imaging.

For general VTI/TTI elastic media, the 6 × 6 matrix is

$$C = A\begin{bmatrix} (1+2\varepsilon) & (1+2\varepsilon)-2(1+2\gamma)f^2 & \left\{\sqrt{\left(1-f^2\right)^2+2\delta(1-f^2)}-f^2\right\} & 0 & 0 & 0 \\ (1+2\varepsilon)-2(1+2\gamma)f^2 & (1+2\varepsilon) & \left\{\sqrt{\left(1-f^2\right)^2+2\delta(1-f^2)}-f^2\right\} & 0 & 0 & 0 \\ \left\{\sqrt{\left(1-f^2\right)^2+2\delta(1-f^2)}-f^2\right\} & \left\{\sqrt{\left(1-f^2\right)^2+2\delta(1-f^2)}-f^2\right\} & 1 & 0 & 0 & 0 \\ 0 & 0 & 0 & f^2 & 0 & 0 \\ 0 & 0 & 0 & 0 & f^2 & 0 \\ 0 & 0 & 0 & 0 & 0 & (1+2\gamma)f^2 \end{bmatrix}, \tag{6.56}$$

where $f = V_{s0} / V_{p0}$ and $A = \rho V_{p0}^2$.

For matrix (6.56), (6.53) reduces to

$$\begin{cases} V_{p0} > 0, V_{s0} \geq 0 \\ \varepsilon \geq \max\left(-\dfrac{1-(1+2\gamma)f^2}{2}, -\dfrac{1}{2}\right) \\ \dfrac{\varepsilon + f^2 + f^2\sqrt{1+2\varepsilon}}{1-f^2} \geq \delta \geq -\dfrac{1}{2}. \\ \gamma \geq -\dfrac{1}{2} \\ 2(\varepsilon-\delta) + f^2 - 2(\gamma-\delta)f^2 + 2f^2\sqrt{(1+2\varepsilon)-(1+2\gamma)f^2} \geq 0 \end{cases} \tag{6.57}$$

For P–SV-wave propagation VTI/TTI elastic media, if we ignore SH-waves, (6.57) becomes

$$\begin{cases} V_{p0} > 0, V_{s0} \geq 0 \\ \varepsilon \geq \max\left(-\dfrac{1}{2}\left(1-f^2\right), -\dfrac{1}{2}\right) \\ \delta \geq -\dfrac{1}{2}\left(1-f^2\right) \\ \delta \leq \dfrac{\varepsilon + f^2 + f^2\sqrt{1+2\varepsilon}}{1-f^2} \end{cases}. \tag{6.58}$$

For orthorhombic acoustic media, the 6 × 6 matrix is replaced by the following 3 × 3 matrix (also see (6.38))

$$\rho V_{p0}^2 \begin{bmatrix} 1+2\varepsilon_2 & (1+2\varepsilon_2)\sqrt{1+2\delta_3} & \sqrt{1+2\delta_2} \\ (1+2\varepsilon_2)\sqrt{1+2\delta_3} & 1+2\varepsilon_1 & \sqrt{1+2\delta_1} \\ \sqrt{1+2\delta_2} & \sqrt{1+2\delta_1} & 1 \end{bmatrix}. \tag{6.59}$$

Constraints in (6.53) reduce to

$$\begin{cases} \varepsilon_1 \geq -\frac{1}{2}, \delta_1 \geq -\frac{1}{2}, \varepsilon_2 \geq -\frac{1}{2}, \delta_2 \geq -\frac{1}{2} \\ \varepsilon_1 \geq \delta_1, \varepsilon_2 \geq \delta_2, \delta_3 \leq \frac{\varepsilon_1 - \varepsilon_2}{1+2\varepsilon_2}. \\ \det(M) \geq 0 \end{cases} \tag{6.60}$$

The last inequality cannot be written in a simple explicit form. Equation (6.60) can be used for orthorhombic model building (Li et al., 2012), but usually one has to use further assumptions to constrain inversion of surface seismic data.

For general orthorhombic elastic media, simple explicit formulation does not exist. During numerical modeling, one has to verify each of the six conditions in (6.53) at each point in the model space.

To show why we need to be concerned about the physical constraints for the anisotropic parameters, we compare the ranges allowed for ε and δ in VTI acoustic media (6.55) and VTI elastic media (6.58). One typical case is shown in Figure 6.15. We assume $V_{p0} = 2000$ m/s and $V_{s0} = 1000$ m/s. Inequality (6.55) gives

$$\varepsilon \geq -0.5, \delta \geq -0.5, \varepsilon \geq \delta. \tag{6.61}$$

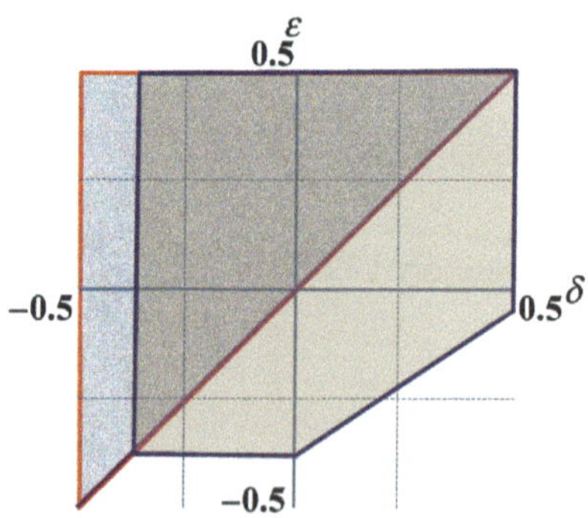

FIGURE 6.15 Comparison between the constraints on ε and δ in VTI acoustic media (area enclosed by red curve) and VTI elastic media (area enclosed by blue curve, the segment at right bottom is not exact).

and (6.58) gives

$$\varepsilon \geq -0.375, \delta \geq -0.375, \frac{\varepsilon + 0.25 + 0.25\sqrt{1+2\varepsilon}}{0.75} \geq \delta. \tag{6.62}$$

The area enclosed by red curve is for VTI acoustic media and the area enclosed by blue curve is for VTI elastic media. Obviously, for practical values of ε and δ, the range allowed by elastic assumption is much larger than that by acoustic assumption. Including shear-wave terms in VTI/TTI, wave equations can increase the stability of the system (Fletcher et al., 2009). Similar conclusion is also valid for orthorhombic media. Generally, including shear-wave terms results in more stable and more realistic equations.

In the next section, we will discuss shear-wave noise in VTI/TTI acoustic media and more stability-related examples will be given.

6.5.3 Artificial and Physical Noises of Reverse-Time Migration in Anisotropic Media

Anisotropy is intrinsically an elastic phenomenon. The widely used acoustic VTI and orthorhombic wave equations result in significant shear-wave presence in both modeling data and RTM images. Figure 6.3 shows the wavefront of both P-wave and SV-wave in a homogeneous VTI model. As discussed previously, the amplitude and travel time of P-wave wavefront changes slightly while the SV wavefront changes dramatically when V_{s0} is close to zero and the SV-wave wavefront becomes a singular point.

The phase and group velocities for P-wave and SV-waves are shown in Figure 6.16. They are drawn at equally spaced phase angles and group angles, respectively. The P-wave phase *slowness* curves remain convex disregarding the changes of V_{s0}. However, SV-wave phase slowness is often concave, especially for strong anisotropy. The concave nature of SV-wave phase slowness leads to the cusps in the SV-wave group velocity or wavefront at 1 s (Cerveny, 2001) and results in amplitude singularity in the amplitude. Figure 6.3 contains both kinematic and dynamic information of the wavefronts for both P-wave and SV-wave.

Shear-wave noise has caused a great deal of attention in seismic modeling and RTM. Zhang et al. (2003) may be the first few who correctly explain the physical reasons. Zhang and Zhang (2009) proposed several simple methods to removal shear-wave noises. Fletcher et al. (2009) added SV-wave terms to the original acoustic VTI/TTI wave equations to make them stable, which basically simulate elastic wave propagation within acoustic wave equations. They also pointed out that shear waves can be initiated by incident P-wave everywhere in the model, not necessarily just in the neighbor of the source location.

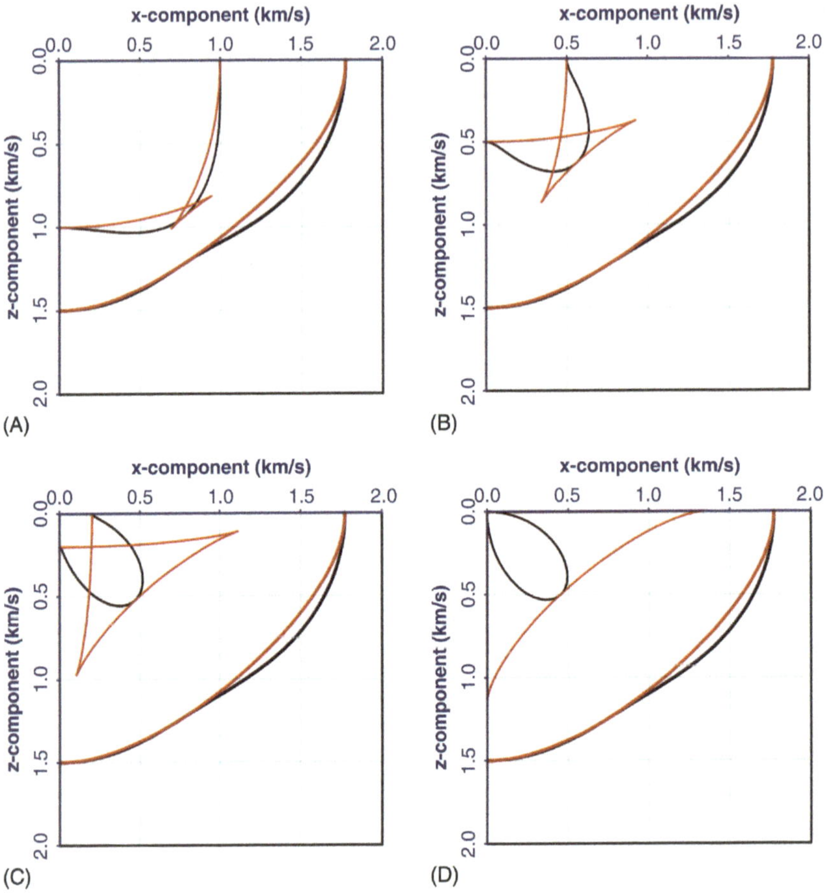

FIGURE 6.16 Phase and group velocity surfaces of P-wave and SV-waves in VTI media. V_{p0} is fixed at 1500 m/s, $\varepsilon = 0.2$, and $\delta = -0.2$. (A) $V_{s0} = 1000$ m/s, (B) $V_{s0} = 500$ m/s, (C) $V_{s0} = 200$ m/s, and (D) $V_{s0} = 1$ m/s. Black curves are for phase velocities (for reference), and red curves are for group velocities (wavefronts). Outer curves are for P-waves and inner curves are for SV-waves.

In real practice, there are several approaches to mitigate the problems caused by shear-wave noise. The simplest way is to clip the ε and δ around the source location to be zeroes, that is, to create an isotropic zone around source location to prevent shear wave from propagating. The small isotropic zone usually does not generate noticeable travel-time errors. This approach is simple and usually yields satisfying results in practice once the anisotropic parameters follow the constraints defined by (6.53).

The second approach is to add SV-wave terms to the original acoustic VTI/TTI wave equations (Fletcher et al., 2009). The wavefield computed contains both P-waves and shear waves. But the amplitude of the shear wave is one magnitude smaller than the P-wave. So shear wave does not produce significant noise into the image. One set of equations is given by

$$\begin{cases} \dfrac{1}{V_{p0}^2}\dfrac{\partial^2 p}{\partial t^2} = (1+2\varepsilon)\,H_2 p + H_1 q + \gamma H_1\,(p-q) \\ \dfrac{1}{V_{p0}^2}\dfrac{\partial^2 q}{\partial t^2} = (1+2\delta)\,H_2 p + H_1 q - \gamma H_2\,(p-q) \end{cases}, \tag{6.63}$$

where $\gamma = V_{s0}^2 / V_{p0}^2$, $H_1 = n_x^2 \frac{\partial^2}{\partial x^2} + n_y^2 \frac{\partial^2}{\partial y^2} + n_z^2 \frac{\partial^2}{\partial z^2} + 2n_x n_y \frac{\partial^2}{\partial x \partial y} + 2n_z n_y \frac{\partial^2}{\partial z \partial y} + 2n_x n_z \frac{\partial^2}{\partial x \partial z}$

and $H_2 = \frac{\partial^2}{\partial x^2} + \frac{\partial^2}{\partial y^2} + \frac{\partial^2}{\partial z^2} - H_1$. (n_x, n_y, n_z) defines the direction of the symmetry axis, $n_x^2 + n_y^2 + n_z^2 = 1$. It is not hard to see that (6.63) is about 50% more expensive than (6.23). Modeling using (6.63) is similar to modeling in elastic anisotropic media with normal shear-wave velocity, which can be seen in Figures 6.3, 6.6, and 6.7; the amplitude of the SV-wave is much smaller compared with that of P-wave.

In (6.63), to ensure the stability of the system, V_{s0} can be chosen (Fletcher et al., 2009).

$$\sigma = \frac{V_{p0}^2}{V_{s0}^2}(\varepsilon - \delta) \le 0.8. \tag{6.64}$$

Condition (6.64) put a lower limit for shear-wave velocity. Under this condition, triplications of the SV wavefronts disappear (Tsvankin, 2001).

In the third approach (Zhang and Zhang, 2009), we still use (6.22) for VTI and (6.23) for TTI, but only save the P-wave wavefield after modeling at each time step. The can be achieved by

$$\tilde{p}(\mathbf{x}) = FT_{\mathbf{k}}^{-1}\left\{\frac{1}{k^2} FT_{\mathbf{x}}\,[H]\right\}, \tag{6.65}$$

where H is the right side of the first equation of (6.22) or (6.23), k is total wavenumber, $FT_{\mathbf{x}}$ is the forward Fourier transform, and $FT_{\mathbf{k}}^{-1}$ is the inverse Fourier transform. The output $\tilde{p}(\mathbf{x})$ is the eigen waveform for P-wave of the total wavefield.

Figure 6.17 shows one example with three layers. The first layer has strong anisotropy with $\varepsilon = 0.25$ and $\delta = 0$. The symmetry axis is tilted 45° to the right. Figure 6.17A is the model. Figure 6.17B shows the image from RTM using wavefield without any processing. Figure 6.17C shows the image based on the wavefield filtered using (6.65). The linear noise in the first layer is contributed from shear-wave noise.

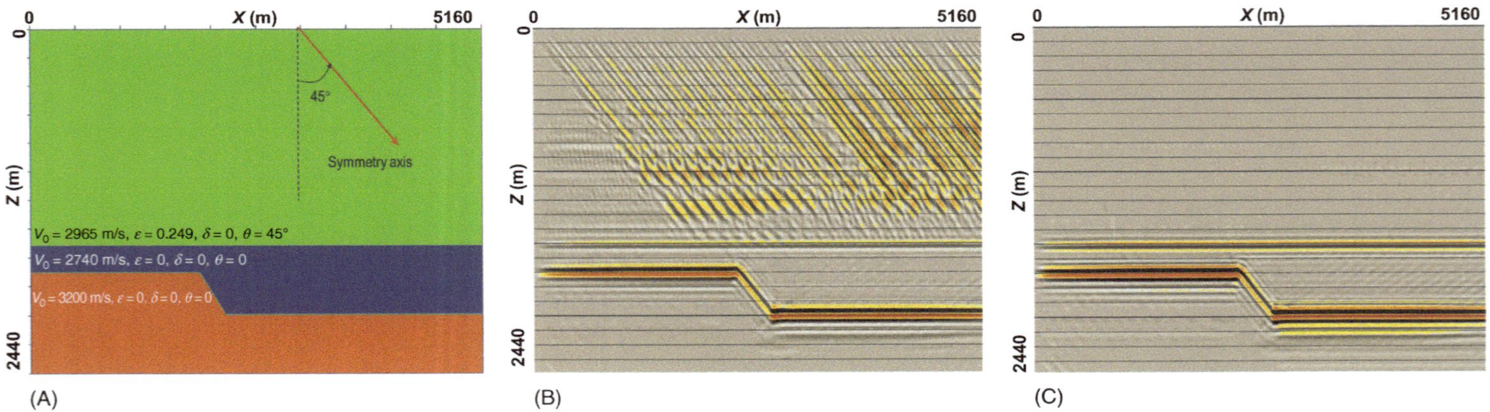

FIGURE 6.17 Shear-wave removal using eigen waveform. (A) The geometry of the 2D TTI model. (B) RTM image without shear-wave removal. (C) RTM image after shear-wave noise removal.

Another important development in order to produce P-wave wavefield only without shear-wave interference during VTI or TTI modeling is to derive so-called pure P-wave wave equations (Du et al., 2010; Chu et al., 2011). Pure-acoustic approximation decouples the P-wave propagation from the elastic wavefield and has looser constraints on anisotropic parameters than that required by pseudo-acoustic approximation ($\varepsilon \geq \delta$).

Each of the previously discussed methods has its own advantages and shortcomings. The first approach is simple and efficient and works for most cases, but may not be able to remove all the SV noises at places with very strong discontinuities of ε, δ, θ, and ϕ. The second method simulates elastic wave propagation in acoustic media and is closer to the actual wave propagation, but is more expensive than the other two. The third approach needs two 3D Fourier transforms, but provides an in-depth analysis for the causes of the shear waves.

It should be emphasized that all the methods discussed here can be equally applied on orthorhombic modeling and imaging with necessary modifications to the wave equations.

6.5.4 Illumination Issues in Complicated Areas

The quality of image in a particular area depends on how well the energy can penetrate into that area and how much the angular coverage input data can provide for that area. If the energy is weak, such as shadow zones in subsalt area, illumination holes will exist and noise will be dominate the image. If there is not enough angular coverage, effective events will not be well collapsed; artificial structures such as swings will be present. Subsequent geological interpretation may be misled. For subsalt areas in the Gulf of Mexico, due to the presence of salt bodies, illumination is a big issue. The complicated geometrical shapes of salt bodies and high-velocity contrasts between salt and sediments cause part of subsalt areas unable to be illuminated and therefore, the data acquired on the surface may not contain all information needed. Offsets that can provide sufficiently large angle coverage for shallow targets or simple structures may result in very limited angle coverage for subsalt targets due to the sharp velocity contrasts. This is illuminated in Figure 6.18 for simple salt geometry. The high velocity in salt causes the energy to be bent twice, one at top of salt and one at base of salt, from the surface to the target image point. A large portion of rays has been restricted inside the salt body. At the top of salt, due to the large contrast of velocity, incident waves with moderate angles may lead to postcritical reflections and head waves. As a result, only small angle incidence can penetrate into the subsalt area. However, sufficiently large angle coverage is critical for subsalt imaging (Bleistein et al., 2001).

Figure 6.19 shows one typical salt model (A) and the stacked illumination field (B) (the denominator of (6.67)). The penetrating energy is unevenly distributed under the salt flanks and below the salt body. Energy is weak in most subsalt areas. Some areas have strong energy due to the focusing effects of the salt body. In real-data imaging, due to the weak energy of the real reflections,

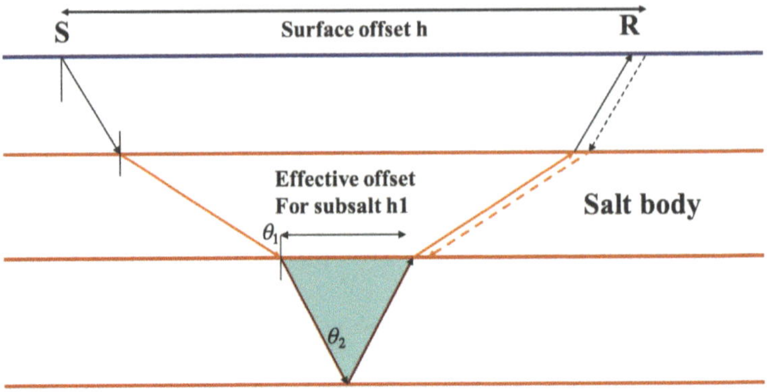

FIGURE 6.18 Large offsets in the surface may only provide small angle coverage for subsalt area.

noises will dominate the image. Figure 6.19C shows the ray field for a shot at one location and Figure 6.19D shows the ray field for a shot at second location. They clearly show that the energy penetrating to the subsurface depends on both shot positions and local incidence and scattering angles. Therefore, in order to fully compensate the energy losses, the compensation must be a function of both space location and spatial directions. The ray fields also show that most of the energy is reflected back to the surface at the top of salt and only a small portion of rays can penetrate into subsalt. As a result, the subsalt energy is usually weak and the subsalt information contained in the data is limited.

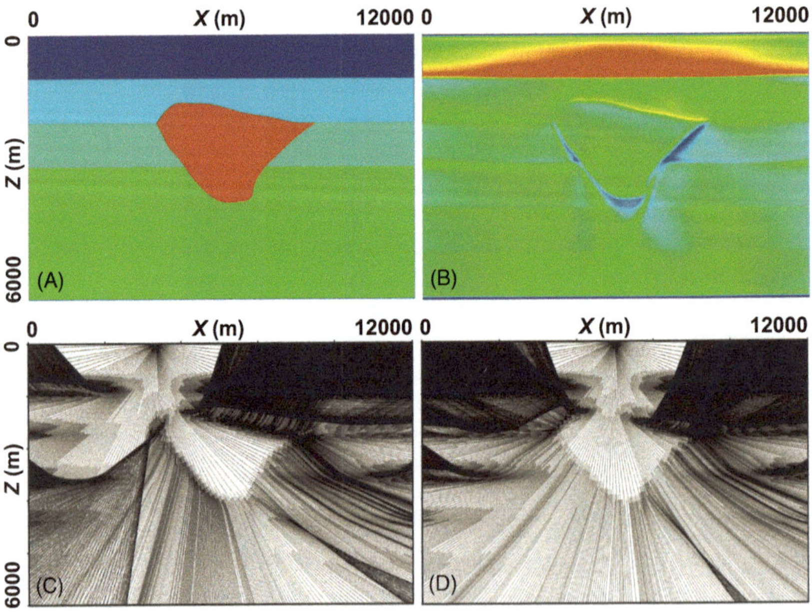

FIGURE 6.19 (A) Velocity model. (B) Illumination field. (C) Ray field for shot 1. (D) Ray field for shot 2.

Unfortunately, there are no systematic and uniform approaches or workflows that can be used for completely solving subsalt illumination problem. However, various methods have been proposed to mitigate the issue and here we give a brief review for those methods that are usually used in real-data processing.

One of the most often used methods is to use source wavefield energy to compensate the cross-correlation image produced by (6.3)

$$I(\mathbf{x}) = \frac{I_0(\mathbf{x})}{\sum\limits_{\mathbf{x}_s,t} p_s^2(\mathbf{x},t;\mathbf{x}_s)} = \frac{\sum\limits_{\mathbf{x}_s} p_s(\mathbf{x},t;\mathbf{x}_s) \otimes p_r(\mathbf{x},t;\mathbf{x}_s)}{\sum\limits_{\mathbf{x}_s,t} p_s^2(\mathbf{x},t;\mathbf{x}_s)}. \tag{6.66}$$

The denominator is the total source wavefield energy for all the time and for all the shots, thus this illumination compensation can be called global illumination compensation. Numerical implementation of (6.66) is straightforward. However, practice showed that the illumination field computed this way is too smooth and does not provide enough compensation for subsalt imaging.

Cogan et al. (2011) proposed shot-domain illumination compensation scheme

$$I(\mathbf{x}) = \sum_{\mathbf{x}_s} \frac{p_s(\mathbf{x},t;\mathbf{x}_s) \otimes p_r(\mathbf{x},t;\mathbf{x}_s)}{\sum\limits_{t} p_s^2(\mathbf{x},t;\mathbf{x}_s)}. \tag{6.67}$$

Note the difference between (6.67) and (6.66). One is to apply illumination compensation after summation over shots and the other is to apply illumination compensation before summation over shots. Real-data examples in the Gulf of Mexico showed that results produced by shot domain illumination compensation are significantly better than the global illumination compensation. Figure 6.20 showed one example from Chazalnoel et al. (2011).

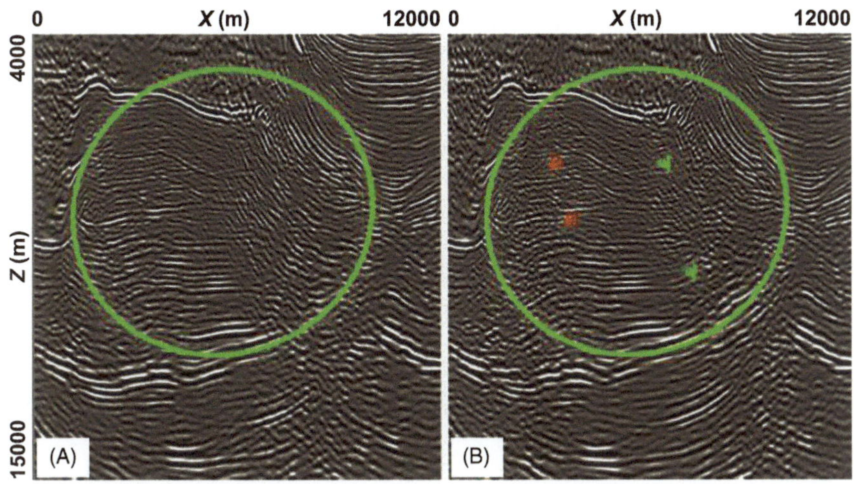

FIGURE 6.20 Comparison of images. (A) Conventional RTM stacked image by (6.6). (B) Image using shot domain illumination compensation (6.23) *(Chazalnoel et al., 2011).*

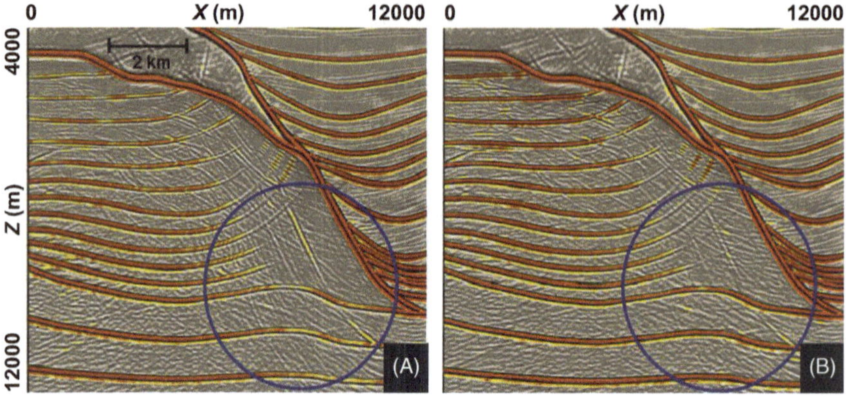

FIGURE 6.21 Removal of crossing events in subsalt using 3D-dip gathers *(Li et al., 2012)*. A synthetic data example. (A) Without noise removal. (B) After noise removal using 3D-dip gather.

The reason why shot domain illumination compensation can produce better image than conventional stacked RTM image is fully documented in Chattopadhyay and McMechan (2008). Shot domain illumination compensation can preserve the angle-dependent reflections. This angle-dependent information is destroyed during stacking in global illumination compensation. Therefore, with shot-domain illumination compensation, when applied to structures such as three-way traps, sediments below salt keels and below salt flanks, image quality can be significantly improved.

Illumination compensation can boost the reflections in critical areas, but it also boosts the noise. So, noise attenuation or removal is equally important to improve the S/N of image. In terms of geological interpretation, reflection gather does not have direct connection with geological contexts. It is the dip gather that has direct connection with geological structures, and geological structures can be preinterpreted from previous stacked image. On the basis of this thinking, Li et al. (2012) proposed using 3D-dip gathers to remove coherent noise from RTM image. Events with same dip angles are preserved during stacking and will be geologically consistent with the final stacked image. Figure 6.21 shows one synthetic example without and with noise removal using 3D-dip gathers. If the coherent noise in the image has opposite or different dips compared with the desired events, they can be muted in the 3D-dip gathers. Figure 6.22 shows one real-data example from Walker Ridge. Because of the removal of the interfering noise, the effective events are much more coherent in the subsalt areas.

A complete illumination analysis can be carried out in local-angle (scattering and dip angles) domain for each point in the image space (Xie and Wu, 2002; Xie et al., 2006; Wu and Toksöz, 1987). The directional illumination field is computed by

$$I(\mathbf{r}_T, z, \theta, \omega; \mathbf{r}_s) = \left|u(\mathbf{r}_T, z, \theta, \omega; \mathbf{r}_s)\right|^2 = \left|FT\{w(\mathbf{r}_T)u(\mathbf{r}_T, z, \omega; \mathbf{r}_s)\}\right|^2. \quad (6.68)$$

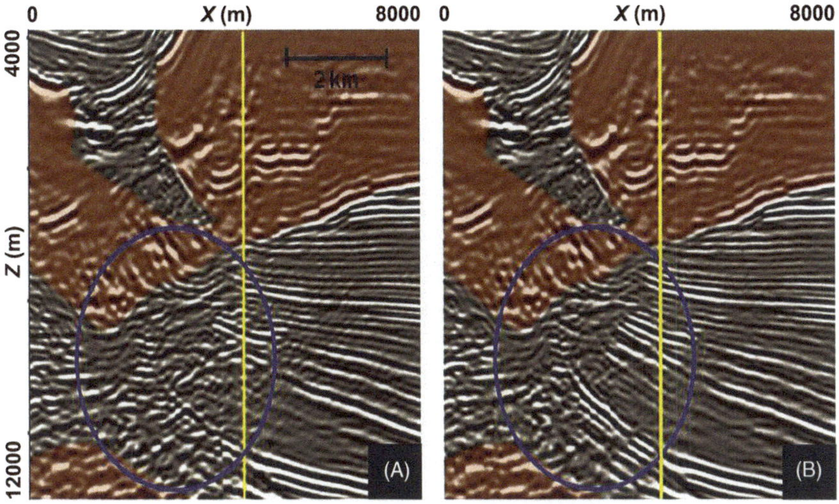

FIGURE 6.22 Removal of crossing events in subsalt using 3D-dip gathers *(Li et al., 2012)*. Real-data example from Walker Ridge. (A) Without noise removal. (B) After noise removal using 3D-dip gather.

Here $\mathbf{r}_T$ is the horizontal coordinate in the image space, $w(\mathbf{r}_T)$ is the window around $\mathbf{r}_T$, and FT is the Fourier transform of the wavefield defined by window $w(\mathbf{r}_T)$. Equation (6.68) implicitly assumes a conversion between the Fourier wavenumber vector $\mathbf{K}_T$ and angle θ. $\mathbf{r}_s$ in (6.68) is the input source or trace location. By combining both source side and receiver side, we can define the dip response

$$I\left(\mathbf{r}_T, z, \theta, \omega; \mathbf{r}_s, \mathbf{r}_g\right) = I\left(\mathbf{r}_T, z, \theta_s, \omega; \mathbf{r}_s\right) I\left(\mathbf{r}_T, z, \theta_r, \omega; \mathbf{r}_g\right), \qquad (6.69)$$

where $\mathbf{r}_s$ and $\mathbf{r}_g$ are the source and receiver location, respectively, θ_s is the propagation angle of the source wavefield at image location $(\mathbf{r}_T, z)$. θ_g is the propagation angle of the receiver wavefield at image location $(\mathbf{r}_T, z)$. θ is the bisector angle between θ_s and θ_r. For common-shot gather, its dip response is

$$I\left(\mathbf{r}_T, z, \theta, \omega; \mathbf{r}_s\right) = \sum_{\mathbf{r}_g} I\left(\mathbf{r}_T, z, \theta_s, \omega; \mathbf{r}_s\right) I\left(\mathbf{r}_T, z, \theta_r, \omega; \mathbf{r}_g\right). \qquad (6.70)$$

For multiple common-source gathers, we have

$$I\left(\mathbf{r}_T, z, \theta, \omega\right) = \sum_{\mathbf{r}_s, \mathbf{r}_g} I\left(\mathbf{r}_T, z, \theta_s, \omega; \mathbf{r}_s\right) I\left(\mathbf{r}_T, z, \theta_r, \omega; \mathbf{r}_g\right). \qquad (6.71)$$

$I\left(\mathbf{r}_T, z, \theta, \omega\right)$ in (6.71) is the same as the 3D-dip gathers used by Li et al. (2012). Equations (6.70) and (6.71) can be used for illumination compensation for dip-angle gathers.

Figure 6.23A shows the subsalt illumination function for different locations of SEG/EAGE salt model (Xie et al., 2006). The illumination is clearly much weaker in the subsalt and the spectrum band is narrower. Figure 6.23B shows

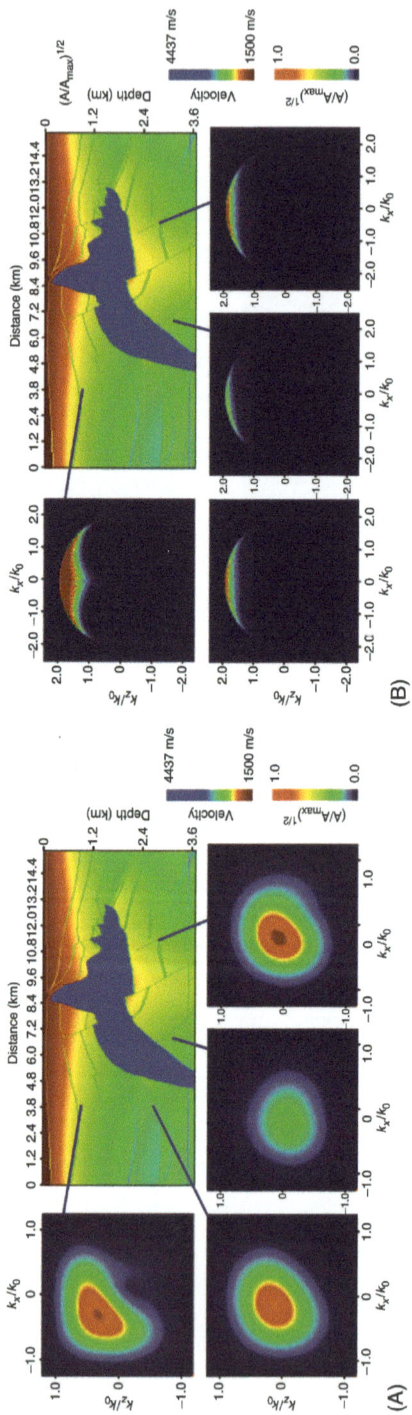

FIGURE 6.23 (A) Local illumination field for different locations in the SEG/EAGE salt model. (B) Wavenumber domain illumination for different locations in the SEG/EAGE salt model.

the wavenumber domain illumination. In the shallow part, illumination is well defined for wide-angle coverage. The dip-angles coverage becomes worse in the subsalt area, and the angle spectrum band is narrower. In the subsalt area, both horizontal and vertical resolutions are very poor.

6.6 IMPACTS OF LONG OFFSETS AND FULL AZIMUTHS ON REVERSE-TIME MIGRATION

Broad frequency band and sufficient subsurface angular coverage are key factors for model building and subsalt imaging in the Gulf of Mexico. Although conventional WAZ has brought significant improvement for multiple attenuation and crossline direction illumination; however, it still suffers from lack of low frequency, limited azimuth coverage, and limited offsets.

In this section, we will give a brief summary on how these factors affect subsalt imaging. Examples from data acquired by recently developed acquisition technologies will be given and will be compared with those using conventional WAZ data.

The subsurface is an absorptive system. The energy dissipation is frequency dependent: the amplitude of high-frequency signal decays much faster than that of low frequency. The subsurface is also a heterogeneous system. The overall impact of the heterogeneities depends on the relative scales of seismic wavelength and dimensions of the scatters. When the seismic wavelength is shorter compared with the size of the scatters, these seismic waves show dispersive behavior. When the seismic wavelength is comparable with the size of the scatters, seismic scattering will be dominated. Seismic waves will be scattered into different directions and there will be multiple scattering between scatters. Therefore, the amplitude of the data received at the surface is weak and the multiple scattering cannot be correctly positioned by RTM. Because of either absorption or scattering, the high-frequency components of the signal experience more significant losses than the low-frequency component. As a result, only low-frequency data can penetrate into the subsalt area. Therefore, collection and preservation of low-frequency data is important for subsalt imaging. The low-frequency component enables the final stacked image to have less rings and reflections to be clearly defined. During model building using full waveform inversion (FWI), the availability of low-frequency data is critical to prevent cycle skipping (Virieux and Operto, 2009).

As shown in Section 6.5.4, for simple geological subsurface with slow variation of velocity, long offset usually implies large angle coverage for subsurface targets. By integrating the contributions from all illumination angles, RTM or any other imaging methods are able to produce well-focused image (see Figure 6.23). On the contrary, for areas with strong velocity contrasts such as between salt body and sediments, large surface offsets may only be able to illuminate a small portion of angles for subsalt targets (Figure 6.18). Even worse, small-angle incidence may not correspond to small offset. The lack of

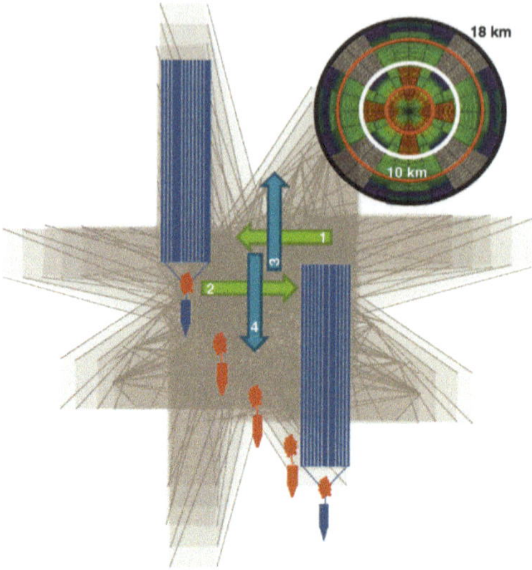

FIGURE 6.24 Acquisition geometry and rose diagram for WAZ (inside the white circle) and StagSeis.

near-angle and sufficiently large-angle spectrum in common image gathers imposes great challenge for model building and subsalt imaging. Therefore, in order to have large subsurface angle coverage for those areas, large offsets are needed.

Figures 6.24–6.27 show some examples to demonstrate the impact of offsets and azimuths on velocity model building and imaging. Figure 6.24 shows the acquisition patterns and rose diagram for a particular acquisition technology, StagSeis, designed by CGG (Mandroux et al., 2013; Grenie et al., 2014). The acquisition is antiparallel and orthogonal to provide a variety of azimuth directions in the super-shot gather. This particular staggered geometry can provide very long offsets, up to 18 km, and full-azimuth distribution up to 9 km (assuming source–receiver reciprocity). Moreover, the straight-line acquisition geometries provide uniform bin coverage that facilitates efficient processing when compared with other geometries. Conventional WAZ data is within the white circle (about 10 km) in the rose diagram, but with less uniform coverage. Extra offsets, from 10 to 18 km) and extra azimuth coverage are additional information beyond that provided by conventional WAZ.

Figure 6.25 shows the velocity models inverted using FWI from different offsets (Mothi et al., 2013). The offset of WAZ is 7.5 km in the inline direction and 4 km in the crossline direction. The offset of the StagSeis data is 18 km in both inline and crossline directions. The velocity model from WAZ is highly oscillatory in the deep portion and carbonate is incorrectly inverted. The inverted model from long-offset and full-azimuth data is more characteristic of the geology and does not have the oscillatory nature.

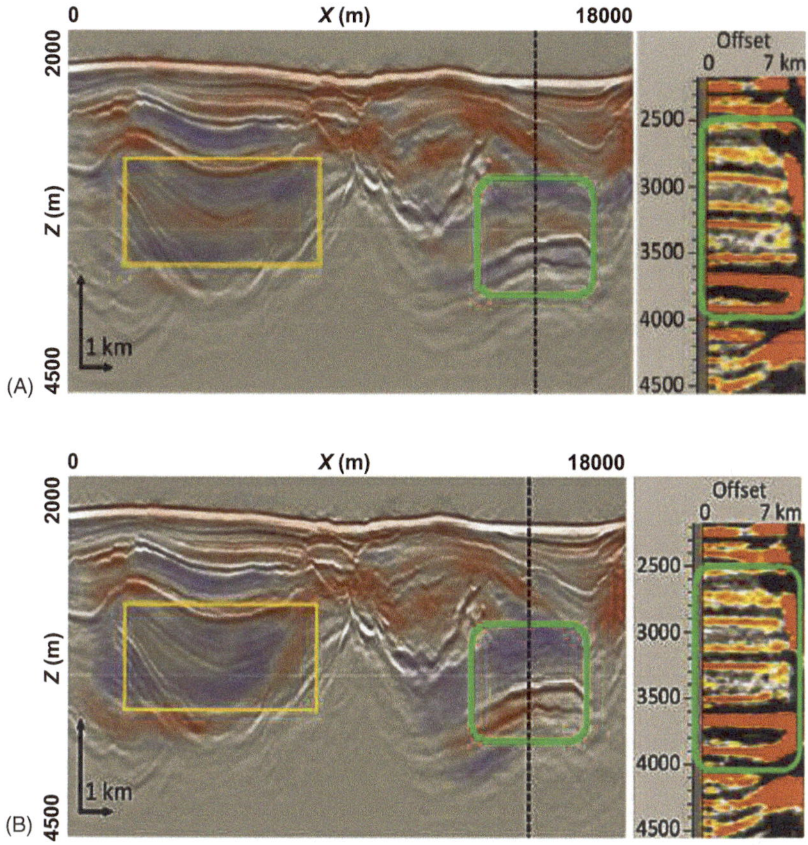

FIGURE 6.25 Comparison of velocity update based on FWI using (A) WAZ data and (B) StagSeis data *(Mothi et al., 2013).*

Figure 6.26 shows the depth slices of the velocity model inverted using only north–south shot lines (Figure 6.26A) and using both north-south and east–west shot lines (Figure 6.26B) (Mothi et al., 2013). With limited azimuth coverage, the velocity model suffers from stripping patterns, which is the acquisition footprint. The sampling in the crossline direction is sparser than that in the inline direction.

Figure 6.27 shows the image produced by RTM using conventional WAZ data and the image using long-offset and full-azimuth data. Because of the limited angular coverage of the WAZ data, subsalt area suffers from illumination issues and this lead to incoherent and broken reflections in the subsalt.

6.7 SUMMARY AND CONCLUSIONS

RTM has quickly become one of the major imaging methods, since it was first used in industry, and practically the dominant method for imaging complicated areas such as subsalt in the Gulf of Mexico and offshore West Africa where

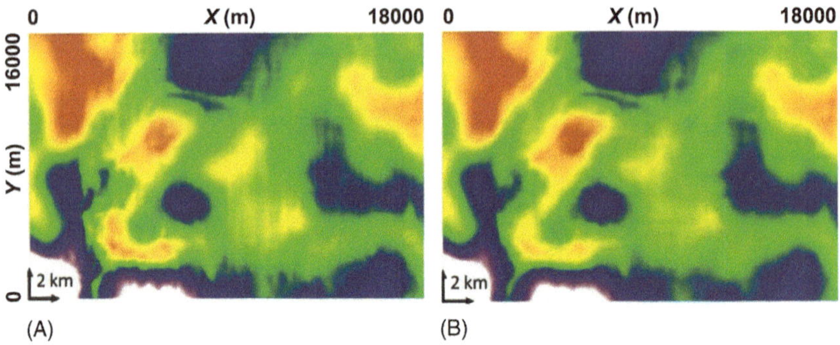

FIGURE 6.26 Comparison of velocity update based on FWI using (A) North–south shot lines and (B) Both north–south and east–west shot lines *(Mothi et al., 2013)*.

strong velocity contrasts are present. The uniform framework of RTM allows it to be implemented in arbitrary media such as isotropic, VTI, TTI, and orthorhombic media. Imaging condition is the most critical part of RTM. Different gathers can be generated from RTM in addition to the basic stacked image. In this overview, we also discussed various practical issues in RTM. The mathematical aspects of RTM is becoming mature, the focus is now on how to build more accurate and more realistic models and how to improve the quality of image. Improving image quality from inversion point of view is also an important direction. Examples show that data collected using modern acquisition technologies contains more low-frequency components and long offsets. These acquisition improvements are able to build better models and improve RTM image quality, especially in areas with poor illuminations.

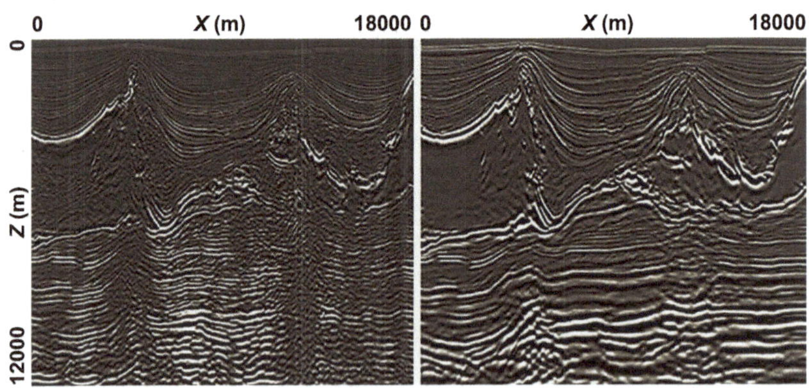

FIGURE 6.27 Comparison of image using WAZ data (left) and StagSeis data (right).

REFERENCES

Aki, K., Richards, P., 2009. Quantitative Seismology. University Science Books Oxford, UK, Sausalito, CA.

Alkhalifah, T., 2000. An acoustic wave equation for anisotropic. Geophysics 65, 1239–1250.

Alkhalifah, T., 2003. An acoustic wave equation for orthorhombic anisotropy. Geophysics 65, 1169–1172.

Backus, G.E., 1962. Lon-wave elastic anisotropy produce by horizontal layering. J. Geophys. Res. 67, 4427–4440.

Baysal, E., Kosloff, D.D., Sherwood, J.W.C., 1983. Reverse-time migration. Geophysics 48, 1514–1524.

Billette, F.J., Brandsberg-Dahl, S., 2005. The 2004 BP velocity benchmark, 67th Meeting, European Association of Geoscientists and Engineers, Expanded Abstracts, B035.

Bleistein, N., Cohen, J.K., Stockwell, J.W., Jr., 2001. Mathematics of multidimensional seismic imaging, migration, and inversion. Springer, New York.

Canuto, C., Hussaini, M.Y., Quarteroni, A., Zang, T.A., 2006. Spectral Methods, Fundamentals in Single Domains. Springer, New York.

Carcione, J.M., 2001. Wave Fields in Real Media: Wave Propagation in Anisotropic, Anelastic, and Porous Media. Pergamon Press, New York.

Cerveny, V., 2001. Seismic Ray Theory. Cambridge University Press, Cambridge.

Chattopadhyay, S., McMechan, G.A., 2008. Imaging conditions for prestack reverse time migration. Geophysics 73, S81–S89.

Chazalnoel, N., Ong, B., Zhao, W., 2011. Imaging 3-way closures by combining a deconvolution imaging condition with vector offset output RTM. 81st Annual International Meeting, SEG, Expanded Abstracts.

Chu, C., Macy, B., Anno, A., 2011. Approximation of pure acoustic seismic wave propagation in TTI media. Geophysics 76, WB97–WB107.

Claerbout, J., 1971. Toward a unified theory of reflector mapping. Geophysics 36 (3), 467–481.

Claerbout, J., 1985. Imaging the Earth's Interior. Blackwell Scientific Publishers, Hoboken, NJ.

Cogan, M., Fletcher, R., King, R., Nichlos, D., 2011. Normalization strategies for reverse time migration. 81st Annual International Meeting, SEG, Expanded Abstracts.

Dong, Z., McMechan, G.A., 1991. Numerical modeling of seismic waves with a 3D anisotropic scalar wave equation. Bull. Seis. Soc. Am. 81, 769–780.

Douma, H., Yingst, D., Vasconcelos, I., Tromp, J., 2010. On the connection between artifact filtering in reverse-time migration and adjoint tomography. Geophysics 75, S219–S223.

Du, X., Fletcher, R.P., Fowler, P.J., 2010. Pure P-wave propagators versus pseudo-acoustic propagators for RTM in VTI media. 72nd European Association of Geoscientists and Engineers, Expanded Abstracts.

Duveneck, E., Bakker, P.M., 2011. Stable P-wave modeling for reverse-time migration in tilted TI media. Geophysics 76, S65–S75.

Duveneck, E., Milcik, P., Bakker, P., 2008. Acoustic VTI wave equations and their application for anisotropic reverse-time migration. 78th Annual International Meeting, SEG, Expanded Abstracts.

Fletcher, R.P., Du, X., Fowler, P.J., 2009. Reverse-time migration in tilted transversely isotropic (TTI) media. Geophysics WCA179–WCA187.

Fowler, P.J., Du, X., Fletcher, R.P., 2010. Coupled equations for reverse time migration in transversely isotropic media. Geophysics 75, S11–S22.

Grenie, D., Siliqi, R., Mandroux, F., Ting, C., 2014. Subsalt imaging through the acquisition of full-azimuth broadband data with extra-long offsets. 76th European Association of Geoscientists and Engineers, Expanded Abstracts.

Helbig, K., 1994. Foundations of Anisotropy for Exploration. Pergamon Press, New York.
Hu, J., McMechan, G.A., Guan, H., 2014. Comparison of methods for extracting ADCIGs from RTM. Geophysics 79, S89–S103.
Huang, T., Zhang, Y., Zhang, H., 2009. The benefit of TTI reverse time migration for subsalt imaging, gulf of Mexico. 71st European Association of Geoscientists and Engineers, Expanded Abstracts.
Huang, T., Zhang, Y., Zhang, H., Young, J., 2009. Subsalt imaging using TTI reverse time migration. Lead. Edge 28, 448–452.
Khalil, A., Sun, J., Zhang, Y., Gordon, P., 2013. RTM noise attenuation and image enhancement using time-shift gathers. 83rd Annual International Meeting, SEG, Expanded Abstracts.
Leveille, J.P., Jones, I.F., Zhou, Z., Wang, B., Liu, F., 2011. Subsalt imaging for exploration, production, and development: a review. Geophysics 76, WB3–WB20.
Li, Y., Han, W., Chen, C., Huang, T., 2012. Velocity model building for tilted orthorhombic depth imaging. 82nd Annual International Meeting, SEG, Expanded Abstracts.
Li, Z., Tang, B., Ji, S., 2012. Subsalt illumination analysis using RTM 3D dip gathers. 82nd Annual International Meeting, SEG, Expanded Abstracts.
Mandroux, F., Ong, B.S., Ting, C., Mothi, S., Huang, T., Li, Y., 2013. Broadband, long-offset, full-azimuth, staggered marine acquisition in the Gulf of Mexico. First Break 31, 125–132.
McMechan, G.A., 1983. Migration by extrapolation of time-dependent boundary values. Geophys. Prospect. 31, 413–420.
Mothi, S., Schwarz, K., Zhu, H., 2013. Impact of full-azimuth and long-offset acquisition on Full Waveform Inversion in deep water Gulf of Mexico. 83rd Annual International Meeting, SEG, Expanded Abstracts.
Sava, P., Fomel, S., 2006. Time-shift imaging condition in seismic migration. Geophysics 71 (6), S209–S217.
Schoenberg, M., Helbig, K., 1997. Orthorhombic media: Modeling elastic wave behavior in a vertically fractured earth. Geophysics 62, 1954–1974.
Stork, C., 1992. Reflection tomography in the postmigrated domain. Geophysics 57, 680–692.
Thomsen, L., 1986. Weak elastic anisotropy. Geophysics 51, 1954–1966.
Thomsen, L., 2002. Understanding seismic anisotropy in exploration and exploitation, SEG DISC No. 5, p. 248.
Tsvankin, I., 1997. Anisotropic parameters and P-wave velocity for orthorhombic media. Geophysics 62, 1292–1309.
Tsvankin, I., 2001. Seismic Signatures and Analysis of Reflection Data in Anisotropic Media. Pergamon, New York.
Virieux, J., Operto, S., 2009. An overview of full-waveform inversion in exploration geophysics. Geophysics 74, WCC127–WCC152.
Whitmore, N.D., 1983. Iterative depth migration by backward time propagation. 53rd Annual International Meeting, SEG, Expanded Abstracts, Session S10.1.
Wu, Q., Li, Y., Li, Z., 2013. Improved subsalt imaging of full azimuth data with tilted orthorhombic PSDM. 83rd Annual International Meeting, SEG, Expanded Abstracts.
Wu, Q., Li, Y., 2014. Benefit of ultra-long offset data for subsalt imaging in deep water Gulf of Mexico. 74th European Association of Geoscientists and Engineers, Expanded Abstracts.
Wu, R., Toksöz, M.N., 1987. Diffraction tomography and multisource holography applied to seismic imaging. Geophysics 52, 11–25.
Xie, X., Jin, S., Wu, R., 2006. Wave-equation based seismic illumination analysis. Geophysics 71, S169–S177.
Xie, X., Wu, R., 2002. Extracting angle domain information from migrated field. 72nd Annual International Meeting, SEG, Expanded Abstracts.

Xu, S., Zhang, Y., Tang, B., 2011. 3D angle gathers from reverse time migration. Geophysics 76, S77–S92.

Zhu, H., Luo, Y., Nissen-Meyer, T., Morency, C., Tromp, J., 2009. Elastic imaging and time-lapse migration based on adjoint methods. Geophysics 74, WCA167–WCA177.

Zhang, L., Rector, J.W., Hoversten, G.M., 2003. An acoustic wave equation for modeling in tilted TI media. 73rd Annual International Meeting, SEG, Expanded Abstracts.

Zhang, Y., Sun, J., 2008. Practical issues of reverse time migration: true-amplitude gathers, noise removal and harmonic-source encoding. 70th European Association of Geoscientists and Engineers, Expanded Abstracts.

Zhang, H., Zhang, Y., 2008. Reverse time migration in 3D heterogeneous TTI media. 78th Annual International Meeting, SEG, Expanded Abstracts.

Zhang, H., Zhang, Y., 2009. Attenuating S-waves in TTI reverse time migration. 79th Annual International Meeting, SEG, Expanded Abstracts.

Zhang, H., Zhang, Y., 2011. Reverse time migration in vertical and tilted orthorhombic media. 81st Annual International Meeting, SEG, Expanded Abstracts.

Zhou, H., Zhang, G., Bloor, R., 2006. An anisotropic acoustic wave equation for modeling and migration in 2D TTI media. 76th Annual International Meeting, SEG, Expanded Abstracts.

Chapter 7

Wave-Propagation Operators for True-Amplitude Reverse-Time Migration

Qizhen Du[1], Gang Fang[2], Xufei Gong[1], Mingqiang Zhang[1]
[1]*School of Geosciences, China University of Petroleum (East China), Qingdao, China;*
[2]*Qingdao Institute of Marine Geology, Qingdao, China*

Chapter Outline

7.1 INTRODUCTION

Seismic migration is a major processing step of imaging in hydrocarbon seismic exploration (Yilmaz and Doherty, 2001). It not only produces interior structure maps of the earth but is also a prerequisite for seismic attribute analysis. Reflection coefficients are one of the most important attributes for reservoir description. The amplitude variation with offset (AVO) or amplitude variation with

Seismic Exploration of Hydrocarbons in Heterogeneous Reservoirs. http://dx.doi.org/10.1016/B978-0-12-420151-4.00006-4

angle (AVA) based on accurate reflection coefficient's amplitude can indicate the reservoir and nonreservoir distributions (Aki and Richards, 1980). True-amplitude migration, which recovers reflection coefficient maps from recorded data, provides a solid physical basis for the quantitative analysis of AVO/AVA. Because migration – AVO – inversion produces are more accurate than the produces of AVO – inversion – migration (Beydoun et al., 1994), true amplitude migration is an essential requirement for reliable AVO/AVA analysis.

Most of the proposed prestack depth migration techniques can produce accurate image locations but hardly accurate amplitudes because they do not compensate the propagation effects of geometric spreading or transmission losses. These methods often choose the imaging conditions that are stable at the cost of accurate amplitude. Claerbout (1971) proposed a unifying imaging principle for common-shot prestack depth migration. According to this principle, the reflection-coefficient maps are obtained by dividing the reflected wavefield from receivers by the incident wavefield from sources. With the unifying imaging principle, migration can be defined as an adjoint operator of the linear single scattering approximation to wave-propagation operators in inhomogeneous media (Leveille et al., 2011). The migration includes two independent steps: reconstruction of two wavefields – the incident and reflected wavefields, and the application of an imaging condition at each subsurface point. This approach has no specific requirement for propagation operators. So it can be used for all currently developed prestack depth migrations, such as Kirchhoff depth migration (KDM), one-way wave-equation migration (WEM), reverse-time migration (RTM), and beam migration (BM) (Leveille et al., 2011). The difference among the above four methods is how the source and receiver wavefields are reconstructed. The propagator of KDM is based on high-frequency approximation that makes KDM hard to handle wavefield caustics, resulting in multiple arrivals at many subsurface locations. One necessity for KDM is to employ a smoothed velocity. Because of this smoothing, the KDM cannot achieve the accurate image amplitude that WEM and RTM can bring, especially when truly sharp and rogues velocity boundaries are encountered as in subsalt imaging. BM uses beam decomposition to treat wavefield as a superposition of Gaussian beams, thereby reducing the likelihood to develop caustic. The WEM-based one-way wavefield extrapolation has no difficulty in constructing complicated wave-front caustics, but cannot handle high-angle wave propagation. The amplitude of one-way wave extrapolation needs to be corrected when used for true-amplitude migration (Zhang et al., 2003a, 2005a). RTM using two-way wave equation as its propagator is consider as the leading migration technology because of its no limitation of image angle and accurate wavefield amplitude (Leveille et al., 2011).

True amplitude migration requires proper handling of wavefield propagation as well as proper implementation of imaging condition. The correlation-based imaging condition is often used for its computational stability, but it can only produce the correct kinematic imaging rather than the true amplitude. The true KDM (Bleistein, 1987; Docherty, 1991; Hanitzsch et al., 1994;

Hanitzsch, 1997; Vanelle and Gajewski, 2000; Vanelle et al., 2006) is proposed based on inverse theory that produces correct amplitude in proportion to the reflection coefficient at the image point. For WEM, Zhang et al. (2005b) proposed a true-amplitude migration method by using a new one-way wave equation that compensates the propagation effect and a deconvolution-type image condition (Kelly and Ren, 2004; Zhang et al., 2005b). For RTM, the unify imaging condition (Claerbout, 1971) is true amplitude. But it is hard to implement because of its instability. In practice, several approximations of the unifying imaging principle are introduced for reasons of stability (Valenciano and Biondi, 2003; Kaelin and Guitton, 2006; Guitton et al., 2007). Kaelin and Guitton (2006) showed that using smoothed source illumination in the unifying imaging condition improves the stability of the computation and preserves the amplitude information. To remove the zero divide, Guitton et al. (2006) included a damping factor in the numerator. Chattopadhyay and McMechan (2008) and Schleicher et al. (2008) compared the output amplitudes of these imaging conditions. They concluded that the imaging condition, which divides the cross-correlation of incident and reflected wavefields by autocorrelation of the incident wavefield, is stable and produces image with stable, reliable amplitudes and fewer artifacts.

Instead of relying on the unifying imaging principle, true-amplitude migration can be also formulated under the principle of inversion. There are two main categories of such inversion approaches. One category estimates the perturbations of the medium parameters based on the least-square principle (Tarantola, 1984) or the generalized inverse Radon transform (Beylkin, 1985). The least-square migration has the potential to obtain accurate reflection coefficients distributions and to deal with irregular illumination. It involves an iterative approach based on loops of a linearized modeling problem, usually in the frequency domain, and an inversion procedure. Lailly (1983) derived the connection between the gradient of least-squares objective function and migration algorithm. For a true-amplitude migration, however, the gradient needs to be premultiplied by the inverse of the Hessian matrix of the least-square misfit, whose size and nondiagonal structure make it unfeasible to evaluate its inverse for real seismic data. Several methods have been proposed to approximate the Hessian with a diagonal matrix (Chavent and Plessix, 1999; Shin et al., 2001; Plessix and Mulder 2004). To handle sparse acquisition geometry, the iterative migration is formulated after linearization around a background model (Nemeth et al., 1999; Sevink 1996; Pica et al., 1990).

The other category of inversion methods expresses the reflected wavefield in terms of the linear relationship of reflection coefficients or perturbations of model parameters. Then the reflection coefficients or the perturbations of model parameters are estimated from recorded data by an inversion approach. For example, Born inversion (Clayton and Stolt, 1981) estimates the perturbation of angle-independent medium parameters, whereas Kirchhoff inversion (Bleistein et al., 2001; Bleistein, 1987) estimates angle-dependent reflection coefficients from seismic data with high-frequency asymptotics. In fact, high-frequency asymptotics is a powerful tool to derive various migration weights for true-amplitude ray-based migration methods (Bleistein, 1987). However, the implementations of these migrations, which rely on ray tracing, may break down for complicated medium.

Most recently, high-frequency asymptotics is used for deriving true-amplitude wave-equation migration methods, which are based on numerical solutions of one- or two-way wave equations. These approaches are fully established for depth imaging of complex structures. Based on the high-frequency asymptotic theory, these approaches convert the Kirchhoff migration weights into partial differential operators and also modify initial or boundary conditions to develop true-amplitude one-way WEM (Zhang et al., 2005b, 2007) or RTM (Costa et al., 2009; Zhang and Sun, 2008). The main task for true-amplitude one-way WEM is to develop a dynamically correct propagation operator. As proved by Zhang et al. (2003b), the original one-way wave equations used for WEM (Claerbout, 1971) are only kinematically correct, which means they were designed to produce accurate travel times but their geometric spreading was inaccurate because of using approximate pseudodifferential operators. Zhang et al. (2003b) provided a term of residual correction for one-way wave-equation propagation operator to correct the geometric spreading behavior. Based on this new propagation operator, several true-amplitude WEMs (Vivas and Pestana, 2010; Zhang et al., 2005b, 2012) are developed. This correction term plays the same role as the weights that give the full geometric spreading in true-amplitude Kirchhoff migration.

In addition to the geometric spreading, transmission losses, seismic Q attenuation and illumination are additional factors that can prevent WEM from obtaining true-amplitude output. Most current true-amplitude migration methods introduced transmission compensation terms (Sun and Fu, 2012; Zhang et al., 2003b) and Q-compensation (McDonal et al., 1981; Widmaier et al., 1995; Deng and McMechan, 2008; Valenciano et al., 2011) when the propagation operator is based on one-way wave equation.

The common-images gathers (CIGs) is a set of images at the same image location, which are commonly used for depth-domain AVO/AVA analysis (Beydoun et al., 1993; Tura et al., 1998; Dong and Ponton, 1999). With a correct velocity model, all of the images at the same image location should focus at the same depth and the reflection events on the CIGs is flat, which can be direct used for AVO/AVA inversion if the amplitude is reliable. For KDM, there is no significant additional cost to compute common-offset CIGs (COCIGs). But for wave-equation-based migration such as WEM or RTM, the computational cost is extremely expensive (Ehinger et al., 1996). The migration artifacts caused by multiple paths of wave propagation will limit the quality of COCIGs, whether or not the propagator can handle multipath of wavefield accurately (Nolan and Symes, 1996). Considering the computational cost of CIGs in the surface-offset domain obtained by wave-equation migration, other types of gather such as ray-parameter gathers (de Bruin et al., 1990; Prucha et al., 1999; Mosher and Foster 2000; Kuehl and Sacchi 2001, 2002), plane-wave parameters gathers (Duquet and Lailly 2006; Soubaras, 2003, 2006), and angle gathers (Weglein and Stolt 1999; Sava and Fomel, 2003; Wu et al., 2004) are also proposed. Xu et al. (2011) proposed a condition for artifact-free CIGs in a subsurface angle domain, and they show that an angle-domain common-image gather (ADCIG), in which the migrated contributions are sorted by reflection angle, reduces artifacts related to multipaths. ADCIGs can be obtained by KDM, BM, one-way WEM, or RTM. Compared with other migration method, RTM implemented by the two-way wave equation provides the most kinematically and dynamically accurate image. The motivation described above for calculating ADCIGs from RTM is given in more detail by Zhang et al. (2010) and Xu et al. (2011). There are three main methods for exacting ADCIGs from RTM: the direction-vector-based methods, the local plane-wave decomposition methods, and the local-shift image-condition methods. The direction-vector-based methods use Poynting vector (Yoon et al., 2004) or polarization-vector-based method (Zhang and McMechan, 2011a, 2011b) to extract ADCIGs directly from the propagated wavefield. These methods are fast and accurate for amplitude, but are not stable for complicated wavefields. The local plane-wave decomposition methods decompose spherical wavefronts into local planar waves by slant stack transform (Xie and Yang, 2008; Yan and Xie, 2011, 2012) or from the horizontal and vertical wavenumbers obtained by 2D Fourier transform (Xie and Wu, 2002; Xu et al., 2011). For the local-shift imaging condition method, which is an indirect method, Sava and Fomel (2003) proposed to output local subsurface-offset CIGs by applying local space-shift imaging condition, and then transforming into the ADCIGs. Sava and Fomel (2006) proposed an alternative approach to produce ADCIGs, which is called time-shift image condition. The time-shift image condition is similar to the local space-shift imaging condition, but shifting is applied in the time domain between the source and receiver wavefields, rather than in the space domain. The local-shift imaging condition methods also involve slant stacking (Sava and Fomel, 2003; Biondi and Symes, 2004) or Fourier transforms (Sava and Fomel, 2003; Sava and Fomel, 2005; Sava et al., 2005) to transform subsurface-offset domain CIGs to the ADCIGs. Because the transformation is done only once for each CIG, the method is fast. Comparing time-shift CIGs with space-shift CIGs, time-shift CIGs are very efficient in 2D, but the angle resolution of ADCIGs obtained from the method is lower than that from the space-shift imaging condition. The time-shift ADCIGs cannot provide sufficient azimuthal information in 3D (Xu et al., 2011), which makes time-shift ADCIGs less popular than the other method. Joncour et al. (2011) considered the amplitude of angle gather and presented preserved amplitude migration formulas in the aperture angle domain based on the strategy of the extended imaging condition.

RTM was derived in the 1980s (Baysal et al., 1983; McMechan, 1983; Whitmore, 1983) and was originally developed for poststack migration. It is based on the concept of exploding reflectors, which means that the image can be extracted at zero time of the reverse-time extrapolated wavefield. RTM has also been extended to prestack form (Chang and McMechan, 1986, 1987). Currently, this method has been well developed and widely used in industry. Because of its power of imaging without dip limitations and handling various types of waves with different propagation paths (e.g., reflected waves, diffracted waves, prismatic waves, and evanescent waves), RTM can obtain accurate images of complex medium such as subsalt area (Zhang and Sun, 2008; Liu et al., 2011) and anisotropic medium (Kang and Cheng, 2011; Zhang et al., 2011). Based on the elastic wave equation, RTM is also developed to image multicomponent seismic data (Chang and McMechan, 1987, 1994; Du et al., 2012a). Also several researches were carried out to provide more accurate medium parameters (Du et al., 2012b; Sava et al., 2005). Chang and McMechan (1986) first proposed to use excitation-time imaging condition in prestack RTM. In excitation-time imaging condition, images are extracted from the extrapolated receiver wavefield at the excitation time, which is the one-way travel time from the source to each image point. The excitation time can be calculated by ray tracing (Chang and McMechan, 1986) or by solving the two-way wave equation with the maximum amplitude criterion or the minimum arrival-time criterion (Lowenthal and Hu, 1991). In the excitation-time imaging condition, only the travel-time information of the source wavefield is used, which is one of special cases of the cross-correlation imaging condition (Claerbout, 1971). Compared to the one-way WEMs, the propagation operators of RTM are built based on two-way wave equation. This gives RTM inherent-amplitude accuracy over one-way wavefield extrapolation method. The migration weights involved in true-amplitude Kirchhoff migration or the residual correction derived by Zhang et al. (2005b) are no longer explicit in RTM. To compensate the transmission losses, Zhang et al. (2003b) estimated reflection coefficients with a frequency-space domain-migration method; Deng and McMechan (2007) present a prestack depth-migration method that uses the framework of reverse-time migration to compensate for geometric spreading, intrinsic Q losses, and transmission losses. In their method, they compensate the transmission losses for RTM with plane-wave transmission coefficients. They also extend the compensation procedure to elastic RTM to compensate for transmission and elastic attenuation losses in an isotropic medium (Deng and McMechan, 2008). However, plane wave is a kind of approximation of the wavefield in real media. The theoretical proof is necessary for supporting these compensation methods with plane-wave transmission coefficients.

In this chapter, we propose theoretical derivations about geometric spreading and transmission losses in RTM for variable density and velocity media via a high-frequency asymptotic analysis. With stationary phase theory, we analyze how geometric spreading and transmission losses of propagation operators affect the imaging amplitude. The analyses provide a mathematical base for current methods of transmission compensation.

This chapter is organized as follows. First, we introduce the wave propagators used in the forward and backward wave propagation of RTM. Second, we derive a true-amplitude migration formula for variable density and velocity media. Third, we make a theoretical analysis about geometric spreading and

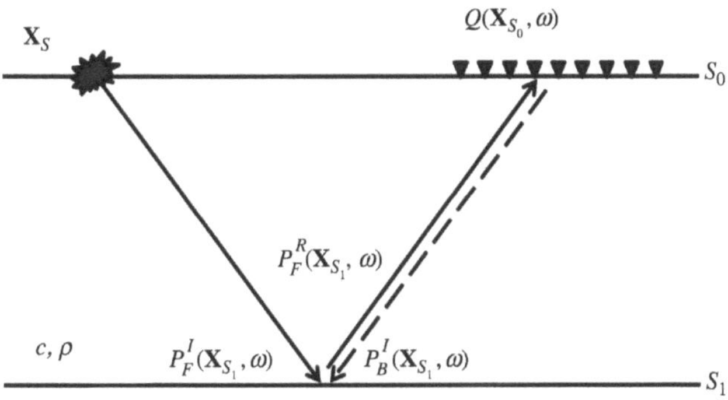

FIGURE 7.1 Sketch of a wavefield propagation path in two-layer media. The solid line denotes the forward-propagation path. The dashed line denotes the backward-propagation path.

transmission losses for wave-propagation operators under the high-frequency approximation. Then we proposed a procedure to compensate for the transmission losses in RTM. Finally we provide several numerical examples to verify the theoretical derivations and the compensation procedure in RTM.

7.2 THEORY

The layered medium is a reasonable simplification of the strata generated by geological sediments. The influence of transmission losses on the migration amplitude is significant in a layered medium. Within each layer, the medium is homogenous, which simplifies the theoretical analysis compared with that for more general heterogeneous media. We begin by considering wavefield propagation in the layered medium shown in Figure 7.1. The interfaces are caused by variations in velocity, density, or both. Our purpose is to obtain the true-amplitude image of interface S_2 with RTM. A point source is fired on the surface at the point $\mathbf{x}_S$. $P_F^I(\mathbf{x}_{S_1}, \omega)$ and $P_F^R(\mathbf{x}_{S_1}, \omega)$ are the incident and reflected wavefields at point $\mathbf{x}_{S_1}$, respectively. The reflected wavefield $P_F^R(\mathbf{x}_{S_1}, \omega)$ propagates upward along the path denoted by the solid line, and is recorded at the observation surface S_0 as $Q(\mathbf{x}_{S_0}, \omega)$. The reflection coefficient at point $\mathbf{x}_{S_1}$ is obtained by dividing the reflected wavefield by the incident wavefield (Claerbout, 1971) at this point

$$R(\mathbf{x}_{S_1}) = \int \frac{P_F^R(\mathbf{x}_{S_1}, \omega)}{P_F^I(\mathbf{x}_{S_1}, \omega)} d\omega. \tag{7.1}$$

As indicated by migration/inversion theory (Bleistein et al., 2001), if we use eq. (7.1) as an imaging condition, the migration will produce a true-amplitude output. Its peak amplitude is proportional to the ray-theoretic reflection coefficient.

To obtain the reflectivity of S_1 with RTM includes three steps: forward propagating the source wavefield, backward propagating the receiver wavefield, and applying an imaging condition. The main task of wavefield propagation is to construct the source (forward) and receiver (backward) wavefields. For isotropic RTM, if mode conversion, elastic effects, and attenuation are ignored, the source wavefield can be constructed with a two-way acoustic wave equation, whose solution is both kinematically and dynamically correct. The geometric spreading and transmission losses of the source wavefield will be incurred automatically (Haney et al., 2005). Therefore, if the exact velocity model is used in RTM, the forward-propagation operator can correctly recover the incident wavefield $P_F^I(\mathbf{x}_{S_1},\omega)$. However, the reflected wavefield $P_F^R(\mathbf{x}_{S_1},\omega)$ is difficult to be obtained directly. In RTM, the receiver wavefield is used as the reflected wavefield for imaging. The receiver wavefield is generated by backward propagation of the recorded data along the dashed line in Figure 7.1. The image at point $\mathbf{x}_{S_1}$ is given by the imaging condition (Claerbout, 1971)

$$I(\mathbf{x}_{S_1}) = \int \frac{P_B^I(\mathbf{x}_{S_1},\omega)}{P_F^I(\mathbf{x}_{S_1},\omega)} d\omega. \tag{7.2}$$

Comparing eqs. (7.1) and (7.2), it is obvious that, when the amplitudes of the receiver wavefield $P_B^I(\mathbf{x}_{S_1},\omega)$ and the reflected wavefield $P_F^R(\mathbf{x}_{S_1},\omega)$ are equal at the imaging point $\mathbf{x}_{S_1}$, RTM with the imaging condition of eq. (7.2) will produce the true-amplitude migration image of eq. (7.1). However, backward propagation of the wave equation is not the inverse of its forward propagation. The energy of the backward-propagating wave will also partition when the wavefront encounters the interface. When the media become complex, the receiver wavefield cannot be used as the reflected wavefield. We next analyze the influence of wave-propagation operator on the amplitude of RTM.

7.2.1 Wave-Propagation Operators in Reverse-Time Migration

We begin from the forward-propagation operator in RTM. We define the temporal Fourier transform of a space- and time-dependent quantity $P(\mathbf{x},t)$ as

$$P(x,\omega) = \int_{-\infty}^{\infty} P(x,t)\exp(-j\omega t)dt. \tag{7.3}$$

In frequency domain, the acoustic pressure and particle velocity in a lossless arbitrary inhomogeneous acoustic medium obey the velocity–stress equations as follows (Wapenaar and Fokkema, 2006):

$$\begin{cases} j\omega\rho\upsilon_i + \partial_i P = 0 \\ j\omega P / K + \partial_i \upsilon_i = F(\omega) \end{cases}. \tag{7.4}$$

Here υ_i is the particle velocity vector in the x_i-direction, and ∂_i denotes the partial derivate in the x_i-direction, in which Einstein's summation convention applies to repeated subscripts. ρ is the mass density of the medium. $K = \rho c^2$ is compressibility (Berkhout, 1985). $F(\omega)$ is a source term.

According to $v_i = -(j\omega\rho)^{-1}\partial_i p$, we can get the acoustic wave equation with variable density. If the media is homogeneous, we can give the spherical wave solution as (Berkhout, 1985)

$$P(r,\omega) = -j\omega\rho F(\omega)\frac{e^{-jkr}}{r} = -j\omega\rho F(\omega)\frac{e^{-j\omega r/c}}{r}, \tag{7.5}$$

and the cylindrical wave solution as

$$P(r,\omega) = -j\omega\sqrt{2\pi}\rho F(\omega)\frac{e^{-j\left(kr+\frac{\pi}{4}\right)}}{\sqrt{kr}}, \tag{7.6}$$

where $\mathbf{r} = |\mathbf{x} - \mathbf{x}_s|$ is the propagation distance.

The acoustic wave equation is often used in its second-order form, which can easily be expressed explicitly with its asymptotic expression. For variable velocity and density media, the second-order acoustic wave equation is

$$\left[\rho\nabla\cdot\frac{1}{\rho}\nabla + \frac{\omega^2}{c^2}\right]P(\mathbf{x},\omega) = -F(\omega)\delta(\mathbf{x}-\mathbf{x}_S), \tag{7.7}$$

where $P(\mathbf{x},\omega)$ is the wavefield at location $\mathbf{x}$ with frequency ω, which is the response to a source at $\mathbf{x}_S$. $F(\omega)$ is the source signal, and ρ, c are the density and velocity of the media, respectively. For forward propagation, the causal WKBJ Green's function is

$$G(\mathbf{x},\mathbf{x}_s,\omega) \sim B(\mathbf{x},\mathbf{x}_s)e^{j\omega\tau(\mathbf{x},\mathbf{x}_s)}, \tag{7.8}$$

where $G(\mathbf{x},\mathbf{x}_S,\omega)$ represents the response at $\mathbf{x}$ due to a point source at $\mathbf{x}_S$. The anticausal WKBJ Green's function of the backward propagation is

$$G^*(\mathbf{x},x_s,\omega) \sim B(x,x_s)e^{-j\omega\tau(\mathbf{x},\mathbf{x}_s)}. \tag{7.9}$$

where * denotes the complex conjugate. In eqs. (7.8) and (7.9), the phase τ is a solution of the eikonal equation, as in the constant-density case. $B(\mathbf{x},\mathbf{x}_S)$ are the WKBJ leading-order amplitudes for the variable-density wave equation, and these satisfy the variable-density transport equation (Bleistein, 1986). As indicated by Bleistein et al. (2001), $B(\mathbf{x},\mathbf{x}_S)$ are related to the WKBJ leading-order amplitudes $A(\mathbf{x},\mathbf{x}_S)$ for the constant density case by

$$B(\mathbf{x},\mathbf{x}_r) = A(\mathbf{x},\mathbf{x}_r)\sqrt{\frac{\rho(\mathbf{x})}{\rho(\mathbf{x}_r)}}. \tag{7.10}$$

7.2.2 True-Amplitude Migration Formula

Many literatures evaluated the Green's function's representation of the wavefield propagating in the single-layer media (Bleistein et al., 2001; Docherty, 1991; Zhang et al., 2003b). Figure 7.1 shows the wavefield propagation path in RTM. S_0 is observation surface. S_1 is a horizontal reflector. The solid line denotes the forward-propagation path of the source wavefield, and the dashed line denotes the backward-propagation path of the receiver wavefield. P_F^I is the incident wavefield on S_1; excited by a point source at $\mathbf{x}_S$. $Q(\mathbf{x}_{S_0}, \mathbf{x}_S, \omega)$ is the recorded data on $\mathbf{x}_{S_0}$. Using Green's function (7.8), the incident wavefield at $\mathbf{x}$ is

$$P_F(\mathbf{x},\omega) \sim A(\mathbf{x},\mathbf{x}_s)\sqrt{\frac{\rho(\mathbf{x})}{\rho(\mathbf{x}_s)}}e^{i\omega\tau(\mathbf{x},\mathbf{x}_s)}, \tag{7.11}$$

Under the high-frequency approximation, the reflected wavefield can be expressed as an integral of WKBJ Green's function along the reflector S_1 (Docherty, 1991)

$$\begin{gathered}Q(\mathbf{x}_0,\omega) \sim 2i\omega\int R(\mathbf{x})P_F(\mathbf{x},\omega)A(\mathbf{x},\mathbf{x}_0)\sqrt{\frac{\rho(\mathbf{x}_0)}{\rho(\mathbf{x})}}, \\ \mathbf{n}\cdot\nabla\tau(\mathbf{x},\mathbf{x}_0)e^{i\omega\tau(\mathbf{x},\mathbf{x}_0)}d\mathbf{x}\end{gathered} \tag{7.12}$$

where $R(\mathbf{x})$ is the reflection coefficient. Similarly, using the anticausal Green's function, we can get the backward-propagation wavefield (Docherty, 1991)

$$\begin{gathered}P_B^I(\mathbf{x},\omega) \sim 2i\omega\int Q(\mathbf{x}_0,\omega)A(\mathbf{x}_0,\mathbf{x})\sqrt{\frac{\rho(\mathbf{x})}{\rho(\mathbf{x}_0)}}, \\ \mathbf{n}_0\cdot\nabla\tau(\mathbf{x}_0,\mathbf{x})e^{i\omega\tau(\mathbf{x}_0,\mathbf{x})}d\mathbf{x}_0\end{gathered} \tag{7.13}$$

where $Q(\mathbf{x}_0,\omega)$ is the recorded data originating by a point source at $\mathbf{x}_S$. Equations (7.11) and (7.13) give the high-frequency asymptotic expressions of incident and scattering wavefields, respectively. True-amplitude RTM needs the unifying imaging principle (Claerbout, 1971), which can be described as

$$I(\mathbf{x}) = \frac{1}{2\pi}\int\frac{P_B(\mathbf{x},\omega)}{P_F(\mathbf{x},\omega)}d\omega. \tag{7.14}$$

Replacing P_F^I and P_B^I with their WKBJ approximations produces (Docherty, 1991)

$$\begin{gathered}I(\mathbf{x}) = \frac{1}{\pi}\int i\omega d\omega\int\sqrt{\frac{\rho(\mathbf{x}_s)}{\rho(\mathbf{x}_0)}}\frac{A(\mathbf{x}_0,\mathbf{x})}{A(\mathbf{x},\mathbf{x}_s)}\mathbf{n}_0\cdot\nabla\tau(\mathbf{x}_0,\mathbf{x}) \\ e^{-i\omega[\tau(\mathbf{x}_0,\mathbf{x})+\tau(\mathbf{x},\mathbf{x}_s)]}Q(\mathbf{x}_0,\omega)d\mathbf{x}_0\end{gathered}. \tag{7.15}$$

Note that under the high-frequency approximation, the common-source inversion and prestack migration are identical (Bleistein et al., 2001)

$$I(\mathbf{x}) \sim A_f R(\mathbf{x}), \tag{7.16}$$

where $R(\mathbf{x})$ is the reflection coefficient of variable-density media. $A_f = (1/2\pi)\int F(\omega)d\omega$ is an integral of $F(\omega)$ that indicates the band-limited nature of the source.

The migration-formula eq. (7.15) achieves true-amplitude image under the high-frequency approximation. It can be applied directly to Kirchhoff or beam migration. For RTM, the propagation operators have no approximation. Xu et al. (2011) replaced the asymptotic expression for the Green's function with their exact counterpart to obtain a true-amplitude RTM formula. However, according to the true-amplitude RTM theory (Xu et al. 2011; Zhang et al., 2007), even using true-amplitude propagation operator, the RTM formula is still asymptotic in terms of amplitude. Since the asymptotic expression form has an explicit amplitude item, which makes it easy to compare the amplitudes, we keep using the asymptotic expression form of Green's function in the following derivation. The conclusions achieved in this chapter are under the high-frequency approximation.

In RTM, the source wavefield is propagated forward and stored, which costs large memory and I/O operations to implement RTM on high-performance computational (HPC) platforms, such as RTM implemented on the GPU. The source wavefield can also be constructed in the reverse-time direction to match with the reverse-time propagated receiver wavefield using the previously computed forward time results as boundary values (Symes 2007; Clapp et al., 2010).

7.2.3 Wave-Propagation Operators for Reverse-Time Migration

The main tasks of forward- and backward-propagation operators are to construct the source and receiver wavefields, respectively. In RTM, the source wavefield is used as incident wavefield, whereas the backward wavefield is used as reflected wavefield. The migration-formula eq. (7.15) will provide a true-amplitude imaging, if only both the source and receiver wavefields have correct amplitude at imaging points. To obtain explicit representations for the amplitude of wavefield, we use exact model parameters in following derivations.

The forward-propagation operators in RTM are based on two-way wave equation. Since we use exact velocity model, the solution of the two-way acoustic wave equation is both kinematically and dynamically correct. If we ignore the mode conversion and elastic effects, geometric spreading and transmission losses of source wavefield will be incurred automatically, which correctly describe the amplitude behaviors of the seismic wavefield. Therefore, the forward-propagation operator can correctly recover the amplitude of the incident wavefield at imaging points without any compensation.

However, this is not true for the receiver wavefield, even the backward-propagation operator is also built with two-way wave equation. The backward propagation of the receiver wavefield is anticausal. It is not a physical process in the real world. There are few studies on theoretical analysis about the geometric spreading and transmission losses of backward-propagation operator of RTM. In this section, we first investigate geometric spreading and transmission losses of backward-propagation operator. Then, we analyze the effects of transmission losses on imaging amplitude. All of these derivations involve the effects of variable density. These derivations are the main developments of this chapter.

7.2.3.1 Geometric Spreading of Backward-Propagation Operator

Consider the sketch of the propagation path shown in Figure 7.1. The subscripts of $\mathbf{x}$ denote its space location. $P_F^I(\mathbf{x}_{S_1},\omega)$ is the incident wavefield on the reflector S_1 originated by a point source at $\mathbf{x}_S$. $P_F^R(\mathbf{x}_{S_1},\omega)$ is the reflected wavefield at the same point. In RTM, the reflected wavefield is reconstructed by backward propagation of the recorded data

$$P_B(\mathbf{x},\mathbf{x}_s,\omega) \sim 2j\omega \int_{S_0} Q(\mathbf{x}_{s_0},\mathbf{x}_s,\omega)\sqrt{\frac{\rho(\mathbf{x}_s)}{\rho(\mathbf{x}_r)}}A(\mathbf{x}_{s_0},\mathbf{x}_r) \\ \mathbf{n}\cdot\nabla\tau(\mathbf{x}_{s_0},\mathbf{x}_\mathbf{r})e^{-j\omega\tau(\mathbf{x}_{s_0},\mathbf{x}_r)}d\mathbf{x}_{s_0}, \tag{7.17}$$

where $Q(\mathbf{x}_{S_0},\mathbf{x}_S,\omega)$ is the reflected data recorded on S_0. Because of geometric spreading, the amplitudes of reflected wave decay during their forward propagation. According to eq. (7.12), the recorded wavefield can be expressed in terms of a causal Green's function

$$Q(\mathbf{x}_{s_0},\omega) \sim 2i\omega \int_{S_1} P_F^R(\mathbf{x}_{s_1},\omega)A(\mathbf{x}_{s_1},\mathbf{x}_{s_0})\sqrt{\frac{\rho(\mathbf{x}_{s_0})}{\rho(\mathbf{x}_{s_1})}}. \\ \mathbf{n}_{s_0}\cdot\nabla\tau(\mathbf{x}_{s_1},\mathbf{x}_{s_0})e^{i\omega\tau(\mathbf{x}_{s_1},\mathbf{x}_{s_0})}d\mathbf{x}_{s_1} \tag{7.18}$$

Submitting eq. (7.18) into eq. (7.17) produces

$$P_B^I(\mathbf{x},\omega) \sim (2i\omega)^2 \int_{S_0}\int_{S_1} P_F^T(\mathbf{x}_{s_1},\omega)\sqrt{\frac{\rho(\mathbf{x})}{\rho(\mathbf{x}_{s_1})}}A(\mathbf{x}_{s_1},\mathbf{x}_{s_0})A\mathbf{x}_{s_0},\mathbf{x})\mathbf{n}_{s_1} \\ \cdot\nabla\tau(\mathbf{x}_{s_1},\mathbf{x}_{s_0})\mathbf{n}_{s_0}\cdot\nabla\tau(\mathbf{x}_{s_0},\mathbf{x})e^{i\omega[\tau(\mathbf{x}_{s_1},\mathbf{x}_{s_0})-\tau(\mathbf{x}_{s_0},\mathbf{x})]}d\mathbf{x}_{s_1}d\mathbf{x}_{s_0}. \tag{7.19}$$

Applying the stationary-phase method, we prove that eq. (7.19) reduces to

$$P_B^I(\mathbf{x}_{S_1},\omega) \sim P_F^R(\mathbf{x}_{S_1},\omega). \tag{7.20}$$

Since we use WKBJ Green's function in the derivation, which is the first-order asymptotic solution of two-way wave equation, eq. (7.20) just gives a high-frequency approximation between the backward wavefield and reflected wavefield at the interface. Equation (7.20) indicates that, under the high-frequency approximation, the backward-propagation wavefield will automatically compensate the geometric spreading that occurred during the original upward propagation from the reflector to the receivers. However, it is hard to obtain a theoretical derivation such as eq. (7.20) without using WKBJ Green's function. In Section 3.1, we provide a numerical example to verify that with an acceptable accuracy eq. (7.20) can apply to two-way wave-equation propagator, which is solved with the finite-difference method.

Further, we provide a derivation of eq. (7.20). Each layer shown in Figure 7.1 is homogenous. We use a homogenous Green's function to deal with the internal wavefields of each homogenous layer. For the wavefields on the interface, we implement stationary phase analysis to find suitable expressions. As indicated by Bleistein (1984), the amplitude and phase of the Green's function in eq. (7.19) are

$$A(\mathbf{x}_r,\mathbf{x}_S)=\frac{1}{4\pi\left|\mathbf{x}_r-\mathbf{x}_S\right|},\quad \tau(\mathbf{x}_r,\mathbf{x}_S)=\frac{\left|\mathbf{x}_r-\mathbf{x}_S\right|}{c},\quad \mathbf{n}\cdot\nabla\tau(\mathbf{x}_r,\mathbf{x}_S)=\frac{\cos\theta}{c},$$

where c is the velocity of the media and θ is the incident angle. Thus, eq. (7.19) can be rewritten as

$$P_B^I(\mathbf{x}_{S_0},\omega)\sim\left(\frac{i\omega}{2\pi}\right)^2\int_{S_0}\int_{S_1}P_F^R(\mathbf{x}_{S_1},\omega)\sqrt{\frac{\rho(\mathbf{x}_{S_1})}{\rho(\mathbf{x})}}\frac{1}{\left|\mathbf{x}_{S_0}-\mathbf{x}_{S_1}\right|}\frac{1}{\left|\mathbf{x}-\mathbf{x}_{S_0}\right|}\frac{\cos\theta_{S_1}}{c}\frac{\cos\theta_{S_0}}{c}e^{i\omega\left[\frac{\left|\mathbf{x}_{S_0}-\mathbf{x}_{S_1}\right|}{c}+\frac{\left|\mathbf{x}-\mathbf{x}_{S_0}\right|}{c}\right]}d\mathbf{x}_{S_1}d\mathbf{x}_{S_0},\tag{7.21}$$

where (x, y, z) denotes the three Cartesian coordinates of $\mathbf{x}$; θ_{S_0} and θ_{S_2} are the angles between the wavefield-propagation direction and the normal direction to S_0 and S_2. According to Green's theorem, the normal directions of S_0 and S_2 are vertically upward and vertically downward, respectively.

The stationary-phase method (Bleistein, 1984) evaluates an n-dimensional oscillatory integral such as

$$I(\omega)=\int_D f(\mathbf{x})\mathrm{e}^{j\omega\phi(\mathbf{x})}\,\mathrm{d}\mathbf{x},\quad \mathbf{x}=(\mathrm{x}_1,\ldots,\mathrm{x}_\mathrm{m}).\tag{7.22}$$

For large ω, its asymptotic expansion depends on the nature of the integrand in the neighborhood of stationary phase points as

$$I(\omega)\sim\left[\frac{2}{\omega}\right]^{\frac{\mathrm{m}}{2}}\frac{f(\bar{\mathbf{x}})}{\sqrt{|\det\mathrm{A}|}}\mathrm{e}^{j\omega\phi(\bar{\mathbf{x}})+j}\frac{\pi}{4}\mathrm{sgnA},\tag{7.23}$$

where $\bar{\mathbf{x}}$ is a simple stationary point satisfying the stationary phase condition

$$\nabla\ \phi(\bar{\mathbf{x}}) = 0, \tag{7.24}$$

and

$$\det A \neq 0, A_{ik} = \left[\frac{\partial^2 \phi(\bar{\mathbf{x}})}{\partial \mathrm{x}_i \partial \mathrm{x}_k}\right], \quad i,k = 1,2,\ldots,m. \tag{7.25}$$

Here, A is the Hessian matrix for ϕ, and sgn$A$ denotes the signature of the matrix A, which is defined by the number of positive eigenvalues minus the number of negative ones.

Applying the stationary-phase method to eq. (7.21) produces

$$P_B^I(\mathbf{x},\omega) \sim \left(\frac{j\omega}{2\pi}\right)^2 \left(\frac{2\pi}{\omega}\right)^2 \frac{P_F^R(\mathbf{x}_{S_1},\omega)}{\sqrt{|\det A|}} \frac{1}{\left|\mathbf{x}_{S_0} - \mathbf{x}_{S_1}\right|} \frac{1}{\left|\mathbf{x} - \mathbf{x}_{S_0}\right|} \frac{\cos\theta_{S_1}}{c} \frac{\cos\theta_{S_0}}{c} \sqrt{\frac{\rho(\mathbf{x}_{S_1})}{\rho(\mathbf{x})}} e^{j\omega\left(\frac{\left|\mathbf{x}_{S_1} - \mathbf{x}_{S_0}\right|}{c} - \frac{\left|\mathbf{x} - \mathbf{x}_{S_0}\right|}{c}\right) + j\frac{\pi}{4}\operatorname{sgn} A}, \tag{7.26}$$

where $\mathbf{x}_{S_0}$ and $\mathbf{x}_{S_1}$ are the stationary phase points on S_1 and S_2. Next, we will determine the stationary phase points by applying the stationary phase condition of eqs. (7.24) and (7.25).

We can define the phase function in eq. (7.26) as

$$\phi\left(\mathbf{x}_{S_0}, \mathbf{x}_{S_1}\right) = \frac{\left|\mathbf{x}_{S_1} - \mathbf{x}_{S_0}\right| + \left|\mathbf{x} - \mathbf{x}_{S_0}\right|}{c}. \tag{7.27}$$

The derivatives of ϕ are

$$\frac{\partial \phi\left(\mathbf{X}_{S_0}, \mathbf{X}_{S_1}\right)}{\partial x_{S_0}} = \frac{1}{c}\left(\frac{x_{S_0} - x_{S_0}}{\left|\mathbf{X}_{S_0} - \mathbf{X}_{S_1}\right|} + \frac{x - x_{S_0}}{\left|\mathbf{X} - \mathbf{X}_{S_0}\right|}\right), \tag{7.28}$$

$$\frac{\partial \phi\left(\mathbf{X}_{S_0}, \mathbf{X}_{S_1}\right)}{\partial x_{S_1}} = \frac{1}{c} \frac{x_{S_0} - x_{S_1}}{\left|\mathbf{X}_{S_0} - \mathbf{X}_{S_1}\right|}, \tag{7.29}$$

$$\frac{\partial \phi\left(\mathbf{X}_{S_0}, \mathbf{X}_{S_1}\right)}{\partial y_{S_0}} = \frac{1}{c}\left(\frac{y_{S_0} - y_{S_1}}{\left|\mathbf{X}_{S_0} - \mathbf{X}_{S_1}\right|} + \frac{y - y_{S_0}}{\left|\mathbf{X} - \mathbf{X}_{S_0}\right|}\right), \tag{7.30}$$

$$\frac{\partial \phi\left(\mathbf{X}_{S_0}, \mathbf{X}_{S_1}\right)}{\partial y_{S_1}} = \frac{1}{c} \frac{y_{S_0} - y_{S_1}}{\left|\mathbf{X}_{S_0} - \mathbf{X}_{S_1}\right|}. \tag{7.31}$$

According to the stationary phase conditions in eqs. (7.27), (7.28), and (7.29), the stationary phase points in the x_{S_0} and x_{S_1} integrals are at the same point as x

$$x_{S_0} = x_{S_1} = x. \tag{7.32}$$

Similarly, from eqs. (7.30) and (7.31), we obtain

$$y_{S_0} = y_{S_1} = y. \tag{7.33}$$

Thus, the angle between the wavefield propagation direction and the normal directions of S_0 and S_1 are

$$\theta_{S_0} = \pi, \qquad \theta_{S_1} = 0. \tag{7.34}$$

The Hessian matrix of phase function ϕ has the following nonzero components at the stationary points

$$\frac{\partial^2 \phi(\mathbf{X}_{S_0}, \mathbf{X}_{S_1})}{\partial^2 x_{S_0}} = \frac{1}{c}\left(\frac{1}{|\mathbf{X}_{S_0} - \mathbf{X}_{S_1}|} + \frac{1}{|\mathbf{X} - \mathbf{X}_{S_0}|}\right), \tag{7.35}$$

$$\frac{\partial^2 \phi(\mathbf{X}_{S_0}, \mathbf{X}_{S_1})}{\partial^2 x_{S_1}} = \frac{1}{c_1|\mathbf{X}_{S_0} - \mathbf{X}_{S_1}|}, \tag{7.36}$$

$$\frac{\partial^2 \phi(\mathbf{X}_{S_0}, \mathbf{X}_{S_1})}{\partial^2 y_{S_0}} = \frac{1}{c}\left(\frac{1}{|\mathbf{X}_{S_0} - \mathbf{X}_{S_1}|} + \frac{1}{|\mathbf{X} - \mathbf{X}_{S_0}|}\right), \tag{7.37}$$

$$\frac{\partial^2 \phi(\mathbf{X}_{S_0}, \mathbf{X}_{S_1})}{\partial^2 y_{S_1}} = \frac{1}{c}\frac{1}{|\mathbf{X}_{S_0} - \mathbf{X}_{S_1}|}, \tag{7.38}$$

$$\frac{\partial^2 \phi(\mathbf{X}_{S_0}, \mathbf{X}_{S_1})}{\partial x_{S_0} \partial_{S_1}} = -\frac{1}{c}\frac{1}{|\mathbf{X}_{S_0} - \mathbf{X}_{S_1}|}, \tag{7.39}$$

and

$$\frac{\partial^2 \phi(\mathbf{X}_{S_0}, \mathbf{X}_{S_1})}{\partial y_{S_0} \partial_{y_{S_1}}} = -\frac{1}{c}\frac{1}{|\mathbf{X}_{S_0} - \mathbf{X}_{S_1}|}. \tag{7.40}$$

Therefore, at the stationary points, the Hessian determinant is

$$\sqrt{|\det A|} = \frac{1}{c^2}\frac{1}{|\mathbf{x}_{s_0} - \mathbf{x}_{s_1}|}\frac{1}{|\mathbf{x} - \mathbf{x}_{s_0}|}. \tag{7.41}$$

The signature of the Hessian can be determined by analyzing its eigenvalues. The Hessian matrix A is 4×4 and can be decomposed into four 2×2 submatrices

$$A = \begin{pmatrix} A'_{11} & A'_{12} \\ A'_{21} & A'_{22} \end{pmatrix}, \tag{7.42}$$

where

$$A'_{11} = \begin{pmatrix} \dfrac{\partial^2 \phi(\mathbf{x}_{s_0}, \mathbf{x}_{s_1})}{\partial^2 x_{s_0}} & \dfrac{\partial^2 \phi(\mathbf{x}_{s_0}, \mathbf{x}_{s_1})}{\partial x_{s_0} \partial x_{s_1}} \\ \dfrac{\partial^2 \phi(\mathbf{x}_{s_0}, \mathbf{x}_{s_1})}{\partial x_{s_0} \partial x_{s_1}} & \dfrac{\partial^2 \phi(\mathbf{x}_{s_0}, \mathbf{x}_{s_1})}{\partial^2 x_{s_1}} \end{pmatrix}, \tag{7.43}$$

$$A'_{22} = \begin{pmatrix} \dfrac{\partial^2 \phi(\mathbf{x}_{s_0}, \mathbf{x}_{s_1})}{\partial^2 y_{s_0}} & \dfrac{\partial^2 \phi(\mathbf{x}_{s_0}, \mathbf{x}_{s_1})}{\partial y_{s_0} \partial y_{s_1}} \\ \dfrac{\partial^2 \phi(\mathbf{x}_{s_0}, \mathbf{x}_{s_1})}{\partial y_{s_0} \partial y_{s_1}} & \dfrac{\partial^2 \phi(\mathbf{x}_{s_0}, \mathbf{x}_{s_1})}{\partial^2 y_{s_1}} \end{pmatrix}, \tag{7.44}$$

$$A'_{12} = A'_{21} = 0. \tag{7.45}$$

Because of its particular structure, the problem reduces to the evaluation of the 2×2 submatrices in eqs. (7.43) and (7.44). From eqs. (7.35) to (7.40), we know $A'_{11} = A'_{22}$. Therefore, the eigenvalues of matrix A are in pairs. Considering the fact that

$$|A'_{11}| = -\frac{1}{C^2} \frac{1}{|\mathbf{x}_{s_0} - \mathbf{x}_{s_1}|} \frac{1}{|\mathbf{x} - \mathbf{x}_{s_0}|} < 0, \tag{7.46}$$

the signs of A_{11}'s two eigenvalues are the same. Thus, the difference in the number of positive and negative eigenvalues in A is

$$\text{sgn } A = 0. \tag{7.47}$$

Substituting eqs. (7.34), (7.41), and (7.47) into (7.26), and using $\mathbf{x}_{S_1}$ and $\mathbf{x}_{S_2}$ to denote the stationary phase points to simplify the notation, we have eq. (7.20).

7.2.3.2 Transmission Losses of Backward-Propagation Operator

Because the backward-propagation operator is built based on two-way wave equation, energy of the backward-propagating wave will also partition when

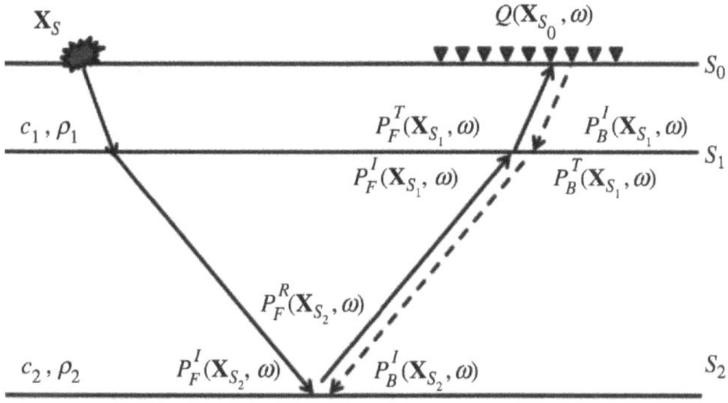

FIGURE 7.2 Sketch of a wavefield propagation path in multilayer media. The solid line denotes the forward-propagation path. The dashed line denotes the backward-propagation path.

the wavefront encounters the interface. When media is complex, it is hard to obtain the exact reflected wavefield at an image point because of transmission losses. In this section, we will derive an RTM formula affected by transmission losses.

To simplify the derivation, we consider wavefield propagation in the layered medium shown in Figure 7.2. The interfaces are caused by variations in velocity, density, or both. Our purpose is to obtain the true-amplitude image of interface S_2 with RTM. A point source is fired on the surface at the point $\mathbf{x}_S$. $P_F^I(\mathbf{x}_{S_2},\omega)$ and $P_F^R(\mathbf{x}_{S_2},\omega)$ are the incident and reflected wavefields at point $\mathbf{x}_{S_2}$, respectively. The reflected wavefield $P_F^R(\mathbf{x}_{S_2},\omega)$ propagates upward along the path denoted by the solid line, and is recorded at the observation surface S_0 as $Q(\mathbf{x}_{S_0},\omega)$. The conclusions are still available for complex media. As shown in Figure 7.2, the reflected source wavefield $P_F^R(\mathbf{x}_{S_2},\omega)$ propagates along the solid line and becomes incident on the interface S_1 from the lower medium. According to Green's theorem, the incident wavefield on S_1 is (Bleistein et al., 2001; Docherty, 1991)

$$P_F^I(\mathbf{x}_{S_1},\omega) \sim 2j\omega\int_{S_2} P_F^R(\mathbf{x}_{S_2},\omega)A(\mathbf{x}_{S_2},\mathbf{x}_{S_1})\sqrt{\frac{\rho(\mathbf{x}_{S_1})}{\rho(\mathbf{x}_{S_2})}}. \\ \mathrm{n}_{S_2}\cdot\nabla\tau(\mathbf{x}_{S_2},\mathbf{x}_{S_1})e^{j\omega\tau(\mathbf{x}_{S_2},\mathbf{x}_{S_1})}d\mathbf{x}_{S_2} \tag{7.48}$$

where $\mathbf{n}_{S_2}$ is the unit normal to S_2. When the upward-traveling wavefield hits surface S_1, the wave energy will partition at the interface. The transmitted source wavefield on surface S_1 is

$$P_F^T\left(\mathbf{x}_{S_1},\omega\right) = \mathrm{T}_\rho\left(\mathbf{x}_{S_1}\right)P_F^I\left(\mathbf{x}_{S_1},\omega\right), \tag{7.49}$$

where $\mathrm{T}_\rho(\mathbf{x}_{S_1})$ is the transmission coefficient for the wavefield incident from the lower medium into the upper medium at point $\mathbf{x}_{S_1}$. $\mathrm{T}_\rho(\mathbf{x}_{S_1})$ depends on density, velocity, and the angle of incidence. Similar to the derivation of eq. (7.48), the recorded wavefield on S_0 is

$$Q\left(\mathbf{x}_{S_0},\omega\right) \sim 2j\omega\int_{S_1} P_F^T\left(\mathbf{x}_{S_1},\omega\right) A\left(\mathbf{x}_{S_1},\mathbf{x}_{S_0}\right)\sqrt{\frac{\rho\left(\mathbf{x}_{S_0}\right)}{\rho\left(\mathbf{x}_{S_1}\right)}}. \qquad (7.50)$$
$$\mathbf{n}_{S_1}\cdot\nabla\omega\left(\mathbf{x}_{S_1},\mathbf{x}_{S_0}\right)e^{j\omega r\left(\mathbf{x}_{S_1},\mathbf{x}_{S_0}\right)}d\mathbf{x}_{S_1}$$

Substituting eqs. (7.48) and (7.49) into eq. (7.50) produces

$$Q(\mathbf{x}_{S_0},\omega) \sim (2j\omega)^2 \int_{S_1}\int_{S_2} T_\rho(\mathbf{x}_{S_1})P_F^R(\mathbf{x}_{S_2},\omega)\sqrt{\frac{\rho(\mathbf{x}_{S_0})}{\rho(\mathbf{x}_{S_2})}}. \qquad (7.51)$$
$$A(\mathbf{x}_{S_1},\mathbf{x}_{S_0})A(\mathbf{x}_{S_2},\mathbf{x}_{S_1})\mathbf{n}_{S_1}\cdot\nabla\tau(\mathbf{x}_{S_1},\mathbf{x}_{S_0})$$
$$\mathbf{n}_{S_2}\cdot\nabla\tau(\mathbf{x}_{S_2},\mathbf{x}_{S_1})e^{j\omega[\tau(\mathbf{x}_{S_1},\mathbf{x}_{S_0})+\tau(\mathbf{x}_{S_2},\mathbf{x}_{S_1})]}d\mathbf{x}_{S_2}d\mathbf{x}_{S_1}$$

From eq. (7.51), it is difficult to estimate the amplitude of the recorded data $Q(\mathbf{x}_{S_0},\omega)$. Applying the stationary phase principle to eq. (7.51) gives an asymptotic approximation of the recorded data at the stationary phase points

$$Q\left(\mathbf{x}_{s_0},\omega\right) \sim T_\rho\left(\mathbf{x}_{s_1}\right)P_F^R\left(\mathbf{x}_{s_2},\omega\right)\sqrt{\frac{\rho\left(\mathbf{x}_{s_0}\right)}{\rho\left(\mathbf{x}_{s_2}\right)}}e^{j\omega\left[\frac{\left|\mathbf{x}_{s_1}-\mathbf{x}_{s0}\right|}{C_1}+\frac{\left|\mathbf{x}_{s_2}-\mathbf{x}_{s_1}\right|}{C_2}\right]}. \qquad (7.52)$$

To build the receiver wavefield, $Q(\mathbf{x}_{S_0},\omega)$ is backward propagated along the dashed line in Figure 7.2.

Further, we derive the Green's function representation of the backward-propagated wavefields. Using the anticausal Green's function, the incident receiver wavefield on S_1 can be represented as (Docherty, 1991)

$$P_B^I\left(\mathbf{x}_{s_1},\omega\right) \sim 2j\omega\int_{S_0} Q\left(\mathbf{x}_{s_0},\omega\right)A\left(\mathbf{x}_{s_0},\mathbf{x}_{s_1}\right)\sqrt{\frac{\rho\left(\mathbf{x}_s\right)}{\rho\left(\mathbf{x}_r\right)}}, \qquad (7.53)$$
$$\mathbf{n}_{s_0}\cdot\nabla\tau\left(\mathbf{x}_{s_0},\mathbf{x}_{s_1}\right)e^{-j\omega\tau\left(\mathbf{x}_{s_0},\mathbf{x}_{s_1}\right)}d\mathbf{x}_{s_0}$$

where $\mathbf{n}_{S_0}$ is the unit normal to S_0. The transmitted receiver wavefield on S_1 is

$$P_B^T\left(\mathbf{x}_{S_1},\omega\right) = T_\rho'\left(\mathbf{x}_{S_1}\right)P_B^T\left(\mathbf{x}_{S_1},\omega\right), \qquad (7.54)$$

where $T_\rho'(\mathbf{x}_{S_1})$ is transmission coefficient for the wavefield incident from the upper medium into the lower medium at point $\mathbf{x}_{S_1}$. Similar to the derivation of eq. (7.18), the incident receiver wavefield on S_2 is

$$P_B^I\left(\mathbf{x}_{S_2},\omega\right)\sim 2j\omega\int_{S_1} P_B^T\left(\mathbf{x}_{S_1},\omega\right)A\left(\mathbf{x}_{S_1},\mathbf{x}_{S_2}\right)\sqrt{\frac{\rho\left(\mathbf{x}_{S_2}\right)}{\rho\left(\mathbf{x}_{S_1}\right)}}\cdot \\ \mathbf{n}_{S_1}\cdot\nabla\tau\left(\mathbf{x}_{S_1},\mathbf{x}_{S_2}\right)e^{-j\omega\tau\left(\mathbf{x}_{S_1},\mathbf{x}_{S_2}\right)}d\mathbf{x}_{S_1} \tag{7.55}$$

Submitting eqs. (7.53)–(7.54) into (7.55), similar to the derivation from (7.19) to (7.20), we obtain

$$P_B^I(\mathbf{x}_{S_2},\omega)\sim(2j\omega)^2\int_{S_1}\int_{S_0} T_\rho'(\mathbf{x}_{S_1})Q(\mathbf{x}_{S_0},\omega)\sqrt{\frac{\rho(\mathbf{x}_{S_2})}{\rho(\mathbf{x}_{S_0})}} \\ A(\mathbf{x}_{S_0},\mathbf{x}_{S_1})A(\mathbf{x}_{S_1},\mathbf{x}_{S_2})\mathbf{n}_{S_0}\cdot\nabla\tau(\mathbf{x}_{S_0},\mathbf{x}_{S_1}) \\ \mathbf{n}_{S_1}\cdot\nabla\tau(\mathbf{x}_{S_1},\mathbf{x}_{S_2})e^{j\omega[\tau(\mathbf{x}_{S_0},\mathbf{x}_{S_1})+\tau(\mathbf{x}_{S_1},\mathbf{x}_{S_2})]}d\mathbf{x}_{S_0}d\mathbf{x}_{S_1} \tag{7.56}$$

Following the derivation of eq. (7.52), we can also apply the stationary phase principle in eq. (7.56) and obtain an asymptotic approximation at the stationary phase points

$$P_B^I\left(\mathbf{x}_{S_2},\omega\right)\sim T_\rho'\left(\mathbf{x}_{S_1}\right)Q\left(\mathbf{x}_{S_0},\omega\right)\sqrt{\frac{\rho\left(\mathbf{x}_{S_2}\right)}{\rho\left(\mathbf{x}_{S_0}\right)}}e^{-j\omega\left[\frac{\left|\mathbf{x}_{S_0}-\mathbf{x}_{S_1}\right|}{c_1}+\frac{\left|\mathbf{x}_{S_1}-\mathbf{x}_{S_2}\right|}{c_2}\right]}. \tag{7.57}$$

Substituting eq. (7.52) into eq. (7.57) produces

$$P_B^I\left(\mathbf{x}_{S_2},\omega\right)\sim T_\rho\left(\mathbf{x}_{S_1}\right)T_\rho'\left(\mathbf{x}_{S_1}\right)P_F^R\left(\mathbf{x}_{S_2},\omega\right). \tag{7.58}$$

This approximation indicates that the amplitude of the receiver wavefield $P_B^I(\mathbf{x}_{S_2},\omega)$ on $\mathbf{x}_{S_2}$ is not equal to the amplitude of the reflected wavefield $P_F^R(\mathbf{x}_{S_2},\omega)$ because of the transmission losses caused by reflector S_1. Substituting eqs. (7.58) into the imaging condition of eq. (7.2), we have

$$I\left(\mathbf{x}_{S_2}\right)\sim\int\frac{T_\rho\left(\mathbf{x}_{S_1}\right)T_\rho'\left(\mathbf{x}_{S_1}\right)P_F^R\left(\mathbf{x}_{S_2},\omega\right)}{P_F^I\left(\mathbf{x}_{S_2},\omega\right)}d\omega. \tag{7.59}$$

This is not a true-amplitude image. The transmission losses affect the amplitudes of RTM.

7.2.4 Compensation of Transmission Losses for the True-Amplitude Reverse-Time Migration

The transmission losses have effects on the amplitudes of RTM and prevent a true-amplitude image. For image points on S_2 in Figure 7.2, considering imaging condition (7.14), and eqs. (7.16) and (7.56), we conclude that the RTM image's amplitude has an asymptotic approximation

$$I\left(\mathbf{x}_{S_2}\right) \sim T_\rho T_\rho' \, R\left(\mathbf{x}_{S_2}\right) A_f. \tag{7.60}$$

A_f is an integral that indicates the band-limited nature of the source. T_ρ, T_ρ', and R_ρ are the density-variable transmissions and reflection coefficient, respectively. Equation (7.60) indicates that the amplitudes of RTM results are not equal to reflection coefficients because of transmission losses. We need to compensate the transmission losses to preserve amplitude. According to eq. (7.58), we can compensate transmission losses with

$$\overline{P}_B^I(\mathbf{x},\omega) = P_B^I(\mathbf{x},\omega) / T(\mathbf{x}), \tag{7.61}$$

where $T(\mathbf{x}) = T_\rho(\mathbf{x}) T_\rho'(\mathbf{x})$ is defined as a transmission compensation factor. $\overline{P}_B^I(\mathbf{x},\omega)$ compensates for the two-way transmission losses during wave propagation. Thus, at point $\mathbf{x}_{S_2}$, the WKBJ leading-order amplitudes of $\overline{P}_B^I$ are equal to those of P_F^R

$$\overline{P}_B^I\left(\mathbf{x}_{S_2},\omega\right) \sim P_F^R\left(\mathbf{x}_{S_2},\omega\right). \tag{7.62}$$

After compensating for the transmission losses, the new imaging condition is given by

$$I_C(\mathbf{x}_{S_2}) = \int \frac{\overline{P}_B^I(\mathbf{x}_{S_2},\omega)}{P_F^I(\mathbf{x}_{S_2},\omega)}. \tag{7.63}$$

According to eq. (7.1), if we ignore other propagation effects, compensation for transmission losses in the receiver wavefield will generate a true-amplitude image for RTM, that is

$$I_C\left(\mathbf{x}_{S_2}\right) \sim R\left(\mathbf{x}_{S_2}\right). \tag{7.64}$$

The previous derivations prove that if we ignore other propagation effects, compensation of transmission losses with plane-wave transmission coefficients in receiver wavefield will generate a true-amplitude image for RTM. For the derivation, we use the exact model. In practice, a smooth model is often used for migration. Deng and McMechan (2007) proposed a more practical procedure to compensate transmission losses for the smooth model. Our derivations support the method of Deng and McMechan (2007) in theory.

It should be mentioned that we obtain eq. (7.60) under the high-frequency approximation, which requires that the wavelength L should be larger than the size of imaging object Λ. According to Rayleigh's principle, they should satisfy

$$L \geq \Lambda / 4. \tag{7.65}$$

When this prerequisite is broken, the image will be distorted. For example, the diffraction wave will deteriorate the migration image. Under this circumstance, compensation of transmission losses according to eq. (7.60) will fail.

To verify the above theoretical derivation about transmission compensation, we use angle-dependent plane-wave reflection coefficients to calculate the transmission factor T_f. In variable-density medium, the angle-dependent transmission coefficient T_ρ in eq. (7.49) is (Červený, 2005)

$$T_\rho(\mathbf{x}) = \frac{2\rho_1(\mathbf{x})c_1(\mathbf{x})\cos\theta_2}{\rho_1(\mathbf{x})c_1(\mathbf{x})\cos\theta_2 + \rho_2(\mathbf{x})c_2(\mathbf{x})\cos\theta_1}, \tag{7.66}$$

where v_1 and v_2 are velocities of the upper and lower medium, respectively. ρ_1 and ρ_2 are densities of the upper and lower medium, respectively. θ_1 is the incident angle in the upper medium and θ_2 is the transmission angle in the lower medium. And the transmission coefficient T'_ρ for pressure wave, which is incident from the lower medium into the upper medium, is

$$T'_\rho(\mathbf{x}) = \frac{2\rho_2(\mathbf{x})c_2(\mathbf{x})\cos\theta_1}{\rho_1(\mathbf{x})c_1(\mathbf{x})\cos\theta_2 + \rho_2(\mathbf{x})c_2(\mathbf{x})\cos\theta_1}. \tag{7.67}$$

The angle-dependent reflection coefficients for a pressure wave is

$$R_\rho(\mathbf{x}) = \frac{\rho_2(\mathbf{x})c_2(\mathbf{x})\cos\theta_1 - \rho_1(\mathbf{x})c_1(\mathbf{x})\cos\theta_2}{\rho_1(\mathbf{x})c_1(\mathbf{x})\cos\theta_2 + \rho_2(\mathbf{x})c_2(\mathbf{x})\cos\theta_1}. \tag{7.68}$$

From eqs. (7.66)–(7.68), we can calculate the transmission-compensation factor with

$$T(\mathbf{x}) = T_\rho(\mathbf{x})T'_\rho(\mathbf{x}) = 1 - R^2_\rho(\mathbf{x}). \tag{7.69}$$

Theoretically, once we obtain the angle-dependent transmission coefficients or parameters of velocity and density near the reflectors, we can compensate the transmission losses accurately.

To determine the incident angle in eq. (7.68), we need to determine the propagation direction and the dip angle. We can calculate the receiver wavefield-propagating direction with its Poynting vector at each point on the computational grid. The Poynting vector can be calculated by (Yoon et al., 2004)

$$\begin{cases} S_x = Pv_x \\ S_z = Pv_z \end{cases}, \tag{7.70}$$

where P represents the pressure, v_x and v_z are components of the particle velocity. The incident angle can be determined by

$$\cos\alpha = \frac{S_z}{\sqrt{s_x^2 + s_z^2}}. \tag{7.71}$$

There are several methods to calculate the reflector dip angle. Here we use plane-wave destruction filter (Fomel, 2002). The incident angle in eq. (7.68) can be calculated from the propagation angle α and the reflector dip angle β as (Zhang and McMechan, 2011b)

$$\theta_1 = \alpha - \beta. \tag{7.72}$$

According to Snell's law, the transmission angle θ_2 in eq. (7.68) is

$$\sin\theta_2 = \frac{v_2}{v_1}\sin\theta_1. \tag{7.73}$$

Calculation of eq. (7.34) needs to know the velocity ratio at each point. For consistency with the theoretical derivations, we use an exact model to compensate for the transmission losses. With this model, it is easy to obtain the velocity ratio directly. In practice, as a macromodel is often used for migration, the velocity ratio is not easy to be obtained directly. Deng and McMechan (2007) proposed a two-pass procedure to estimate the velocity ratio in a macromodel. To process real data, we need the method of Deng and McMechan (2007) to provide a reasonable model. With this angle information and medium parameters, we can obtain the angle-dependent reflection coefficients according to eq. (7.68). Once we have the angle-dependent reflection coefficients, the transmission compensation factor can be calculated by eq. (7.69). Then, the transmission losses of the receiver wavefield can be compensated using eq. (7.61).

It should be mentioned that the aforementioned theoretical derivations and the corresponding procedure to compensate transmission losses are proposed for common-shot RTM rather than angle-gather RTM. According to the true-amplitude angle-gather RTM theory (Zhang et al., 2007, 2012), a cross- correlation imaging condition is needed to achieve the true-amplitude angle-gather image. Thus, it needs to make corresponding modifications in the derivations and procedure before applying them on the true-amplitude angle-gather RTM.

7.3 NUMERICAL EXAMPLES

As indicated by eq. (7.69), the transmission compensation is normally a second-order term of the reflectivity. Generally, reflectivity is quite small. In most cases, RTM can achieve a good image even if the transmission losses are not compensated. The main purpose of compensating transmission losses is to achieve an image with more accurate amplitudes, or to obtain a better propagation operator, which can be used for inversion. In this section, we will use several numerical examples to show how to improve amplitude of common-shot RTM by compensating transmission losses.

First, we use a simple numerical example to verify that eq. (7.20) can apply for the two-way wave propagation operator. Then we use a three-layer model to evaluate the influence of transmission losses on amplitude of RTM. Finally, we move to a numerical test on a part of modified Marmousi2 model (Martin et al., 2002). We use the method presented by Costa et al. (2009) to remove the low wavenumber noises in RTM. This method has less effect on amplitude compared to Laplacian filter (Zhang and Sun, 2008).

7.3.1 Applying Numerical Verification of Eq. (7.20) to Two-Way Wave Propagation Operator

We obtain eq. (7.20) under the high-frequency approximation. In practice, RTM often uses a two-way propagator, which solves the two-way wave equation with finite-difference method. In this section, we use a simple numerical example to verify that without high-frequency approximation, eq. (7.14) can apply to the two-way propagation operator. Figure 7.3A shows the reflected source wavefield from a reflector at the depth of 1800 m. Figure 7.3B shows the backward-propagated wavefield at the same time of Figure 7.3A. Figure 7.3C shows the differences between Figure 7.3A and B. The maximum amplitudes of source wavefield and receiver wavefield are 25.3 and 24.8, respectively. The maximum

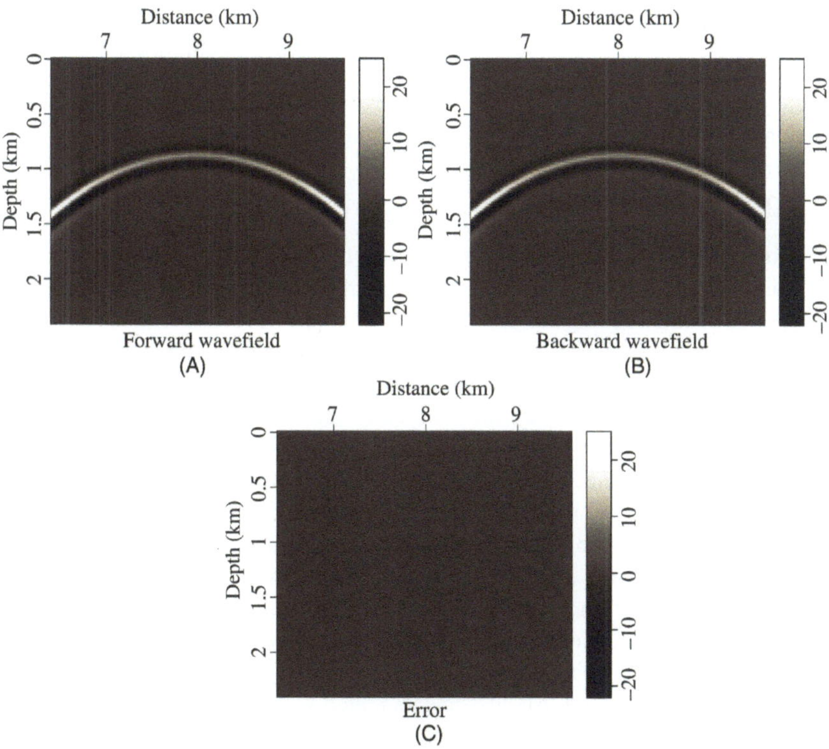

FIGURE 7.3 Numerical verification applying eq. (7.14) to two-way wave equation: (A) forward wavefield, (B) backward wavefield, (C) error between A and B.

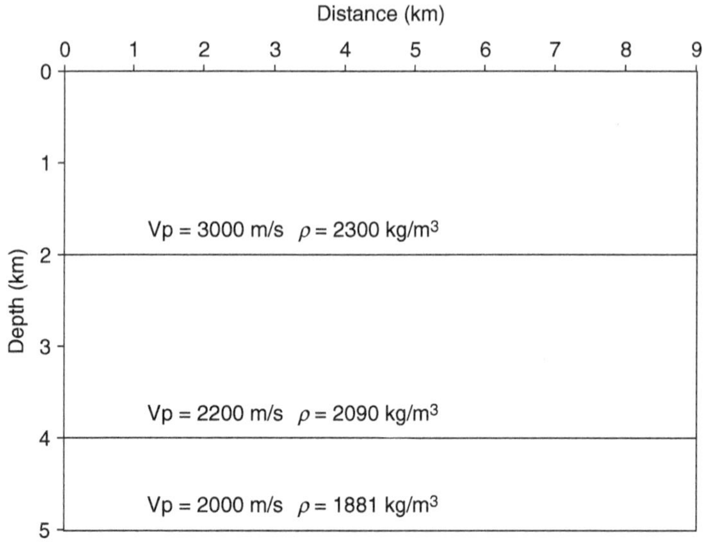

FIGURE 7.4 Model with horizontal reflectors.

amplitude of error in Figure 7.3C is 1.1. The relative errors between the forward and backward wavefield are smaller than 5%. Therefore, with an acceptable accuracy, the reflected wavefield can be approximated by the backward wavefield, which is generated by the two-way wave-equation-propagation operator.

7.3.2 2D Horizontal Reflector Model

The horizontal reflector model is shown in Figure 7.4. The acoustic velocities of each layers are 3000, 2200, and 2000 m/s, respectively. And the densities are 2300, 2090, and 1881 kg/m^3, respectively. Thus the vertical-incident reflection coefficient is 0.2 for the first reflector and 0.1 for the second reflector.

The shot location is at a horizontal distance of 4.5 km. Following the steps introduced in the previous sections, we calculate the incident angle, reflection coefficients, and transmission compensation factor $T(\mathbf{x})$. To illustrate the accuracy of the calculated values, we also obtain the theoretical incident angle, reflection coefficients, and transmission compensation factor according to the geometrical relationship and medium parameters of this model. In Figure 7.5, we compare the calculated incident angle with the theoretical values. Figure 7.6A compares the reflection coefficients obtained by the calculated incident angle with the theoretical incident angle, and Figure 7.6B compares the transmission-compensation factor $T(\mathbf{x})$ obtained by the calculated incident angle with the theoretical incident angle. All coefficients match well except at the edge, where errors increase because of edge effects.

Figure 7.7 compares the results of RTM with and without transmission compensation in the common-shot domain. According to eq. (7.16), the outputs

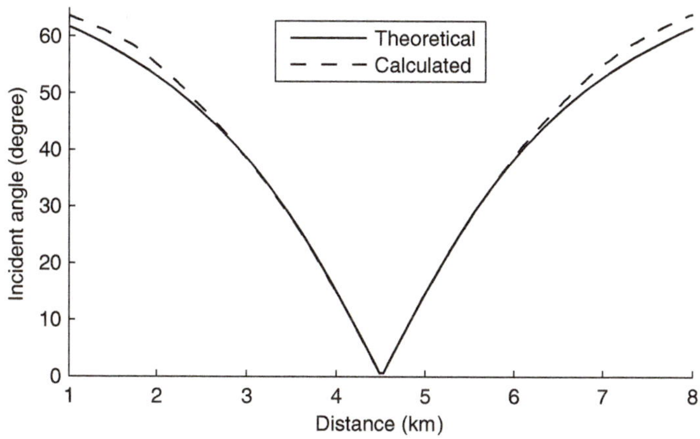

FIGURE 7.5 Comparison of the estimated incident angle and the theoretical values along the first reflector.

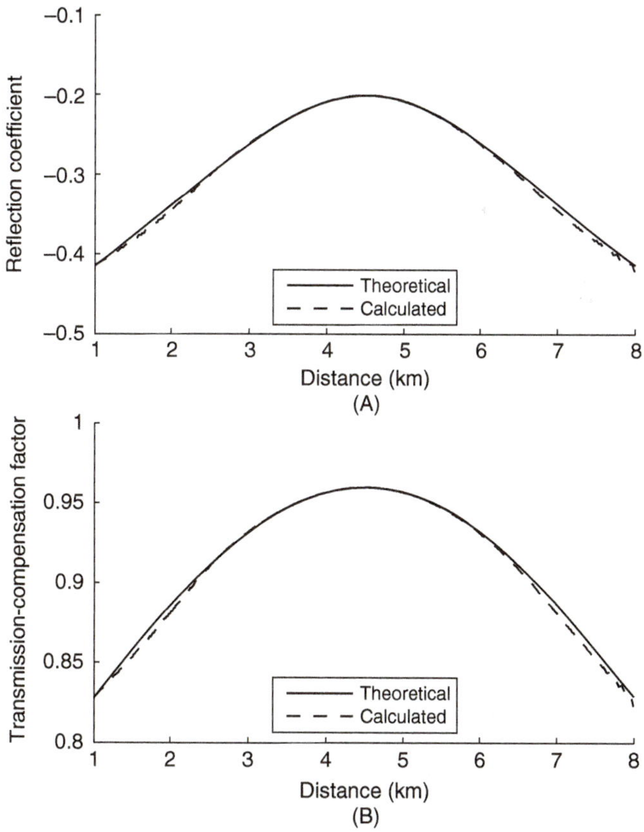

FIGURE 7.6 (A) Comparison of the theoretical and calculated values of the reflection coefficients along the first reflector. (B) Comparison of the theoretical and calculated values of the transmission compensation factor $T(\mathbf{x})$ along the first reflector.

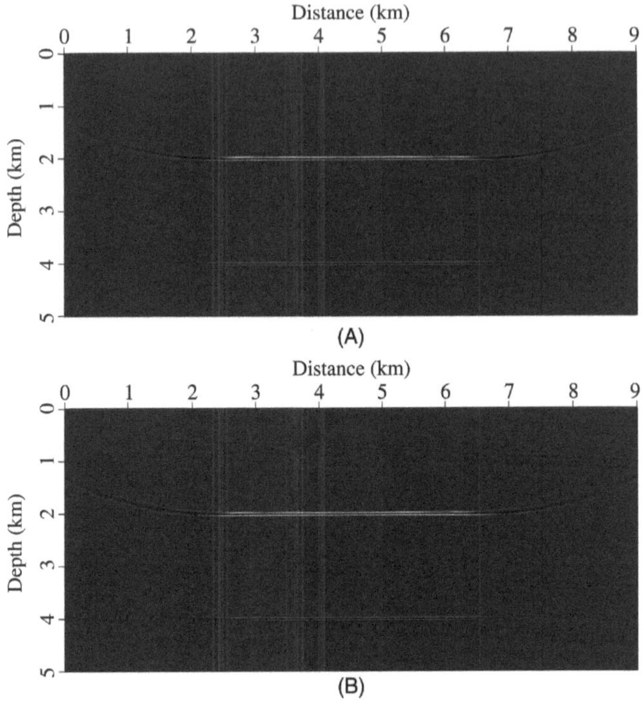

FIGURE 7.7 RTM single-shot results: (A) RTM without transmission compensation. (B) RTM with transmission compensation.

of true-amplitude migration are proportional to the reflection coefficients. To verify this, we pick out the maximum amplitudes along the reflectors from the single-shot migrated results of RTM with and without transmission compensation (see Figure 7.8). Figure 7.8A shows the comparison of maximum amplitudes along the first reflector. Because there is no transmission loss above the first reflector and the geometric spreading is compensated automatically during the propagation of the receiver wavefield, the maximum amplitudes of migration results coincide with the theoretical angle-dependent reflection coefficients, except at two edges (because of the edge effects). The comparison of maximum amplitudes along the second reflector (Figure 7.8B) indicates that the lateral changes of the amplitude of RTM with transmission compensation match better with the correct reflection coefficients than RTM without transmission compensation.

7.3.3 Marmousi2 Model

To illustrate that the transmission-compensation method can improve the imaging quality of complicated media, we use a part of the modified Marmousi2 model (Martin et al., 2002) for RTM. The velocity and density of the model

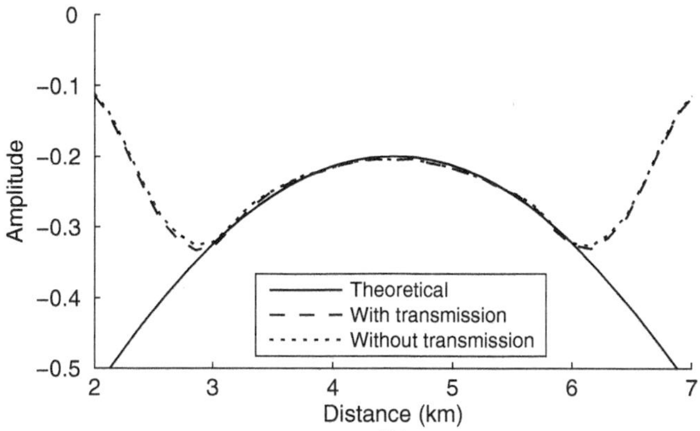

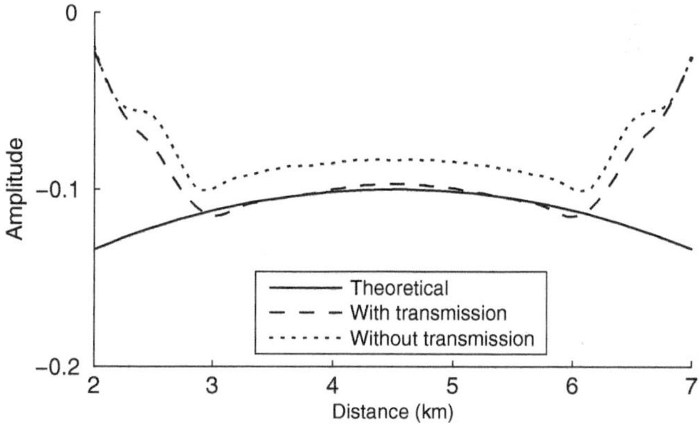

FIGURE 7.8 Comparison of the reflection coefficients along (A) the first reflector and (B) the second reflector in Figure 7.5 with the theoretical values. Correction of geometric spreading is automatically included in all curves.

are shown in Figure 7.9. The section is at the right side of the entire model. We remove the overlying water layer and several top layers slowly with vertical velocity contrast, take every eighth samples in both x- and z-directions, and then redefine grid increments as $dx = dz = 10$ m. We use an acoustic finite-difference code to generate the synthetic seismic data for 51 shots from location 0 to 5000 m with a shot spacing of 100 m. Then, we perform both RTM with and without transmission compensation on those synthetic data in the common-shot domain with true velocities.

Figure 7.10 shows the dip angle of the reflector obtained by applying PWD filter (Fomel, 2002) to traditional RTM imaging (in Figure 7.13A). Figure 7.11 shows how the propagation angle is obtained. Figure 7.11A and B are the snapshots of x and z components of Poynting vector of receiver wavefield of single shot with shot position at horizontal position 2.5 km at the time of 1.3 s. Figure 7.11C shows the propagation angle of the receiver wavefield at the same

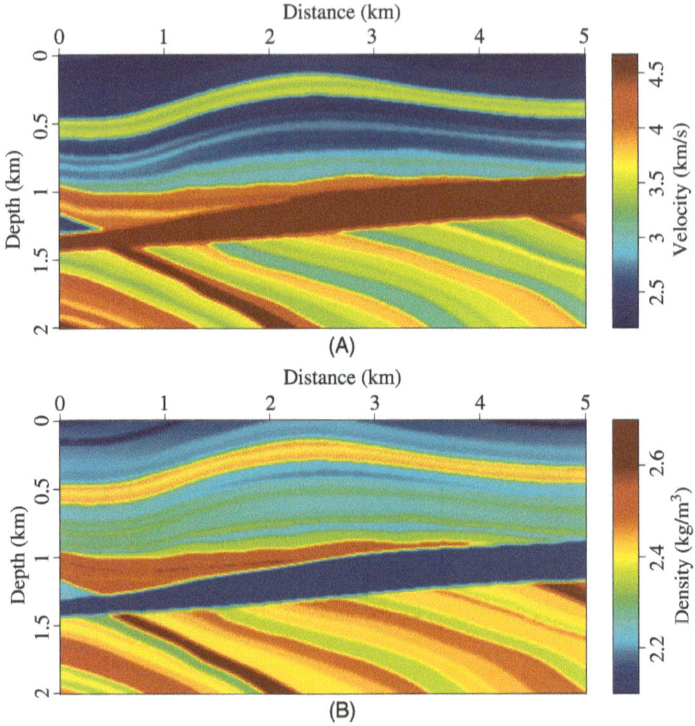

FIGURE 7.9 Portion of the Marmousi2 model (Martin et al., 2002): (A) velocity and (B) density model.

time. It indicates that the method that is used to determine propagation angle works for complex medium.

In Figure 7.12, we show receiver illumination map of single shot with shot position at horizontal position 2.5 km. The illumination map in Figure 7.12A is obtained from the traditional backward-propagation operator without transmission

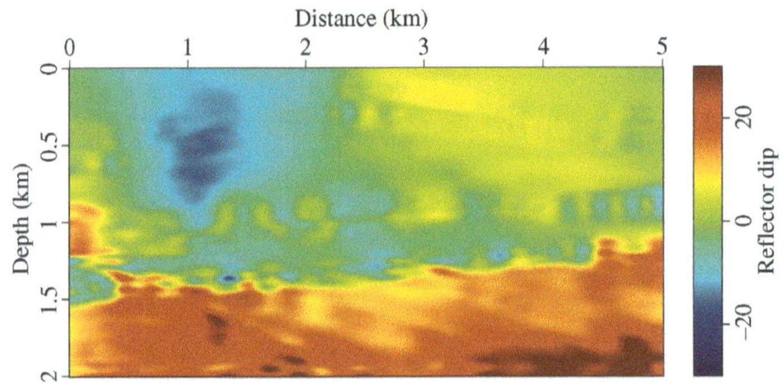

FIGURE 7.10 Reflector dip angle of the model shown in Figure 7.9.

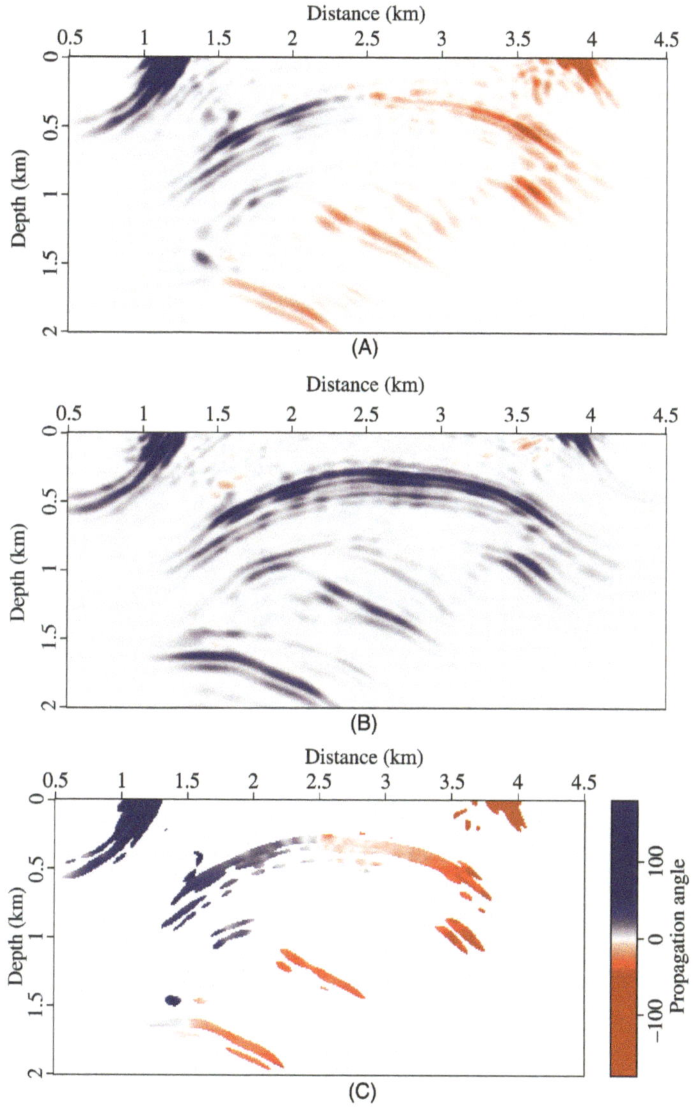

FIGURE 7.11 (A) The x-component and (B) the z-component of the receiver wavefields' Poynting vectors at time 1.3 s after one shot. (C) Corresponding propagation angle of the receiver wavefield.

compensation. The amplitude map is $\sum_t \left|P_U^2(\mathbf{x},t)\right|$. Figure 7.12B is the receiver illumination map of $\sum_t \left|\bar{P}_U^2(\mathbf{x},t)\right|$ calculated by the new backward-propagation operator with transmission compensation (eq. (7.61)). Comparing Figure 7.12A with B, we see that the receiver illumination with transmission compensation has wider illumination range and stronger illumination in deep layers.

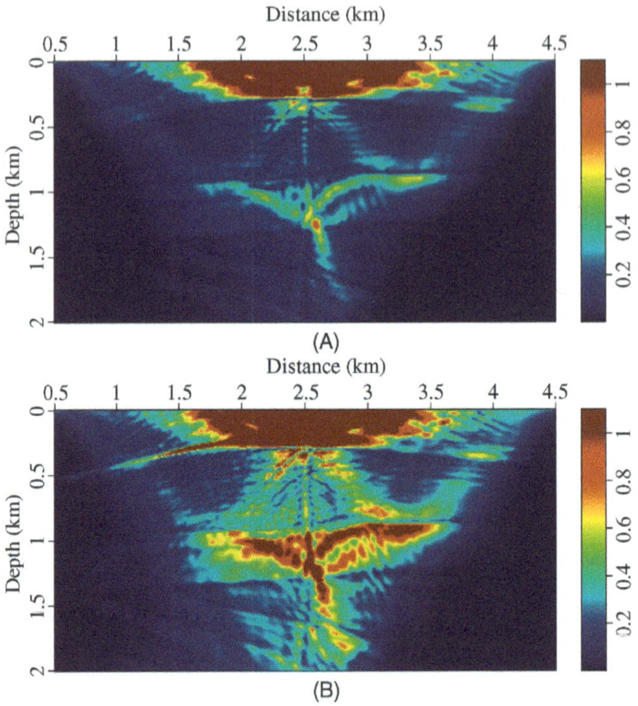

FIGURE 7.12 Receiver illumination map (A) without transmission compensation and (B) with transmission compensation for one shot located at a distance of 2.5 km.

Figure 7.13 shows single-shot migration results using RTM with and without transmission compensation. We can see that with the transmission compensation the migration events below the unconformity are enhanced, but meanwhile the artifacts such as diffraction are also enhanced. Stolk and Symes (2004) showed that artifacts are produced because of incorrect pairing of multipath source and receiver wavefronts in individual prestack migrations; they also show that these false images tend to disappear when the prestack images are stacked. Thus, we stack all shots to suppress these artifacts.

Stacking over all the prestack-migrated common-shot images produces the images in Figure 7.14. With compensation of transmission losses, the new RTM has compensated the transmission losses and gives a fairly good image with clear layer structures and a fairly balanced amplitude in the areas below the unconformity. In Figure 7.15, we compare parts of single traces in Figure 7.14A and B with true reflectivity of the model. The trace compared is at horizontal position 2.4 km from depth 0.8 to 2 km. The solid line is the true reflectivity model. The blue dot line is from traditional RTM image, whereas the cyan dash line is from RTM with compensation of transmission losses. The amplitudes of deep events are compensated, which help to improve the amplitude properties of the final imaging results.

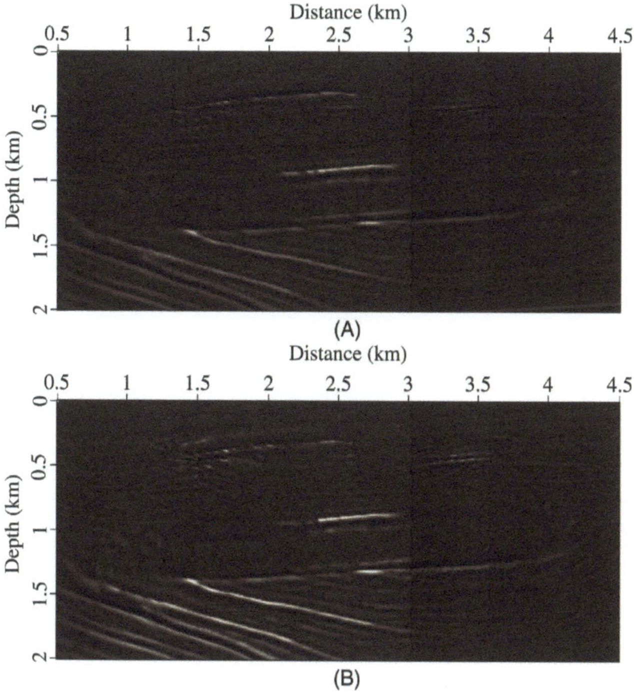

FIGURE 7.13 Single-shot migrated result using (A) RTM without transmission compensation and (B) RTM with transmission compensation. The shot location is at a horizontal distance of 2.5 km.

Computational efficiency is an important issue for RTM. Transmission compensation needs additional computation. In this example, it takes 171.4 s for one-shot migration using traditional RTM, whereas it takes 219.9 s for RTM with transmission loss compensation. This numerical example is used to verify that compensation of transmission losses helps to improve the accuracy of image amplitude. We assume the model information needed is already known. In practice, including the time spent on calculating values needed for transmission compensation, the total time for RTM with transmission compensation should be even longer.

7.4 CONCLUSIONS

In this chapter, we applied the high-frequency asymptotic analysis to analyze the amplitude behaviors of wave-propagation operators used in true-amplitude RTM. Using the unifying imaging condition, the amplitude preservation of RTM is relying on whether or not the forward-source wavefield and backward-receiver wavefield are true amplitude. Based on two-way wave equation, the forward-propagation operator is kinematically and dynamically correct. However, this is not true for backward-propagation operator because of transmission losses.

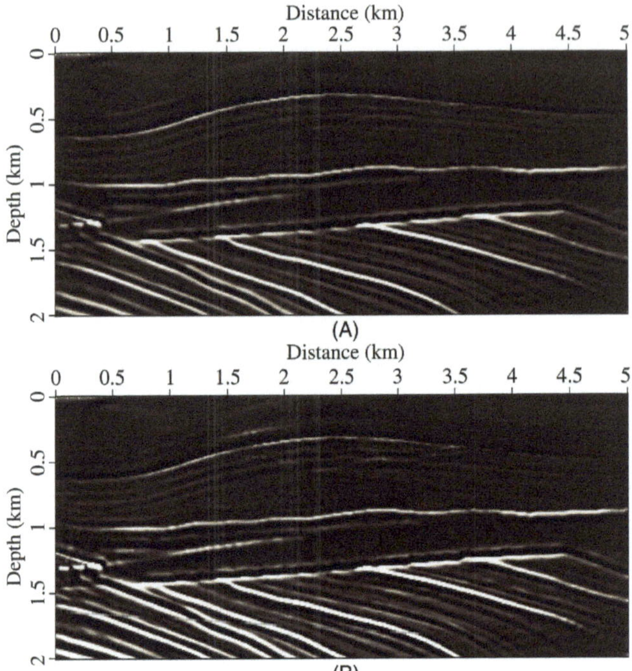

FIGURE 7.14 Stacked RTM image (A) without transmission compensation and (B) with transmission compensation.

With theoretical analysis, we have two conclusions about geometric spreading and transmission losses. We proved that under the high-frequency approximation, the geometric spreading in the forward propagation is corrected automatically in the backward propagation. We use a numerical example to verify that with an acceptable accuracy. This conclusion can also apply to the two-way

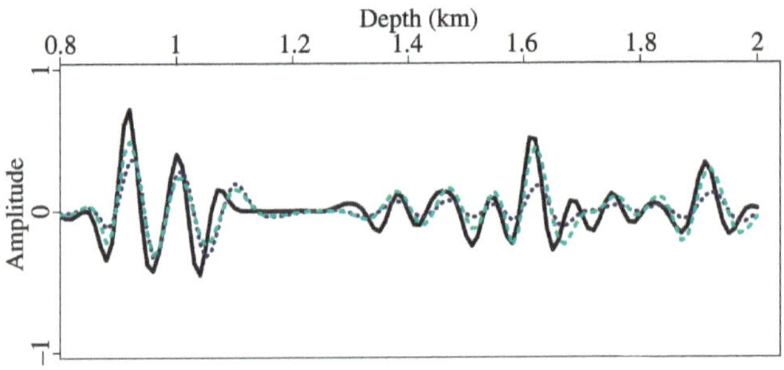

FIGURE 7.15 Comparison of single traces in Figure 7.14 with true reflectivity. The black solid line is the reflectivity of the true model; the blue dotted line is the amplitude of RTM without transmission compensation; and the cyan dashed line is the amplitude of RTM with transmission compensation.

propagation operator without high-frequency approximation, such as solving two-way wave equation with finite-difference method. The second conclusion is that because of the two-way transmission losses in the receiver wavefield, the migration formula contains a transmission-loss item, which prevents RTM from producing true-amplitude image. These theoretical derivations provide a mathematical base for the procedures to compensate transmission losses for RTM. The numerical examples verify that the compensation of transmission losses helps to achieve a better image with more accurate amplitudes.

REFERENCES

Aki, K., Richards, P.G., 1980. Quantitative Seismology. W. H. Freeman and Company, San Francisco, California.

Baysal, E., Kosloff, D.D., Sherwood, J.W.C., 1983. Reverse time migration. Geophysics 48, 1514–1524.

Berkhout, A.J., 1985. Seismic Migration: Imaging of Acoustic Energy by Wave Field Extrapolation, second ed. Elsevier, New York.

Beydoun, W.B., Hanitzsch, C., Jin, S., 1993. Why migrate before AVO? A simple example. 55th Conference & Exhibition, EAGE, Extended Abstracts, B044.

Beydoun, W.B., Jin, S., Hanitzsch, C., 1994. AVO migration and inversion: Are they commutable? 64th Annual International Meeting, SEG, Expanded Abstracts, 952–955.

Beylkin, G., 1985. Imaging of discontinuities in the inverse scattering problem by inversion of a causal generalized Radon transform. J. Math. Phys. 26, 99–108.

Biondi, B., Symes, W.W., 2004. Angle- domain common-image gathers for migration velocity analysis by wavefield-continuation imaging. Geophysics 69, 1283–1298.

Bleistein, N., 1984. Mathematical Methods for Wave Phenomena. Academic Press, New York.

Bleistein, N., 1986. Two-and-one-half dimensional in-plane wave propagation. Geophys. Prosp. 34, 686–703.

Bleistein, N., 1987. On the imaging of reflectors in the earth. Geophysics 52, 931–942.

Bleistein, N., Cohen, J.K., Stockwell, Jr., J.W., 2001. Mathematics of Multidimensional Seismic Imaging, Migration, and Inversion. Springer-Verlag, New York.

Červený, V., 2005. Seismic Ray Theory. Cambridge University Press, New York.

Chang, W., McMechan, G.A., 1986. Reverse-time migration of offset vertical seismic profiling data using the excitation-time imaging condition. Geophysics 51, 67–84.

Chang, W., McMechan, G.A., 1987. Elastic reverse time migration. Geophysics 52, 1365–1375.

Chang, W., McMechan, G.A., 1994. 3-D elastic prestack, reverse-time depth migration. Geophysics 59, 597–609.

Chattopadhyay, S., McMechan, G.A., 2008. Imaging conditions for prestack reverse-time migration. Geophysics 73, S81–S89.

Chavent, G., Plessix, R.E., 1999. An optimal true amplitude least-squares prestack depth migration. Geophysics 64, 508–515.

Claerbout, J.F., 1971. Toward a unified theory of reflector mapping. Geophysics 36, 467–481.

Clapp, R.G., Fu, H., Lindtjorn, O., 2010. Selecting the right hardware for reverse time migration. Lead. Edge 29, 48–58.

Clayton, R.W., Stolt, R.H., 1981. A Born-WKBJ inversion method for acoustic reflection data. Geophysics 46, 1559–1567.

Costa, J., Alcântara, F.A.S., Schleicher, J., Novais, A., 2009. Obliquity-correction imaging condition for reverse time migration. Geophysics 74, S57–S66.

de Bruin, C.G.M., Wapenaar, C.P.A., Berkhout, A.J., 1990. Angledependent reflectivity by means of prestack migration. Geophysics 55, 1223–1234.

Deng, F., McMechan, G.A., 2007. True-amplitude prestack depth migration. Geophysics 72, S155–S166.

Deng, F., McMechan, G.A., 2008. Viscoelastic true-amplitude prestack reverse-time depth migration. Geophysics 73, S143–S155.

Docherty, P., 1991. A brief comparison of some Kirchhoff integral formulas for migration and inversion. Geophysics 56, 1164–1169.

Dong, W., Ponton, M., 1999. AVO inversion and interpretation via localized 3D migrations. 69th Annual International Meeting, SEG, Expanded Abstracts, 816–819.

Du, Q., Li, F., Ba, J., Zhu, Y., Hou, B., 2012a. Multicomponent joint migration velocity analysis in the angle domain for PP-waves and PS-waves. Geophysics 77, U1–U13.

Du, Q., Zhu, Y., Ba, J., 2012b. Polarity reversal correction for elastic reverse time migration. Geophysics 77, S31–S41.

Duquet, B., Lailly, P., 2006. Efficient 3D wave-equation migration using virtual planar sources. Geophysics 71, S185–S197.

Ehinger, A., Lailly, P., Marfurt, K., 1996. Green's function implementation of common-offset wave-equation migration. Geophysics 61, 1813–1821.

Fomel, S., 2002. Applications of plane-wave destructor filters. Geophysics 67, 1946–1960.

Guitton, A., Valenciano, A., Bevc, D., 2006. Robust imaging condition for shot-profile migration. 74th Annual International Meeting and Exposition, SEG, Expanded Abstracts, 2519–2523.

Guitton, A., Valenciano, A., Bevc, D., Claerbout, J., 2007. Smoothing imaging condition for shot-profile migration. Geophysics 72, S149–S154.

Haney, M.M., Bartel, L.C., Aldridge, D.F., Symons, N.P., 2005. Insight into the output of reverse-time migration. What do the amplitudes mean? 75th Annual International Meeting, SEG, Expanded Abstracts, 1950–1953.

Hanitzsch, C., 1997. Comparison of weights in prestack amplitude-preserving Kirchhoff depth migration. Geophysics 62, 1812–1816.

Hanitzsch, C., Schleicher, J., Hubral, P., 1994. True-amplitude migration of 2-D synthetic data. Geophysics 42, 445–462.

Joncour, F., Lambaré, G., Svay, L.J., 2011. Preserved-amplitude angle domain migration by shot-receiver wavefield continuation. Geophys. Prosp. 59 (2), 256–268.

Kang, W., Cheng, J., 2011. Prestack scalar reverse time migration of elastic seismic data in TI media. 81st Annual International Meeting, SEG, Expanded Abstracts, 3367–3371.

Kaelin, B., Guitton, A., 2006. Imaging condition for reverse time migration. 76th Annual International Meeting, SEG, Expanded Abstracts, 2594–2598.

Kelly, S., Ren, J., 2004. Key elements in the recovery of relative amplitude for prestack shot record migration. Lead. Edge 23, 882–886.

Kuehl, H., Sacchi, M., 2001. Generalized least-squares DSR migration using a common angle imaging condition. 71st SEG meeting, San Antonio, Texas, US, Expanded Abstracts, 1025–1028.

Kuehl, H., Sacchi, M., 2002. Robust AVP estimation using leastsquares wave-equation migration. 72nd SEG meeting, Salt Lake City, Utah, US, Expanded Abstracts, 281–284.

Lailly, P., 1983. The seismic inverse problem as a sequence of before stack migrations. In: Conference Proceeding on Inverse Scattering, Theory and Application, SIAM, Philadelphia, 206–220.

Leveille, J.P., Jones, I.F., Zhou, Z.-Z., Wang, B., Liu, F., 2011. Subsalt imaging for exploration, production, and development: a review. Geophysics 76, WB3–WB20.

Liu, F., Zhang, G.S., Morton, A., Leveille, J.P., 2011. An effective imaging condition for reverse-time migration using wavefield decomposition. Geophysics 76, S29–S39.

Lowenthal, D., Hu, L., 1991. Two methods for computing the imaging condition for common-shot prestack migration. Geophysics 56, 378–381.

Martin, G.S., Marfurt, K.J., Larsen, S., 2002. Marmousi-2: An updated model for the investigation of AVO in structurally complex areas. 72nd Annual International Meeting, SEG, Expanded Abstracts, 1979–1982.

McDonal, F.J., Angona, F.A., Mills, R.L., Sengbush, R.L., VanNostrand, R.G., White, J.E., 1981. Attenuation of shear and compressional waves in Pierre shale. In: Toksöz, M.N., Johnston, D.H. (Eds.), Seismic Wave Attenuation. SEG.

McMechan, G.A., 1983. Migration by extrapolation of time-dependent boundary values. Geophys. Prosp. 31, 413–420.

Mosher, C., Foster, D., 2000. Common angle imaging conditions for pre-stack depth migration. 70th SEG meeting, Calgary, Canada, Expanded Abstracts, 830–833.

Nemeth, T., Wu, C., Schuster, G.T., 1999. Least-squares migration of incomplete reflection data. Geophysics 64, 208–221.

Nolan, C., Symes, W.W., 1996. Imaging in complex velocities with general acquisition geometry. In: Annual Report, T.R.I.P.: The Rice Inversion Project.

Pica, A., Diet, J., Tarantola, A., 1990. Nonlinear inversion of seismic reflection data in a laterally invariant medium. Geophysics 55, 284–292.

Plessix, R.E., Mulder, W.A., 2004. Frequency-domain finite difference amplitude-preserving migration. Geophys. J. Int. 157, 975–987.

Prucha, M., Biondi, B., Fomel, S., 1999. Angle domain common image gathers by wave equation migration. 69th SEG meeting, Houston, Texas, US, Expanded Abstracts, 824–827.

Sava, P.C., Biondi, B., Etgen, J., 2005a. Wave-equation migration velocity analysis by focusing diffractions and reflections. Geophysics 70, U19–U27.

Sava, P., Fomel, S., 2003. Angle-domain common-image gathers by wavefield continuation methods. Geophysics 68, 1065–1074.

Sava, P., Fomel, S., 2005b. Wave equation common-angle gathers for converted wave. 75th Annual International Meeting, SEG, Expanded Abstracts, 947–951.

Sava, P., Fomel, S., 2006. Time-shift imaging condition in seismic migration. Geophysics 71 (6), S209–S217.

Schleicher, J., Costa, J.C., Novais, A., 2008. A comparison of imaging conditions for wave-equation shot-profile migration. Geophysics 73, S219–S227.

Sevink, A., 1996. Asymptotic seismic inversion, PhD thesis, Delft University of Technology, The Netherlands.

Shin, C., Jang, S., Min, D., 2001. Improved amplitude preservation for prestack depth migration by inverse scattering theory. Geophys. Prosp. 49, 592–606.

Soubaras, R., 2003. Angle gathers for shot-record migration by local harmonic decomposition. 73rd Annual International Meeting, SEG, Expanded Abstracts, 889–892.

Soubaras, R., 2006. Modulated-shot migration. 76th SEG meeting, New Orleans, Louisiana, US, Expanded Abstracts, 2430–2433.

Stolk, C.C., Symes, W.W., 2004. Kinematic artifacts in prestack depth migration. Geophysics 69, 562–575.

Sun, W., Fu, L.Y., 2012. Compensation for transmission losses based on one-way propagators in the mixed domain. Geophysics 77, S65–S72.

Symes, W.W., 2007. Reverse time migration with optimal checkpointing. Geophysics 72, SM213–SM221.

Tarantola, A., 1984. Inversion of seismic reflection data in the acoustic approximation. Geophysics 49, 1259–1266.

Tura, A., Hanitzsch, C., Calandra, H., 1998. 3-D AVO migration/inversion of field data. Lead. Edge 17, 1578–1583.

Valenciano, A.A., Biondi, B., 2003. 2-D deconvolution imaging condition for shot-profile migration. 73rd Annual International Meeting, SEG, Expanded Abstracts, 1059–1062.

Valenciano, A.A., Chemingui, N., Whitmore, N.D., Brandsberg-Dahl, S., 2011. Wave equation migration with attenuation and anisotropy compensation. 81st Annual International Meeting, SEG, Expanded Abstracts, 232–236.

Vanelle, C., Gajewski, D., 2000. Traveltime based true-amplitude migration. 70th Annual International International Meeting, SEG, Expanded Abstracts, 961–964.

Vanelle, C., Spinner, M., Hertweck, T., Jäger, C., Gajewski, D., 2006. Traveltime-based true-amplitude migration. Geophysics 71 (6), S251–S259.

Vivas, F.A., Pestana, R.C., 2010. True-amplitude one-way wave equation migration in the mixed domain. Geophysics 75, S199–S209.

Wapenaar, K., Fokkema, J., 2006. Green's function representations for seismic interferometry. Geophysics 71, SI33–SI46.

Weglein, A.B., Stolt, R., 1999. Migration-inversion revisited. Lead. Edge 18, 950–953.

Whitmore, N.D., 1983. Iterative depth migration by backward time propagation. 53rd Annual International Meeting, SEG, Expanded Abstracts, 382–385.

Widmaier, M., Müller, T.H., Shapiro, S.A., Hubral, P., 1995. Amplitude-preserving migration and elastic P-wave AVO corrected for thin layering. J. Seismic Explor. 4, 169–177.

Wu, R.S., Luo, M., Chen, S., Xie, X., 2004. Acquisition aperture correction in angle-domain and true-amplitude imaging for wave equation migration. 74th Annual International Meeting, SEG, Expanded Abstracts, 937–940.

Yan, R., Xie, X., 2011. Angle gather extraction for acoustic and isotropic elastic RTM. 81st Annual International Meeting, SEG, Expanded Abstracts, 3141–3146.

Yan, R., Xie, X., 2012. An angle-domain imaging condition for elastic reverse time migration and its application to angle gather extraction. Geophysics 77 (5), S105–S115.

Yilmaz, Ö., Doherty, S.M., 2001. Seismic data analysis: processing, inversion, and interpretation of seismic data, second ed. Society of Exploration Geophysicists, Tulsa, OK.

Yoon, K., Marfurt, K., Starr, E.W., 2004. Challenges in reverse-time migration. 74th Annual International Meeting, SEG, Expanded Abstracts, 1057–1060.

Xie, X., Yang, H., 2008. A full-wave equation based seismic illumination analysis method. 70th Annual International Conference & Exhibition, EAGE, Extended Abstracts, 284.

Xie, X., Wu, R.S., 2002. Extracting angle domain information frommigrated wavefields. 72nd Annual International Meeting, SEG, Expanded Abstracts, 1360–1363.

Xu, S., Zhang, Y., Tang, B., 2011. 3D angle gathers from reverse time migration. Geophysics 49, 1259–1266.

Zhang, L., Hua, B., Calandra, H., 2005a. 3D Fourier finite-difference anisotropic depth migration. 75th Annual International Meeting, SEG, Expanded Abstracts, 1914–1917.

Zhang, H., McMechan, G., Stein, J., 2003a. True amplitude wave equation migration. 73rd Annual International Meeting, SEG, Expanded Abstracts, 913–916.

Zhang, Q., McMechan, G.A., 2011a. Common-image gathers in the incident phase-angle domain from reverse time migration in 2D elastic VTI media. Geophysics 76 (6), S197–S206.

Zhang, Q., McMechan, G.A., 2011b. Direct vector-field method to obtain angle-domain common-image gathers from isotropic acoustic and elastic reverse-time migration. Geophysics 76 (5), WB135–WB149.

Zhang, Y., Roberts, G., Khalil, A., 2012. Compensating for source and receiver ghost effects in reverse time migration. SEG Tech. Prog. Expanded Abstracts, 1–5.

Zhang, Y., Sun, J., 2008. Practical issues of reverse time migration—True-amplitude gathers, noise removal and harmonic-source encoding. 70th Conference and Exhibition, EAGE, Extended Abstracts, F047.

Zhang, Y., Xu, S., Tang, B., Bai, B., Huang, Y., Huang, T., 2010. Angle gathers from reverse time migration. Lead. Edge 29, 1364–1371.

Zhang, Y., Xu, S., Tang, B., Bai, B., Huang, Y., Huang, T., 2005b. Theory of true-amplitude one-way wave equations and true-amplitude common-shot migration. Geophysics 70, E1–E10.

Zhang, Y., Xu, S., Tang, B., Bai, B., Huang, Y., Huang, T., 2007. True-amplitude, angle-domain, common-image gathers from one-way wave-equation migrations. Geophysics 72, S49–S58.

Zhang, Y., Zhang, G., Bleistein, N., 2003b. True amplitude wave equation migration arising from true amplitude one-way wave equations. Inverse Prob. 19, 1113–1134.

Zhang, Y., Zhang, H., Zhang, G., 2011. A stable TTI reverse time migration and its implementation. Geophysics 76, WA3–WA11.

Chapter 8

Rock Physics Models and Quantitative Seismic Prediction of Heterogeneous Gas Reservoirs – A Case Study in Metejan Area of Amu Darya Basin

Jing Ba[1], Hao Yu[2], Jinsong Li[2], Xinfei Yan[2], Xingyang Zhang[2], Xinzhen He[3]

[1]*Department of Earth and Atmospheric Sciences, University of Houston, Houston, Texas, US;*

[2] *Research Institute of Petroleum Exploration and Development, PetroChina, Beijing, China;*

[3]*Amu Darya Petroleum Company Ltd., CNPC, Beijing, China*

Chapter Outline

In recent years, in order to cope with climate change and low-carbon economy development problems, oil companies around the world have attached great importance to unconventional and tight oil/gas resources, especially for natural gas in the reservoirs with strong heterogeneity. The exploration and use of natural

Seismic Exploration of Hydrocarbons in Heterogeneous Reservoirs. http://dx.doi.org/10.1016/B978-0-12-420151-4.00008-8

gas resources is an effective way to relieve the pressure on crude demand of oil and ensure national energy security. In term of China only, more than 80% of natural gas is reserved in the heterogeneous reservoirs; therefore, it is of great significance to carry out systematic studies on quantitative prediction techniques for hydrocarbon in such type of stratums.

With conventional seismic inversion and interpretation techniques, it is difficult to effectively describe and quantitatively predict the complex pore structures and fluid distribution characteristics in heterogeneous reservoir rocks. In this chapter, a case study is shown in the carbonate gas reservoirs of Metejan area on the right bank of the Amu Darya River. By means of computed tomography (CT) scanning, thin section electron microscopy, theoretical modeling, and experimental analysis, the rock physics modeling methods and industrial technical processes that are formed in the actual engineering application of seismic prediction for reservoirs and hydrocarbon are introduced. The related prediction results and the production test reports of wells are presented.

8.1 OVERVIEW ON THE WORK AREA

8.1.1 Amu Darya Basin

8.1.1.1 Regional Tectonic Evolution

On geotectonic location, the Amu Darya Basin is located in the mid-west of the Central Asia Tectonic Zone. Kyzyl Mountains at the northern fringe of the basin divides the Central Asia Tectonic Zone into two parts, including Kazakhstan plate in the north and China–Korea–Tarim plate in the south. The Amu Darya River is located in the northwest of China–Korea–Tarim plate and is also known as Karakum landmass (Li, 1982).

Central Asia and western China are similar in topography and geotectonic setting. Karakum landmass and Tarim landmass are in similar geotectonic locations (Che et al., 1997). According to the plate tectonic theories, the formation of tectonic framework of Central Asia can be divided into 2 stages, namely, the Hercynian and the Alpine Epoch.

1. The Hercynian. For both the Kazakhstan plate in the north and the Karakum plate in the south, the distribution range of Precambrian metamorphic basement was small while that of epimetamorphic rock formed by Hercynian geosynclinal folds was broad. Through the Hercynian tectonic movement, Central Asia, Siberia plate, and Eastern Europe plate were joined to form the vast Eurasian plate, which is also known as the Angara megaplate in paleogeography. In the Palaeozoic era, the African–Arabian–Indian plate in the south was joined with the Eurasian plate to form Gondwana land.
2. The Alpine Epoch. Starting from the Mesozoic era, the Indian plate and the Arabian plate were separated from the Gondwana land and drifted respectively to the north and north east. In that era, the tectonic movements within the Central Asia region were mainly steadily uplift and subsidence. In

the Cenozoic era, the Arabian plate and the Indian plate continued to move northward and finally collided with the Eurasia continent. As a result, the Tethys Trough was extruded into a mountain; a piedmont depression was formed on the southern fringe of Central Asia and a series of intermountain basins were formed in the east.

8.1.1.2 Sedimentary Characteristics

The Amu Darya Basin (Figure 8.1) is a large sedimentary basin on the southeastern fringe of Turan platform and is a Mesozoic basin that is developed on the basis of Hercynian synclines. After experiencing strong folds at the end of the Triassic period, the Amu Darya Basin, together with Turan platform, got to a stable platform development stage. The basin basement was formed before the Jurassic period. The basin experienced sustained and steady subsidences in the Jurassic and Cretaceous periods and the Tertiary uplifting transformation, etc., in its development. The major source of natural gas in the basin is the middle and lower Jurassic continental and transitional coal-bearing clastic rocks, with local unsteady coal fragments, followed by Upper Jurassic carbonate source rocks. Upper Jurassic evaporites are well developed, with widespread gypsum-salt rock of 200–1200 m thick, and in the northwest fault zone, it becomes argillaceous rock, with good cover and storage conditions for gas.

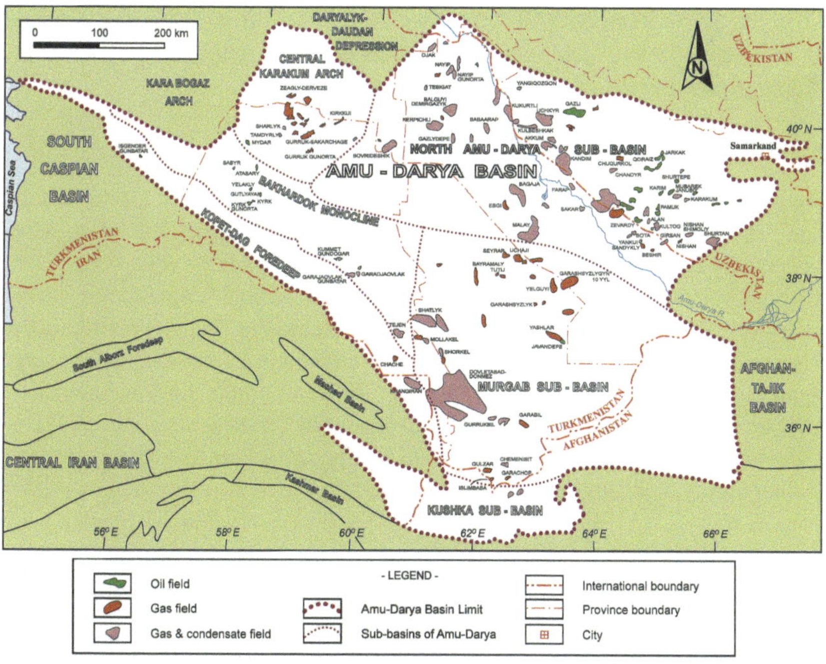

FIGURE 8.1 Amu Darya basin geographic location and oil/gas field distribution.

The strata of the entire basin can be divided into three tectonic layers: Paleozoic basement, Permian-Triassic transition layer, and the Mesozoic and Cenozoic sedimentary cover (Liu and Zhu, 2007). Details of each layer include the following.

1. Basin basement. The basin basement is composed of Permian and Triassic granite, tuffaceous sandstone and shale, with high degree of metamorphism; the bedrock is cut into bedrock rise, graben, and flexure zones by two bedrock fault zones, aligning in the northwest and northeast directions. These control the formation and development of sedimentary cover.
2. Transition layer. The transition layer includes Permian and Triassic strata and is lithologically composed of variegated conglomerate, sandstone, shale, and neutral-acid eruptive rocks. The variation in the thickness of Upper Paleozoic-Triassic bedrock is controlled primarily by Paleozoic tectonic structure. During the Paleozoic era, the Amu Darya Basin, Kopet-Dag paleo-piedmont depression and rudiment of Karakum paleo-uplift had already formed; Paleozoic structure and deep faults played a certain role in controlling the formation and development of the later fracture pattern.
3. Mesozoic and Cenozoic sedimentary cover. In the long geological history from the Mesozoic to the Cenozoic era, the Amu Darya Basin experienced three major tectonic cycle stages of development. The three stages include the early subsidence stage of the basin in the Jurassic period, the steady subsidence stage of the basin in the Cretaceous period, and the stage for intense reformation of the basin and the finalization of the current structure in the Tertiary period. Concurrent with the three tectonic stages, two large-scale transgression and regression events occurred in the basin area, resulting in sedimentation of the corresponding strata.

8.1.2 Overview on Metejan District

Metejan district is located on the Chardzhui terraces in the Amu Darya depression. The drilled stratum from the top to bottom are Neogene, Paleogene, Cretaceous, and Jurassic. The Jurassic system includes the three rock series from the bottom to top. At the bottom, it is a series of transitional coal-bearing clastic rocks from the lower Jurassic to middle Jurassic Bathonian Stage, constituting the main source rock series of the block. In the middle, it is a carbonate sedimentary sequence at the Callovian-Oxfordian Stage, constituting the main accumulation rock series of the block. At the top, it is an evaporite series at the Kimmeridgian-Tithonian Stage, lithologically dominated by gypsum and salt rocks with perfect sealing ability, constituting the main cap rock series for the reservoirs.

8.1.2.1 Structural Characteristics

The Metejan structure is a nearly East-West-trending anticline with multiple culminations, gradually rising from west to east. The structure is cut and complicated by feathery faults. Two of the culminations with larger scales are located close to Well Met22 and Well Met21 (Figure 8.2).

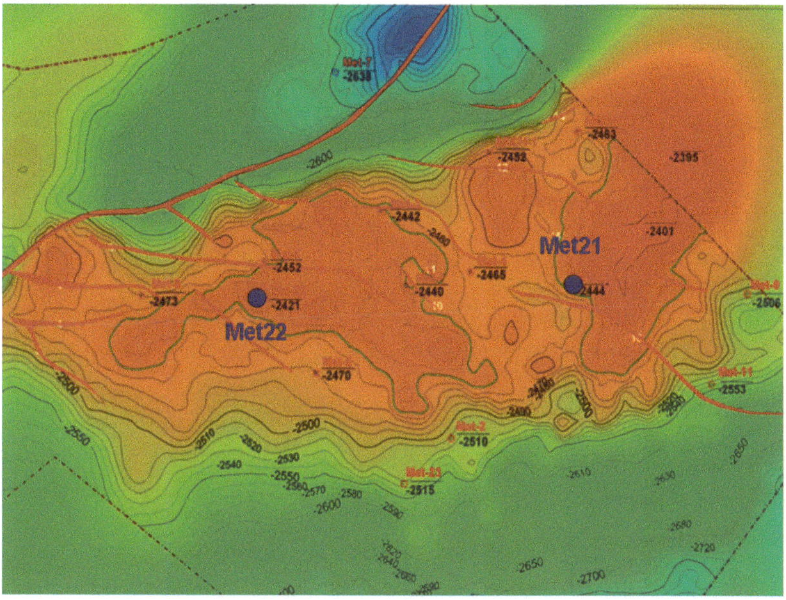

FIGURE 8.2 Structural characteristics in the Metejan district.

8.1.2.2 Reservoir Characteristics

In the sedimentary period of gas reservoirs in the Jurassic Callovian-Oxfordian stage of Metejan district, the stratum was in its platform facies. Carbonate reservoirs in reef-bank facies are mainly developed. The reservoir spaces include pores, karst caves, and fractures. Karst caves with medium-large sizes are observed, which form "honeycomb" structures in some cases. Tectonic fractures dominate the fracture systems. Reservoir porosity is in the range of 5%–12.5%. Effective reservoirs account for 56% in the district. Reservoirs are mainly developed in the XVm layer. Vertically, the reservoir thickness varies significantly. The reservoirs around all the wells are thin except Wells Met3, Met21, Met22, and Met23. Horizontally, the reservoir development is not equal and has got poor continuity. The development degree of reservoirs strongly varies between the different well areas (Figure 8.3).

8.1.2.3 Reservoir Prediction

Based on the analysis of drilling results in the Metejan district, the reservoir development should be first depicted and described before the gas reservoir prediction. Attribute analysis technique is commonly used in reservoir prediction. It is to extract amplitude, frequency, phase, and other seismic attributes from the interpretation results of the tracked target layer along stratum or a certain time window. The reservoir evolution of plane can be identified through coherent analysis; favorable reservoir development area can be identified through

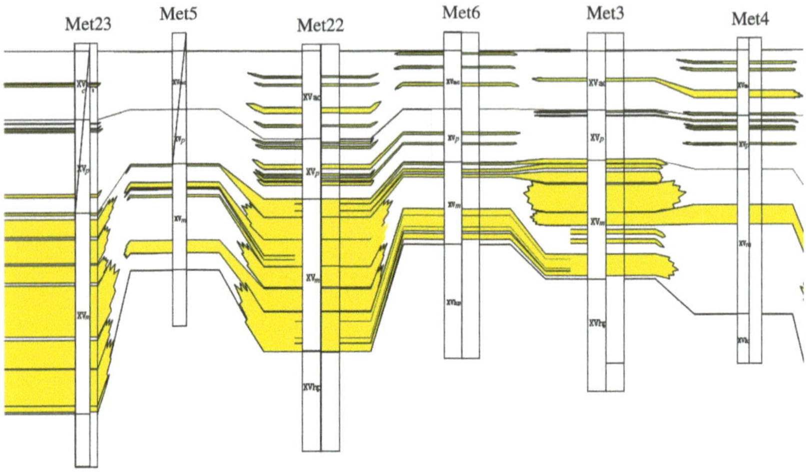

FIGURE 8.3 Reservoir features of the Metejan district.

paleogeomorphic background analysis; and methods of seismic frequency decomposition can help to identify fractures and reservoirs.

Within the effective band of seismic data, the section characteristics of small faults will become clearer with the increasing frequency. Therefore, based on the S-type transformation, reservoir is identified with a data volume at around 45 Hz after frequency decomposition, which is higher than the basic frequency of seismic data (Yu et al., 2013). In the comparison between the conventional seismic profile (Figure 8.4) and the profile after frequency decomposition (Figure 8.5), it can be shown:

1. The depiction of break points on the frequency-decomposition profile is clearer than that on the original profile. On the original seismic profile, some faults are vague and some are shown as two intersecting reflection events. Break points are clearly shown on the frequency-decomposition profile. The

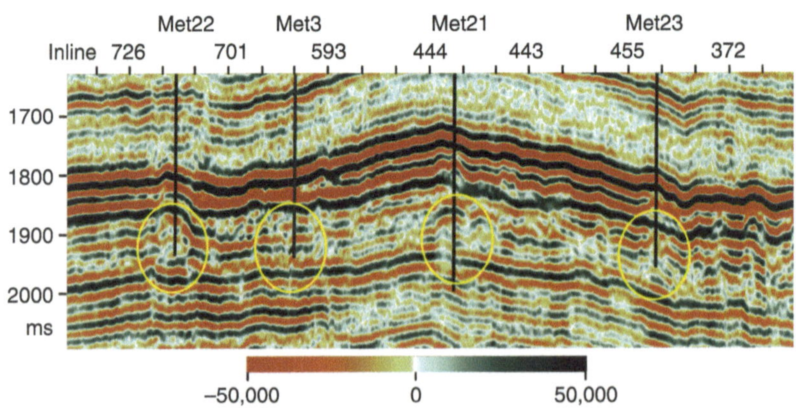

FIGURE 8.4 The original seismic profile.

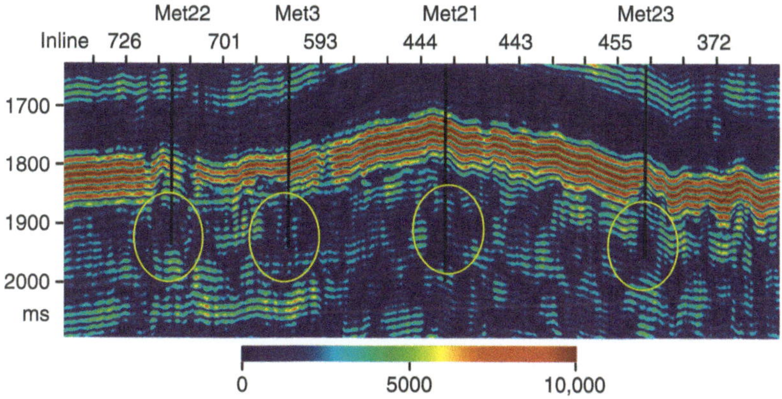

FIGURE 8.5 The frequency-decomposition profile.

interpretation based on frequency-decomposition profiles can efficiently improve the accuracy of fault interpretation.

2. The features of reservoirs on the frequency-decomposition profile are clearer than on those on the original profile. Dissolution reservoirs are mainly developed around Well Met22, Met3, and Met23. On the original seismic profile, reservoirs show disordered weak reflection characteristics, with unclear form and scope; but on the frequency-decomposition profile, reservoirs are located in the deep blue areas of low energy surrounded by yellow areas, with very clear form and scope. It shows frequency composition also help improving the precision of reservoir recognition.

By use of the interpretation results on frequency-decomposition data volume, the disordered weak reflections and low-energy responses from reservoir rocks are considered as reservoir development areas on the profile. Reservoir distribution characteristics can be displayed in a 3D space with an amplitude sculpture technique (Figure 8.6).

Seismic inversion is an imaging process of structures and physical properties of the subsurface formations with the constraint of known geological laws, logging data, and drilling data. Wave impedance inversion is to eliminate the effects of wavelet from seismic profile and then calculate the physical parameter – wave impedance, which reflects the changes in physical properties of the formation rocks, from the reflection coefficient, thereby to obtain the distribution features of rock physical parameters in space. As can be seen from the average wave impedance transection map of reservoirs (Figure 8.7), the wave impedance around Well Met22 and Zen21 is low, possibly with good gas-bearing property; the overall wave impedance in the northeast of the block, such as the area around Well Met21, is high, possibly with less gas-bearing potential. As is shown from the reservoir time-thickness transection map (Figure 8.8), the reservoirs around Well Met22 and Zen21 have got larger time thickness and may be well developed; reservoirs in other areas have got smaller time thickness and are less developed.

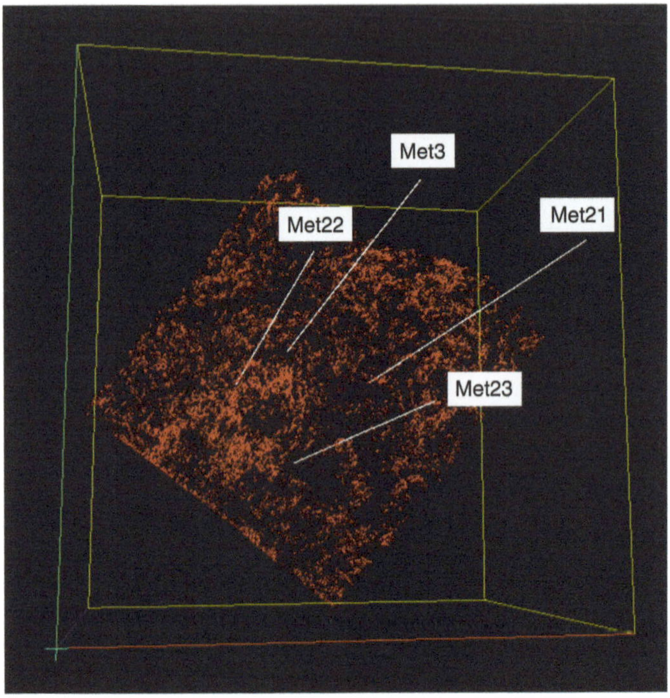

FIGURE 8.6 The 3D amplitude sculpture results of reservoirs.

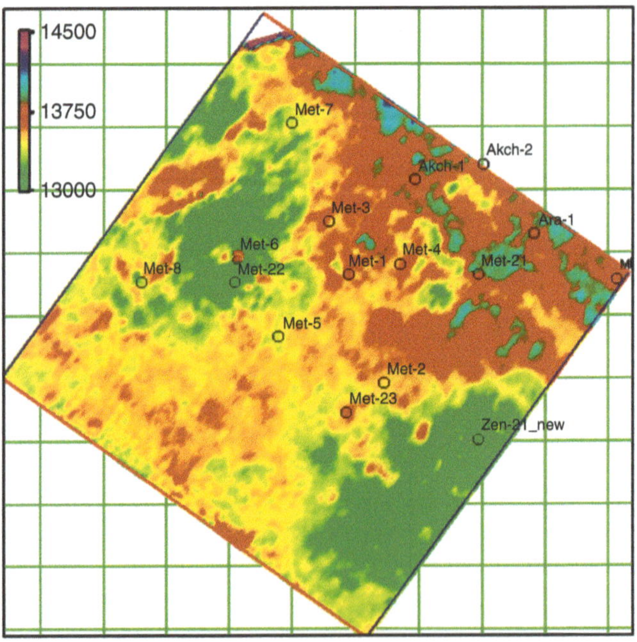

FIGURE 8.7 Average wave impedance transection map of reservoirs.

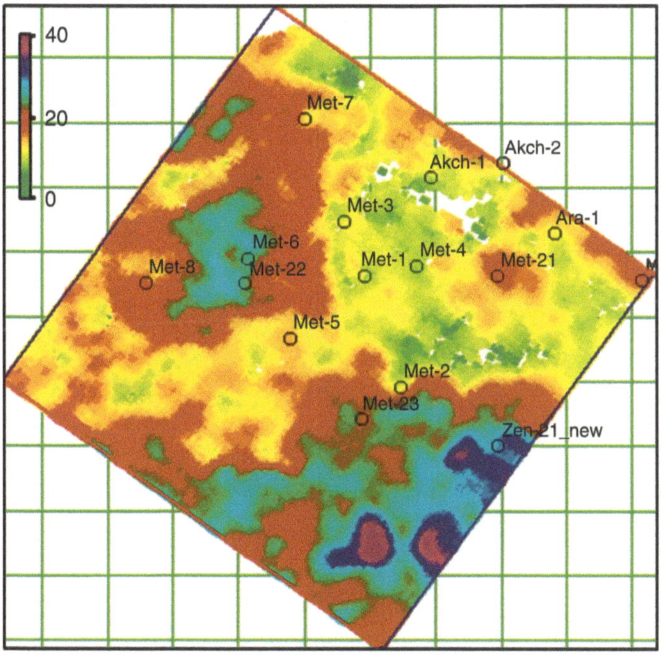

FIGURE 8.8 Reservoir time-thickness transection map.

8.2 EXPERIMENTAL ANALYSIS

8.2.1 CT Scanning Analysis

For the purpose of investigating the pore structure characteristics and fluid distribution features in carbonate rock, industrial CT scans are performed on a carbonate sample that is partially saturated with water and air. The advantage of CT scanning is that it obtains 3D images of the pore spaces and fluid distribution without damaging the internal structure of rock samples (Yu et al., 2014).

The scanned limestone sample is collected from a certain well in the Metejan district. It is a cylinder of 12 mm in diameter and is suitable for identifying immiscible fluids. The CT scans were performed with a "Nano CT" from GE with a resolution of up to 1 μm. The sample was immersed in water for 15 h. The voltage for scanning was 85 kV, with the current of 85 mA. The resolution was 5.4 μm. It took 75 min to obtain 1500 projected 2D images and then 3D reconstruction was performed. Pores, water, and air are recognized by using a mathematical morphology approach and the watershed algorithm. As can be seen from the results of scanning (Figure 8.9), immiscible fluids (air and water) are distributed in patches of different sizes, and gas is mostly surrounded by water; large pores are generally occupied by gas and small ones are generally filled by water. In real rocks, "gas-in-water" patchy saturation is very obvious.

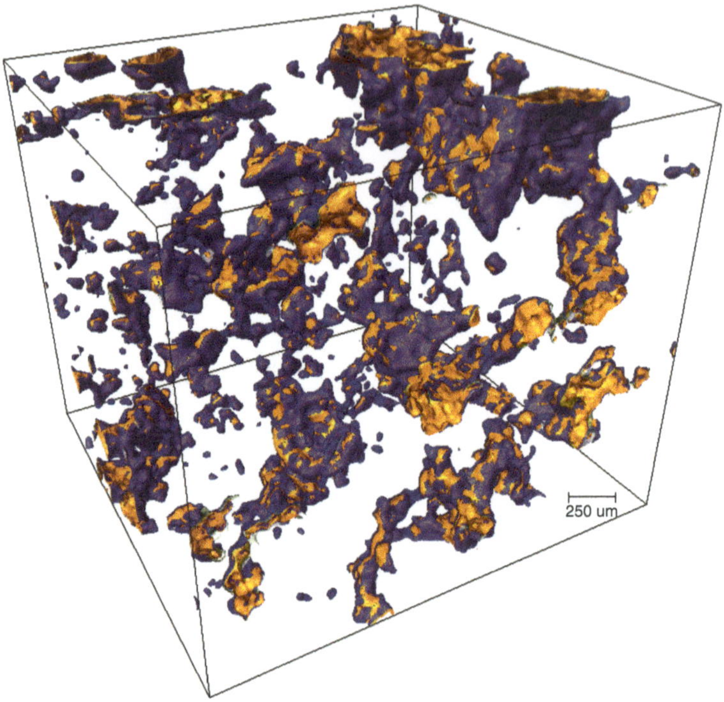

FIGURE 8.9 CT scan of the fluid distribution in limestone (porosity is 6.86% and air saturation is 69.2%).

The size of a single "gas pocket" is larger than that of the grains and pores; however, it is much smaller than the seismic wavelength, defining the mesoscopic heterogeneity.

Figures 8.10 and 8.11 show that the size of most water pockets are in the range of 10^{-7}–10^{-3} mm^3, whereas air pocket is present in the mesoscopic range of 10^{-3}–10^{-1} mm^3, with a volume exceeding the average grain (pore) size. The largest gas pocket has a volume of 0.288 mm^3 with a spherical radius greater than 0.41 mm, if we consider the volume of the adjacent mineral grains. Because the average porosity of the Metejan gas reservoirs is larger than that of the scanned sample (0.0686), it is inferred that the in situ size of the gas patches is larger, that is, in the mm–cm range. Moreover, the comparison of Figures 8.10B and 8.11 shows that although less air pockets are observed as the threshold of the pocket size increases, the volume ratio of the air pockets increases sharply, which means that most of the gas is contained in the mesoscopic patches.

The blue and yellow colors denote water and gas, respectively.

The volume of air pockets are counted in different ranges of sizes and then divided by the total pore volume to calculate the air volume in this range.

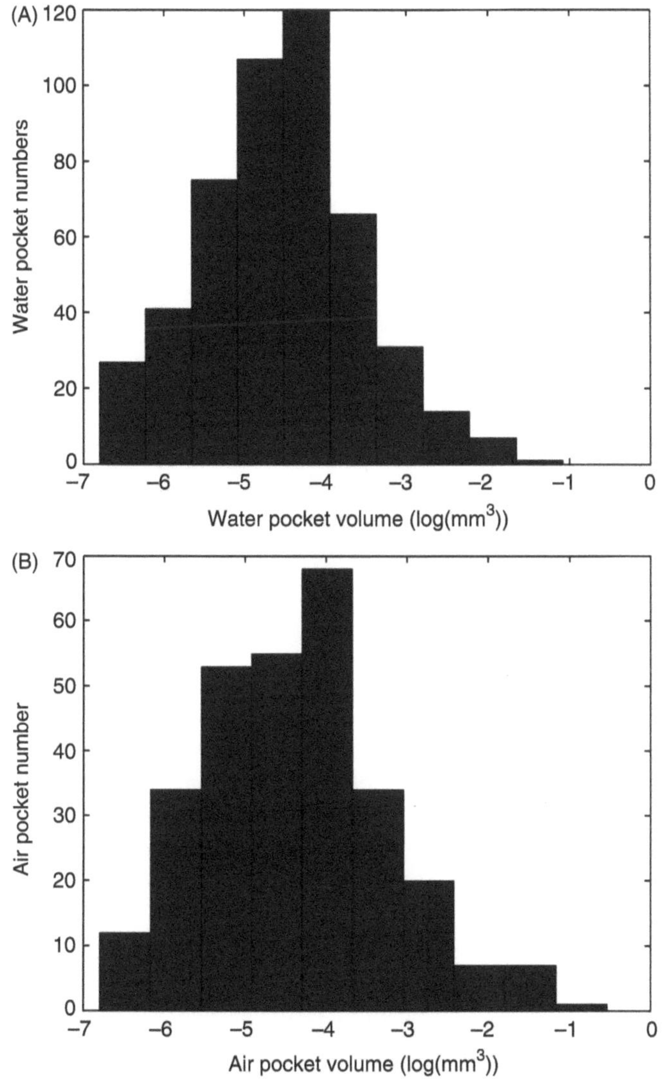

FIGURE 8.10 Histograms of the number of fluid pockets versus pocket size. (A) Water pockets. (B) Air pockets.

8.2.2 Geological Thin Section Analysis for Rocks

The method of thin section analysis can help identify rocks and minerals under a polarizing microscope. It is an important method which has been widely applied in the research of geological exploration for oil and gas. Rock or mineral specimens are ground into thin sections and their structure, composition, crystalline

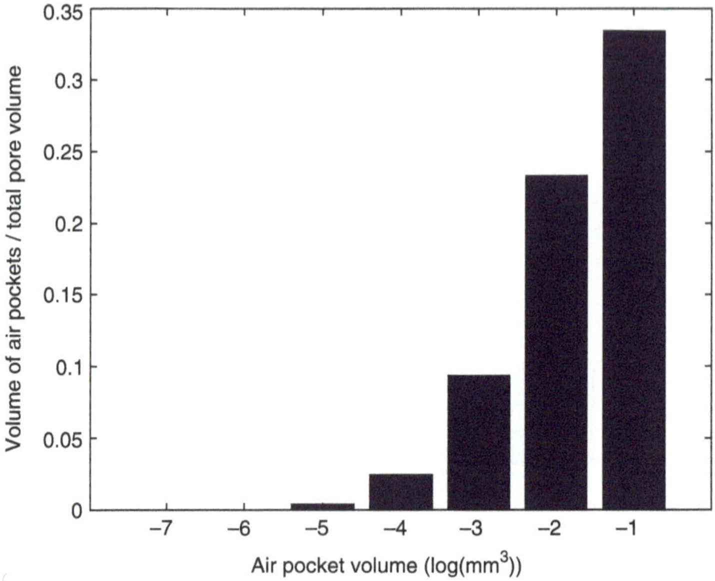

FIGURE 8.11 Histogram of the air volume versus gas pocket size.

characteristics, and optical properties are studied under a polarizing microscope to determine the rock type and genesis features.

Thin sections from the two typical Metejan wells Met21 and Met22 were examined with a microscope. The features about composition, pore structure, and surface porosity of reservoir rocks of the wells are observed. On the basis of a comparison with well log data, geological support is provided for reservoir prediction and fluid detection.

8.2.2.1 Rock Thin Section Analysis of Well Met22

Based on the thin-section analysis on samples from well Met22, it is shown that:

1. Met22-A (Figure 8.12A): sparite calcarenite and calcirudite; calcirudite grain size is in range of 2–11.6 mm and the grains are unevenly distributed; calcarenite grains are generally of small-medium size; the grains are mostly micritized, with unclear internal structures and a small amount of foraminifera (fossil); worm tubes can be observed occasionally; small amount of dolomite and silt are observed near the sutures; and authigenic quartz is observed in grains occasionally. The pores are well developed, including intergranular dissolved pores and intragranular dissolved pores. Some of the pores are partially/fully filled. The plane porosity is about 9%.
2. Met22-B (Figure 8.12B): sparite calcarenite; the grains are generally of small-medium size and unevenly distributed; there are a small amount of rudstone grains, with the maximum diameter of 6 mm; brachiopoda fragments and foraminifera can be observed; and authigenic quartz is observed occasionally.

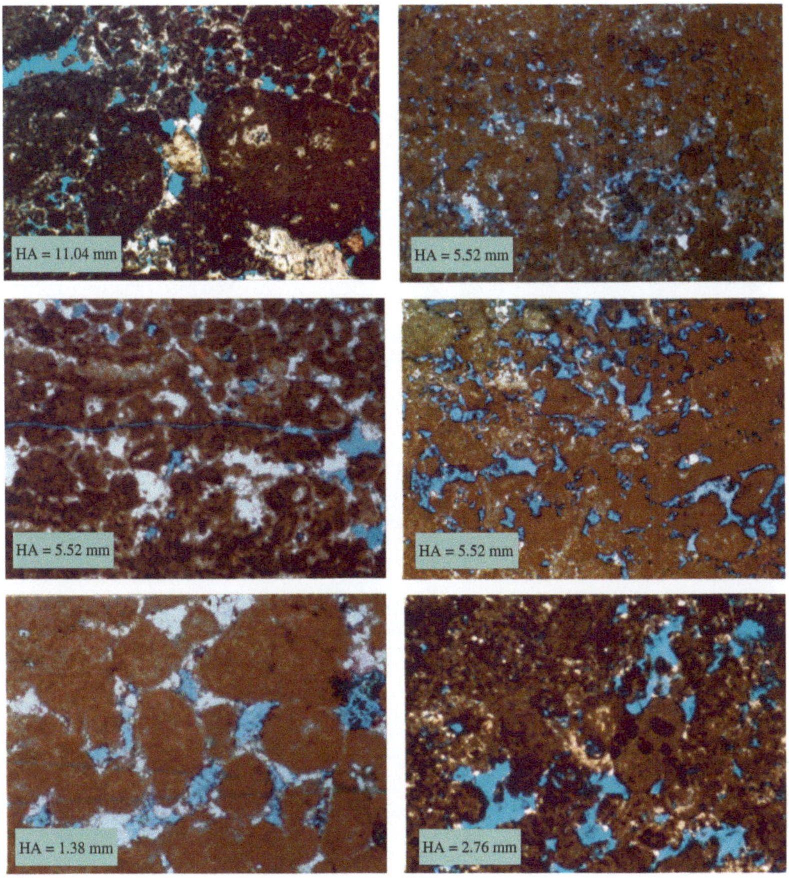

FIGURE 8.12 Core thin sections of Well Met 22. (A) Met22-A. (B) Met22-B.

The pores are mainly intergranular dissolved pores, with a small amount of intragranular dissolved pores; the surface porosity is about 7%.

8.2.2.2 Rock Thin Section Analysis of Well Met21

Based on the thin-section analysis on samples from well Met21, it is shown that:

1. Met21-A (Figure 8.13A): sparite calcarenite; grains are generally of small-medium size and unevenly distributed; and there are mainly fine grains with a few unevenly distributed calcirudite grains. A few bioclasts are observed, generally Caillou bacteria, bivalve, foraminifera, etc.; cement is heterogeneously distributed, with unclear laminae; micro-cross-bedding is observed in some area; and the plane porosity is about 1%.
2. Met21-B (Figure 8.13B): Caillou bacteria sparite calcarenite; there are mainly fine calcarenite grains; it contains many laminer cyanobacteria – Caillou

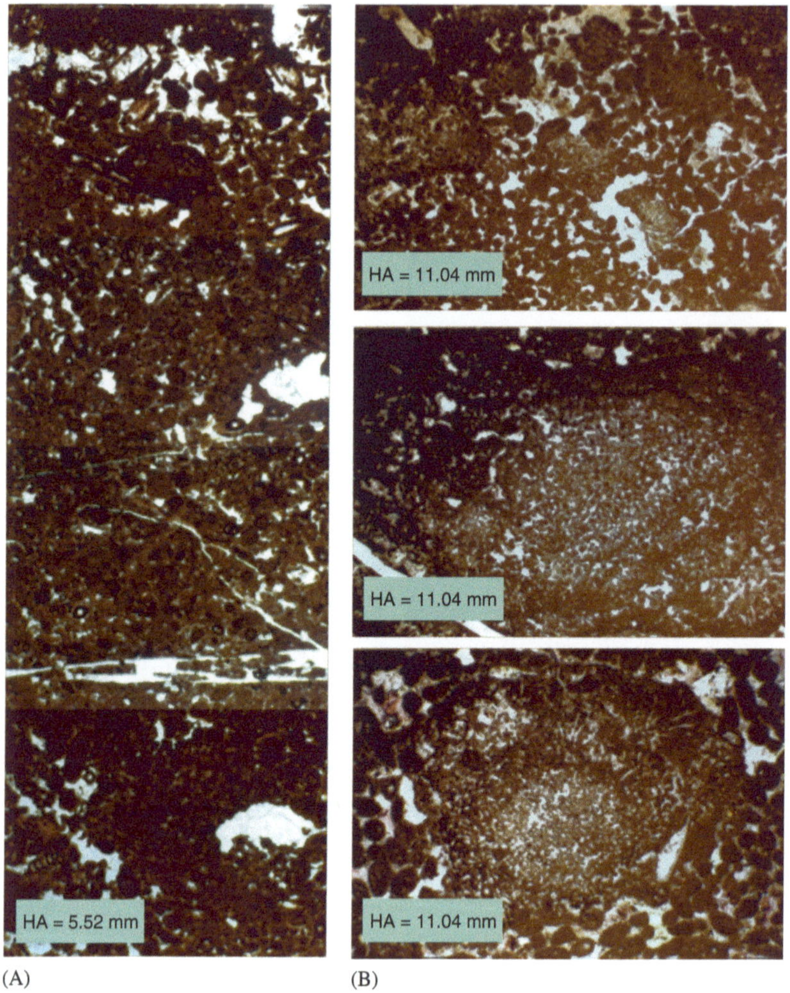

FIGURE 8.13 Core thin sections of Well Met 21. (A) Met21-A. (B) Met21-B.

bacteria (mainly aggregates, most of which are larger than 2 mm); and oncolites are observed, covered by Caillou bacteria, and unevenly distributed. Fenestriform pores are unevenly distributed, with no obvious laminer distribution, filled with crystalline calcite; intergranular pores are not developed; and the plane porosity is 1%.

It is shown from the thin section observation results that reservoir rocks from Well Met22 are mainly sparite calcarenites, dominated by the grains with medium-large sizes, dissolved pores are well developed, rocks have got high porosity, and the reservoir space is conducive to hydrocarbon accumulation. It is comprehensively judged that the sedimentary facies around Met22 is located in open platform facies.

Reservoir rocks from Well Met21 are muddy sparite calcarenites, dominated by fine grains. Pores are poorly developed, and most pores and fractures are filled with calcite. Structural fractures are developed. Rocks have low porosity. Reservoir space is not conducive to hydrocarbon accumulation and it is comprehensively judged that the sedimentary facies around this well is located in the restricted platform facies.

8.2.3 Rock Physics Experimental Measurement

Rock physics experimental measurement can help obtaining more accurate information of reservoir rocks by directly measuring underground core samples. The information includes porosity, permeability, mineral composition, density, P-/S-wave velocity, etc. These are helpful for understanding the relationships between the elastic characteristics and the physical properties of reservoirs to guide the establishment of a theoretical model of target reservoir rocks.

8.2.3.1 Porosity, Permeability, and Density Measurement

The sample is placed in a vacuum oven for drying and weighed with an electronic balance until the weight change is less than 0.01 g. The porosity ϕ and permeability κ of the sample are measured with porosity and permeability measurement system (helium method). Both porosity and permeability of the sample will change with pressure. Since the change in volume of mineral particles is small, the change in volume of the rock under pressure is approximately equal to the change in volume of pores. Namely, when the pore pressure is constant, the changes in porosity with confining pressure can be measured by fluid under pore pressure.

The diameter and length of core sample cylinder are measured with a Vernier caliper to calculate the volume V_b; an electronic balance is used to obtain the weight W_{dry} of the dry sample, so the density of dry rock is

$$\rho_{\text{dry}} = \frac{W_{\text{dry}}}{V_b}. \tag{8.1}$$

The density of water-saturated sample is

$$\rho_{\text{wet}} = \rho_{\text{dry}} + \phi\rho_w. \tag{8.2}$$

where ϕ is the rock porosity and ρ_w is the density of water.

The density of rock matrix is

$$\rho_m = \frac{\rho_{\text{dry}}}{(1-\phi)}. \tag{8.3}$$

As can be seen from the porosity–density relationships (Figure 8.14), the porosities of the two samples from Well Met21 are lower than 2% and the densities are higher than 2.6 g/cm^3, and the porosities of the samples from Well Met22 are

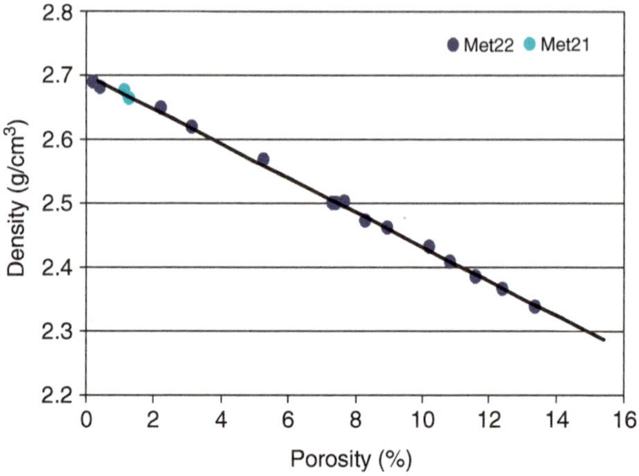

FIGURE 8.14 Porosity–density relationships.

higher than 2%. The densities of limestone samples from Met22 and Met21 decrease with the increasing porosity, with a good linear relationship.

As can be seen from porosity–permeability relationships (Figure 8.15), both samples from Well Met21 have got low permeability. Two samples from Well Met22 have low permeability, and the other samples have got higher porosity and permeability. Measurement data show that the porosity and permeability in this district have good correlations.

8.2.3.2 Single Fluid Saturation and Elastic Wave Experiments

In a single fluid saturation experiment, a rock sample is fully saturated with a single fluid and measured with the changes in P-wave and S-wave velocities with pressure and temperature variation in this state.

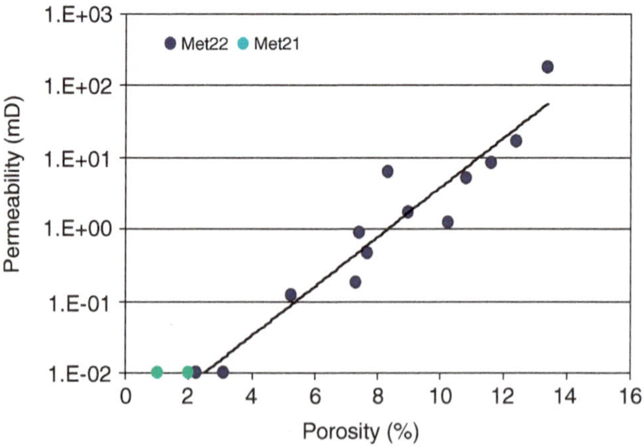

FIGURE 8.15 Porosity–permeability relationships.

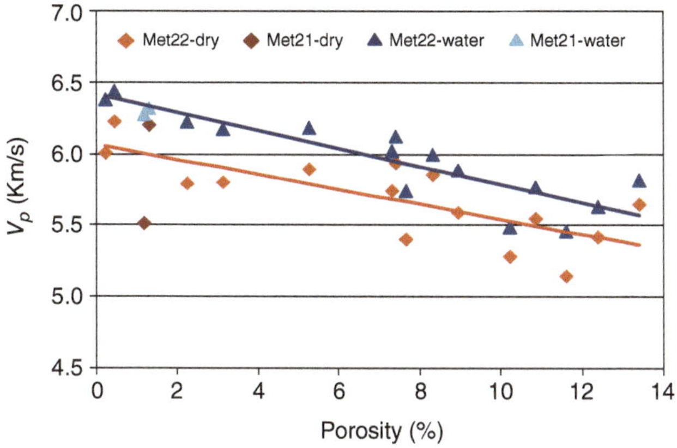

FIGURE 8.16 The relationships between porosities and P-wave velocities.

Pulse transmission technique is used to measure the ultrasonic wave velocity. Electric signals sent by the pulse transmitter are transformed into ultrasonic waves by P-wave or S-wave transducer. The ultrasonic waves transmit through the core and are received by another transducer. The signals are finally sent to the oscilloscope for acquisition. By picking the first arrivals of P-wave and S-wave, the P-wave and S-wave velocities can be obtained by correction and calculation with travel times.

Metejan limestone samples are measured with ultrasonic waves at the temperature of 20°C, pore pressure of 22 MPa and confining pressure of 52 MPa. In the Figure 8.16, orange diamonds denote dry core samples of Well Met22, brown diamonds denote dry core samples of Well Met21, blue triangles denote water-saturated core samples of Well Met22, and blue triangles denote water-saturated core samples of Well Met21. The red line is the fitting line of the measurement data of dry samples, and the blue line is the fitting line of the measurement data of water-saturated samples.

As can be seen from the relationships between P-wave and S-wave velocities and porosity in dry and water-saturation state (Figures 8.16 and 8.17), both P-wave and S-wave velocities decrease with the increasing porosity. In the dry state, the P-wave velocity changes from 6.199 to 5.145 km/s and the S-wave velocity changes from 3.266 to 2.872 km/s. In the water-saturated state, the P-wave velocity changes from 6.323 to 5.458 km/s and the S-wave velocity changes from 3.264 to 2.833 km/s.

As can be seen from the relationships between the P-wave impedances and Vp/Vs ratio (Figure 8.18), with 1.9 as a dividing line, Vp/Vs of dry core samples is generally in range of 1.8–1.9 and that of water-saturated core samples is 1.9–2.0, significantly higher. That indicates that Vp/Vs can be used to effectively distinguish fluid saturation state of rock, thus experimentally proving Vp/Vs can be a sensitive parameter for fluid detection in the Metejan district.

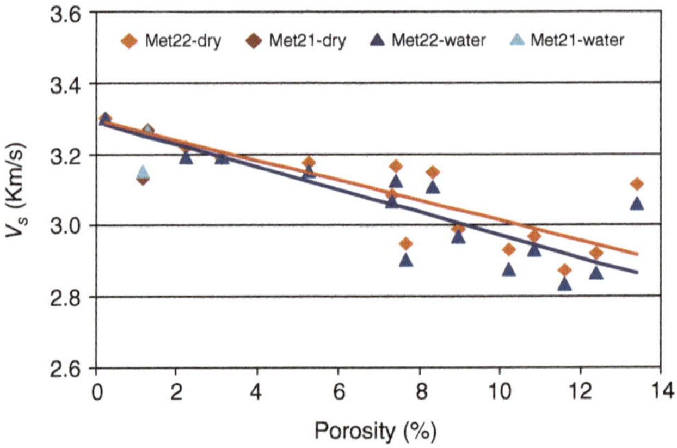

FIGURE 8.17 The relationships between porosities and S-wave velocities.

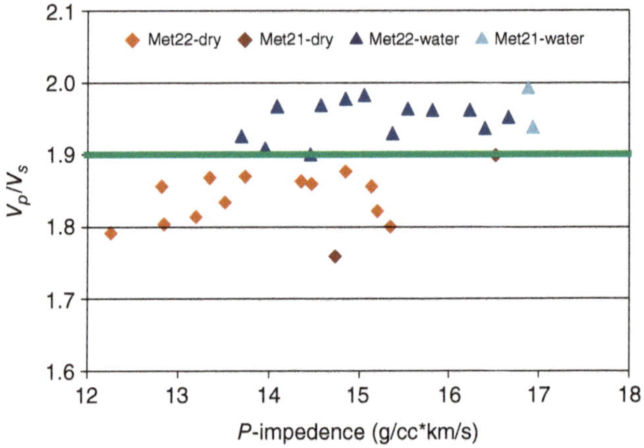

FIGURE 8.18 The relationships between P-wave impedances and VP/VS ratios.

8.3 MULTISCALE ROCK PHYSICAL MODELING

8.3.1 Basic Theories

8.3.1.1 Biot–Rayleigh Equation

Gassmann and Biot were the first to study the mechanism of elastic wave propagation in fluid-saturated porous medium, that is, Biot–Gassmann theory (Gassmann, 1951; Biot, 1956a, 1956b). However, the traditional Biot–Gassmann theory mainly considers the uniform pore structure saturated with a single fluid and cannot describe the complexity of the real reservoirs. White et al. (1975) considered the situation where both gas and water are saturated, where the gas is patchily distributed in the background of fully water-saturated rocks (White, 1975). Dutta and Odé (1979) improved the White theory, so that the P-wave velocity predicted at the low-frequency limit was in good agreement with the analytical results derived from the classical Biot theory (Dutta and Seriff, 1979; Dutta and Odé, 1979). In order to solve the same problem, Johnson proposed a branching function approach to analyze the seismic wave dispersion and attenuation at low and high frequencies and make an approximation at intermediate frequencies, achieving the simulation of wave propagation in partially saturated media at different frequencies (Johnson, 2001). Over the past decade, Müller et al. have conducted simulation of random 1D and 3D partially saturated media and compared the results with the measured data in carbonates and sandstones. They believed that this numerical simulation of random media could also provide a reasonable description and prediction for wave responses of actual rocks (Müller et al., 2010; Gurevivh et al., 1997; Toms et al., 2006).

Nie et al. (2004) considered partial-saturation of fluids in the Biot/Squirt model (Dvorkin and Nur, 1993) (BISQ model). Nie et al. (2012) have considered the influence of both clay-induced viscoelasticity inside the rock and partial-saturation on wave propagation based on the BISQ model (Nie et al., 2012); however, in the treatment of two-phase fluids, it is assumed that gas and water are completely and uniformly mixed. Liu et al. (2009, 2010) studied seismic wave propagation in spherical patchies and alternately layered heterogeneous partially saturated rock model by used of poroelasticity method; however the numerical calculation and implementation process of this approach are complex and not conducive to direct engineering applications.

In 2004, Pride et al. (2004) proposed a theory with the clearer physical significance, double-porosity medium theory, to analyze elastic wave propagation and attenuation in partially saturated medium. In this case, the double-porosity respectively corresponds to the gas-saturated pores and water-saturated pores in a rock saturated with two fluids. Ba et al. (2008a, 2008b, 2008c, 2010) conducted further simulations and analysis on seismic waves in double-porosity medium and compared the prediction results of double-porosity theory with the results of wide-frequency-band experiments on tight sandstone, which is collected from Xujiahe formation from central Sichuan Basin, and proved that the elastic waves in such type of reservoir rocks may exhibit significant dispersion in seismic frequency band. In order to derive the wave propagation equations with succinct formats, less coefficients, and physical realizability of each coefficient, so that they could meet the needs of actual application studies and industrial engineering, Ba et al. (2010, 2011) have derived the Biot–Rayleigh equations (referred to as BR equation) on the basis of Biot theoretical framework that describes the mechanism of seismic wave propagation in partially saturated rocks. By the comparison with previous theories, these equations have been successfully applied in the actual gas exploration project (Ba, 2012). Sun et al. (2012) further compared theoretical results with the experimental data and the scanning results of Casino sandstone from south Australia and discussed the rationality of BR equations aiming at the applications in sandstone reservoirs (Sun et al., 2012).

To analyze the effect of pore fluid on seismic wave propagation in rocks, the traditional Biot theory assumes that the pores inside rocks are uniform. In the 1990s, Berryman and Wang (1995, 2000) considered the situation that there are two types of pores in rocks and derived the basic wave equations of double-porosity medium, that is, an equation set consisting of three formulas. However, the relaxation effect of fluid in the two types of pores that occurs under seismic wave excitation was not considered in the equation set. Therefore, the wave propagation equation in double-porosity medium in that period was only equivalent to the simple sum of two single-porosity Biot equations. The P-wave dispersion and attenuation calculated were in the same magnitude with the traditional Biot theory, much lower than the results of actual measurement. The same problem also occurred in the three-phase (solid, liquid, and gas) theory that was presented by Santos et al. (1990a, 1990b). Zhao (2007) once included viscoelasticity theory into this theory in order to improve the Santos theory's prediction ability of P-wave attenuation in rocks.

Pride and Berryman (2004) improved the double-porosity equations in the three formulas and introduced an additional formula to describe local fluid oscillations under the wave excitation. The method used is similar to Johnson's branch function method (Johnson et al., 1987). The defect of the double-porosity theory is that complex equations and too many coefficients to be determined go against further experimental verifications and engineering applications, and the empirical relationships at the intermediate frequencies have no clear physical meaning. Based on their local flow description method, double-porosity medium wave propagation equations can be changed into an equation set with four formulas. Tang (2011) derived the wave propagation equations in pore-fracture double-porosity medium and applied the equations in the engineering problems of acoustic logging project (Chen and Tang, 2012).

In order to derive a more concise equation to describe the complex fluid oscillation and relaxation process inside heterogeneous rocks under the seismic waves' excitation, the author performed spherical approximation on the heterogeneous body in rocks, thereby to describe isotropic local oscillations of fluid based on the extension of Rayleigh theory (Rayleigh, 1917). The derived BR equations are shown below:

$$\begin{aligned} &N\nabla^2\mathbf{u} + (A+N)\nabla e + Q_1\nabla(\xi^{(1)} + \phi_2\varsigma) + Q_2\nabla(\xi^{(2)} - \phi_1\varsigma) \\ &\quad = \rho_{11}\ddot{\mathbf{u}} + \rho_{12}\ddot{\mathbf{U}}^{(1)} + \rho_{13}\ddot{\mathbf{U}}^{(2)} + b_1(\dot{\mathbf{u}} - \dot{\mathbf{U}}^{(1)}) + b_2(\dot{\mathbf{u}} - \dot{\mathbf{U}}^{(2)}), \end{aligned} \tag{8.4a}$$

$$Q_1\nabla e + R_1\nabla(\xi^{(1)} + \phi_2\varsigma) = \rho_{12}\ddot{\mathbf{u}} + \rho_{22}\ddot{\mathbf{U}}^{(1)} - b_1(\dot{\mathbf{u}} - \dot{\mathbf{U}}^{(1)}), \tag{8.4b}$$

$$Q_2\nabla e + R_2\nabla(\xi^{(2)} - \phi_1\varsigma) = \rho_{13}\ddot{\mathbf{u}} + \rho_{33}\ddot{\mathbf{U}}^{(2)} - b_2(\dot{\mathbf{u}} - \dot{\mathbf{U}}^{(2)}), \tag{8.4c}$$

$$\begin{aligned} &\phi_2(Q_1e + R_1(\xi^{(1)} + \phi_2\varsigma)) - \phi_1(Q_2e + R_2(\xi^{(2)} - \phi_1\varsigma)) \\ &\quad = \frac{1}{3}\rho_{f1}\ddot{\varsigma}R_0^2\frac{\phi_1^2\phi_2\phi_{20}}{\phi_{10}} + \frac{1}{3}\frac{\eta_1\phi_1^2\phi_2\phi_{20}}{\kappa_{10}}\dot{\varsigma}R_0^2, \end{aligned} \tag{8.4d}$$

where $\mathbf{u} = [u_1, u_2, u_3]$, $\mathbf{U}^{(1)} = [U_1^{(1)}, U_2^{(1)}, U_3^{(1)}]$, and $\mathbf{U}^{(2)} = [U_1^{(2)}, U_2^{(2)}, U_3^{(2)}]$, respectively, denote space vectors of the displacements of three components. The suffixes 1, 2, and 3 denote the three directions of vector space, respectively. ς is the local fluid deformation increment generated in the wave excitation, and

$$e_{ij} = \frac{1}{2}\left(\frac{\partial u_i}{\partial x_j} + \frac{\partial u_j}{\partial x_i}\right), \tag{8.5a}$$

$$\xi_{ij}^{(1)} = \frac{1}{2}\left(\frac{\partial U_i^{(1)}}{\partial x_j} + \frac{\partial U_j^{(1)}}{\partial x_i}\right)\delta_{ij}, \tag{8.5b}$$

$$\xi_{ij}^{(2)} = \frac{1}{2}\left(\frac{\partial U_i^{(2)}}{\partial x_j} + \frac{\partial U_j^{(2)}}{\partial x_i}\right)\delta_{ij}, \tag{8.5c}$$

where x_1, x_2, and x_3 denote the coordinates in the three directions, respectively. ϕ_1 and ϕ_2 denote the absolute porosity of the two types of pores. The total porosity of rock is $\phi = \phi_1 + \phi_2$. ϕ_{10} and ϕ_{20} denote the local porosity in the two areas, respectively. If there is only one uniform skeleton in the rock and two fluids are saturated, then $\phi_{10} = \phi_{20} = \phi$. ρ_{f1} and η_1 denote the density and viscosity of the fluid in background-phase; R_0 denotes gas pocket radius and κ_{10} is rock permeability.

8.3.1.2 Numerical Example with Carbonates: A Comparison with Previous Theories

To validate the BR theory, a comparative analysis was performed on the prediction results of BR theory, White theory (reformulated by Dutta), and Johnson theory on the basis of the same limestone properties. The rock physical parameters included matrix bulk modulus of 71.59 GPa, skeleton bulk modulus of 19.77 GPa, skeleton shear modulus of 10.31 GPa, water bulk modulus of 2.25 GPa, air bulk modulus of 0.144 MPa, water viscosity of 0.001 Pa·s, air viscosity of 0.00001 Pa·s, matrix density of 2.7g/cm^3, water density of 0.997 g/cm^3, air density of 0.124 g/cm^3, porosity of 0.21, permeability of 5.5 mD, and the average gas pocket size of 2 cm. In this example, air saturation was set as 0.1.

Figure 8.19 shows the results of comparison between the three theories. In Figure 8.19A, the lines BGW and BGH, respectively, give the Biot–Gassmann–Wood boundary (the two types of fluids are in parallel in mechanics and treated as an equivalent fluid, then the equivalent fluid is substituted into Gassmann equation to estimate the overall modulus of partially saturated rock). The Biot–Gassmann–Hill boundary (the two types of fluid are respectively substituted into Gassmann equation to estimate the two moduli of fully saturated rocks, then Hill average is used on the two moduli to estimate the overall modulus of the rock). The predicted velocities from the three theories are in good

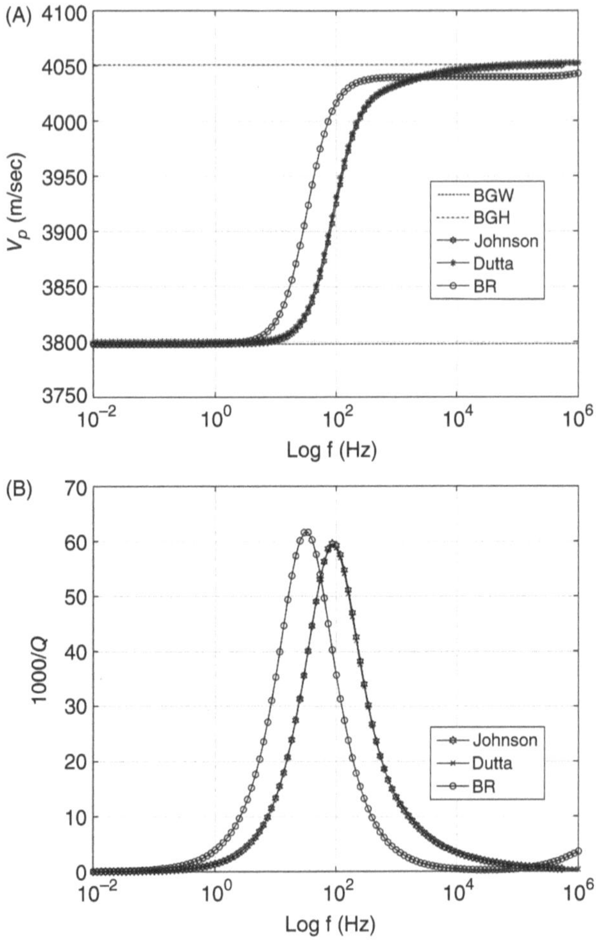

FIGURE 8.19 Comparison between the limestone prediction results of Biot-Rayleigh theory, Johnson theory, and the White theory. (A) P-wave velocity. (B) P-wave attenuation.

agreement with BGW boundary at low-frequency limit. At the high-frequency limit, the prediction results of the White theory and Johnson theory are in good agreement with BGH boundary (4050 m/s), and BR curve also approaches to BGH boundary but is slightly lower.

In comparison with the prediction results of Johnson's theory, the "step" of P-wave velocity dispersion curve and the peak of P-wave attenuation curve in BR prediction results are closer to the left end of frequency axis. As is shown in Figure 8.19B, the predicted attenuation peak by BR theory occurs in the frequency band of 10–100 Hz. Its peak frequency is lower than that of Johnson theory, and the predicted peak attenuation of BR theory is slightly higher than that of Johnson and White theories in amplitude.

8.3.1.3 Comparison of the Theoretical Prediction Results with the Limestone Experimental Data

To verify the applicability of BR theory in the actual gas-saturated limestone reservoirs, the experimental data of limestone wave velocities that were measured by Cadoret et al. (1995) at acoustic and ultrasonic frequencies were adopted. Estaillades limestone is bioclastic limestone, with the solid grain of calcite, porosity of 0.3, permeability of 0.255D, skeleton bulk modulus of 7.45 GPa, and skeleton shear modulus of 6.09 GPa. Cadoret et al. (1995) measured the P-wave and S-wave velocities in rock samples with resonance-bar techniques at the acoustic frequencies (KHz), while the pulse transmission method was used in the masurements at the ultrasonic frequencies (10^2 KHz).

The average gas pocket size was set as 1.05 mm. The comparison between the experimental and theoretical predictions is shown in Figure 8.20. In Figure 8.20A, at the frequency of 1 KHz, the experimental results of the P-wave velocity are distributed close to the BGW boundary and vary with variation of water saturation. The prediction results of three theories are close and consistent with the variation features of the experimental data. Within the range of low water-saturation, the P-wave velocity predicted with the reformulated White theory is higher than other models and its prediction at full gas saturation does not agree with the BG theory. In the range of high water saturation, the prediction of BR theory is closer to the experimental data. According to Figure 8.20B, at the frequency of 500 KHz, the experimental measurements of P-wave velocity are basically distributed close to the BGH boundary. If a gas pocket size is kept unchanged, the prediction of BR theory is in the best agreement with the experimental measurement, followed by that of White theory. The prediction of Johnson theory has a large deviation from the experimental data. The comparison shows that based on the same rock sample and same fluid spatial distribution, BR theory has slight advantages and relatively simple mathematical implementation process in predicting the P-wave velocity variation of rocks in different scales versus gas saturation status of reservoir rocks.

Figure 8.20B also shows that, in the results measured at 500 KHz, some experimental data points exceed the BGH boundary when the rock is nearly full water saturation. This may be related to the measurement error. Besides, only P-wave dispersion caused by gas/water partial saturation is considered in the current theories and the influence of complex pore structures on the waves is neglected, so there is a possibility of underestimated P-wave velocities (especially at the intermediate and high frequencies) in the theoretical prediction when it approaches to full water saturation.

8.3.2 Rock Physical Modeling and Workflow

8.3.2.1 Mineral Grain and Dry Rock Matrix

Rock physics modeling process for the real oil/gas-saturated reservoirs generally involves three steps (Xu and White, 1995; Xu and Payne, 2009): (1) analyzing

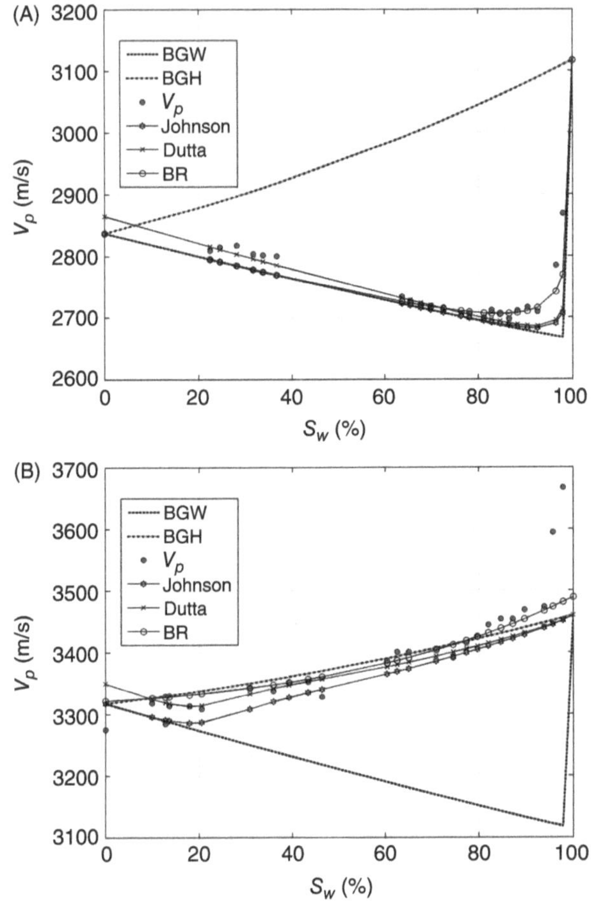

FIGURE 8.20 Comparison of theoretical predictions with experimental observation results of *P*-wave velocities in partially saturated Estaillades limestone. (A) At acoustic frequencies (1 kHz). (B) At ultrasonic frequencies (500 KHz).

the rock mineral components and their volume ratios and calculating the elastic parameters and density of mineral mixture; (2) calculating the elastic parameters and density of dry rock skeleton with theoretical, empirical formula, or other methods on the basis of the rock mineral properties, pore structures of reservoirs, consolidation degree of skeletons, and the other related factors; and (3) analyzing the physical properties of fluid in reservoirs, estimating wave response characteristics of fluid-saturated rocks on the basis of the relationships between fluid distribution characteristics and pore structures, and linking it to practical engineering problems.

The commonly used approaches for modeling rock mineral mixtures include Voigt–Reuss–Hill averaging method, Hashin–Shtrikman–Hill averaging method, and the time average equation (Mavko et al., 2009). In case of complex components and combination, self-consistent theory can be used for modeling. For rocks with relatively less components and compact structures, such as limestone containing more than 90% of calcites and low percentage of clay and other minerals, Voigt–Reuss–Hill averaging method can be used to get a theoretical estimation. The method is used in modeling of this chapter, as shown in the formula below:

$$M_g = \frac{1}{2}\left(\sum_{i=1}^{N} f_i M_i + \frac{1}{\sum_{i=1}^{N} \frac{f_i}{M_i}} \right) \tag{8.6}$$

where M_g denotes elastic modulus of the mineral mixture, f_i and M_i denote volume ratio and elastic modulus of the ith mineral, respectively, and N denotes the number of minerals in the rock.

To obtain the mineral properties of rock, an important aspect of modeling is to estimate the elastic modulus of dry rock skeleton K_b. Seismic wave response characteristic of reservoirs can be obtained by fluid substitution with K_b, so the precision of K_b is directly related to the applicability of the rock physics model. Xu et al. adopted differential effective medium theory (Norris, 1985) and the Kuster–Toksöz theory (Kuster and Toksöz, 1974) for modeling in sandstones and carbonates. In this chapter, modeling for dry rock skeleton is performed with a semiempirical formula proposed by Pride, as shown in the below formula:

$$K_b = \frac{(1-\phi)K_s}{1+c\phi}, \tag{8.7a}$$

$$\mu_b = \frac{(1-\phi)\mu_s}{1+c'\phi}, \tag{8.7b}$$

where μ_b denotes dry rock skeleton's shear modulus, K_s and μ_s denote mineral bulk modulus and shear modulus, respectively, and empirical parameters c and c' are related to the consolidation degree of rock.

Take limestone as an example, if the rock minerals composed mainly of calcite, different parameters c and c' are selected and the variation of dry rock skeleton's bulk and shear moduli versus porosity is shown in Figure 8.21. The elastic parameters of dry rock skeleton decrease with the increase of porosity, and this decreasing trend is more obvious with the decrease of consolidation degree of rocks. The degree of consolidation of the rock can be determined according to the specific conditions of the reservoirs.

8.3.2.2 Fluid Substitution

On the basis of rock mineral mixture and dry skeleton properties, the seismic response characteristics of rock with different pore fluid and at different saturation degrees can be analyzed by substituting fluid in pores. As shown in Figures 8.19

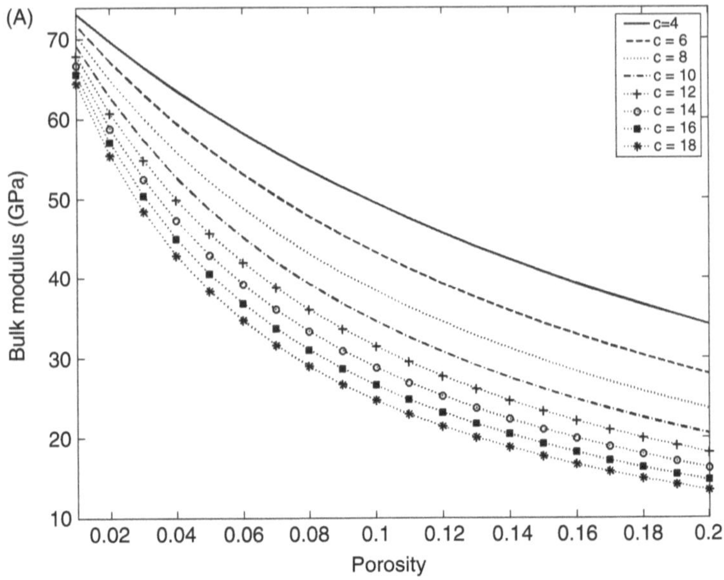

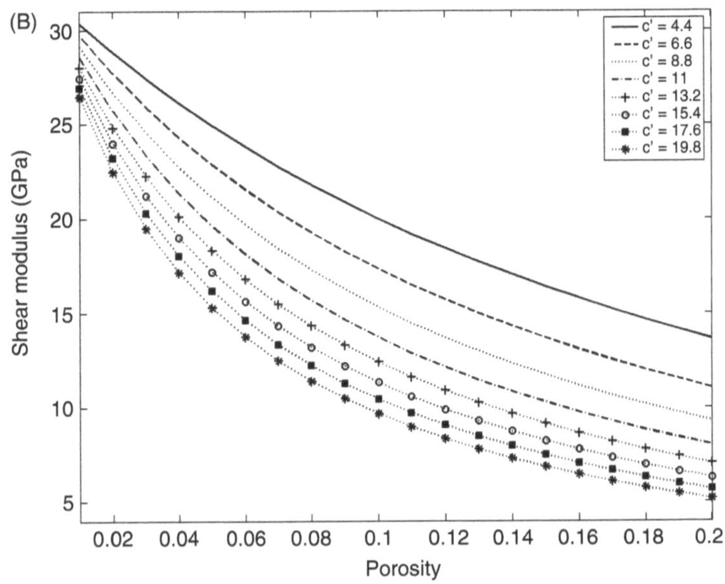

FIGURE 8.21 Elastic moduli of dry limestone matrix versus porosity. (A) Bulk modulus. (B) Shear modulus.

and 8.20, due to the influence of pore fluid, there are significant differences in the P-wave velocities of different frequency bands for the same rock, reflected in the P-wave dispersion effect in multiscale seismic data (Sams et al., 1997). P-wave velocity in different frequency bands in the multiscale waveform data can be predicted with BR equation.

As to a limestonne, dry rock skeleton properties are set as follows: porosity of 0.16, skeleton bulk modulus of 24.81 GPa, skeleton shear modulus of 11.79 GPa, permeability of 30.23 mD, and average gas pocket size of 5 mm. The rock mineral is calcite. Fluid density and elastic parameters at in situ temperature and pressure can be estimated with the van der Waals equation. Figure 8.22 shows the P-wave velocity and attenuation versus water saturation in partially saturated limestone at several frequencies. For the same limestone, P-wave velocity at different frequencies varies significantly with water saturation. At low frequencies, P-wave velocity curve is typically "hook-shaped." At ultrasonic frequencies, P-wave velocity shows a nearly linear monotonic rise with the increase of water saturation. In this example, the strongest P-wave attenuation occurs in the acoustic band and in the water saturation range of 85%–95%. At the intermediate and low frequencies, P-wave attenuation mainly occurs in high water saturation. Besides, the lowest point of velocity curve and the peak of attenuation curve move towards the left end of the water-saturation axis with an increasing frequency.

8.3.2.3 Rock Physics Modeling Workflow

According to the actual reservoir conditions, minerals, dry skeleton, and fluid in pores are analyzed on the basis of theoretical and empirical knowledge and experimental observations to give a quantitative relationship between seismic wave response characteristics and lithology, pores, and fluid. That is rock physics modeling.

Figure 8.23 shows the flowchart of multiscale rock physics modeling. For rock physics modeling and analysis in the actual engineering, the initial model obtained on the basis of theoretical and empirical knowledge certainly needs an iterative debugging. The major basis information for reference in debugging process includes well-log data, well-log interpretation results, prestack, and poststack seismic inversion results and oil/gas production test results of some wells. If petrophysical experiments can be carried out aiming on the target stratum, it will help to improve the accuracy of model prediction. In the situation that sufficient cores and experimental data cannot be obtained, previously published experimental data can be properly consulted on the premise that the rock natures are comparative to the target strata of the block.

8.3.3 Multiscale Rock Physics Template and Its Calibration

Modeling of limestones in Metejan XVm layer is performed with the workflow shown in Figure 8.23. Calcite and a small amount of clay are considered as the mineral mixture and formula (8.7) is used to calculate the elastic moduli of dry rock skeleton at different porosities. At different frequencies, multiscale P-wave velocities can be obtained by plane wave analysis of formula (8.4). In the process of S-wave calculation, the effect of fluid on S-wave dispersion is generally ignored. Finally, rock physics template can be formed on the basis of multiscale

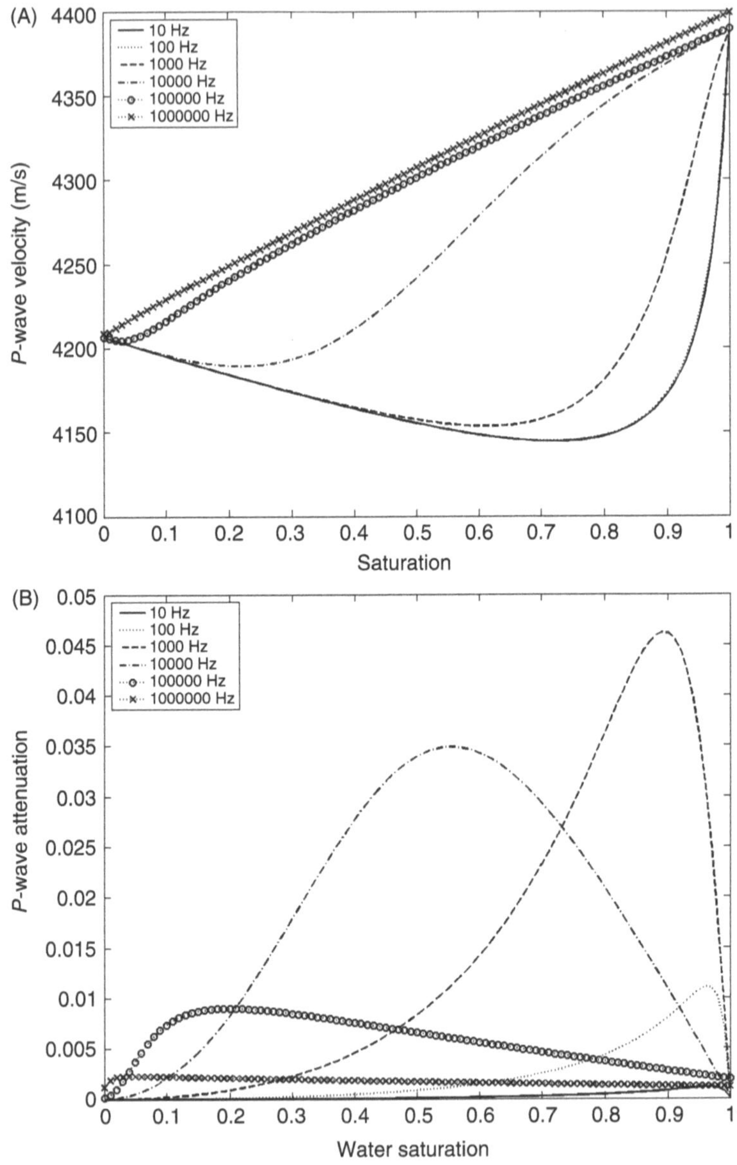

FIGURE 8.22 *P*-wave velocity (A) and attenuation (B) versus water saturation in partially-saturated limestone at different frequencies.

P-wave and S-wave velocities. The calibration of rock physics template is based primarily on experimental data, well-log results, and seismic data.

Figure 8.24 shows the comparative results of rock physics template and experimental data at different frequencies after calibration. There are only a few previous literatures on wave velocity observations in partially saturated

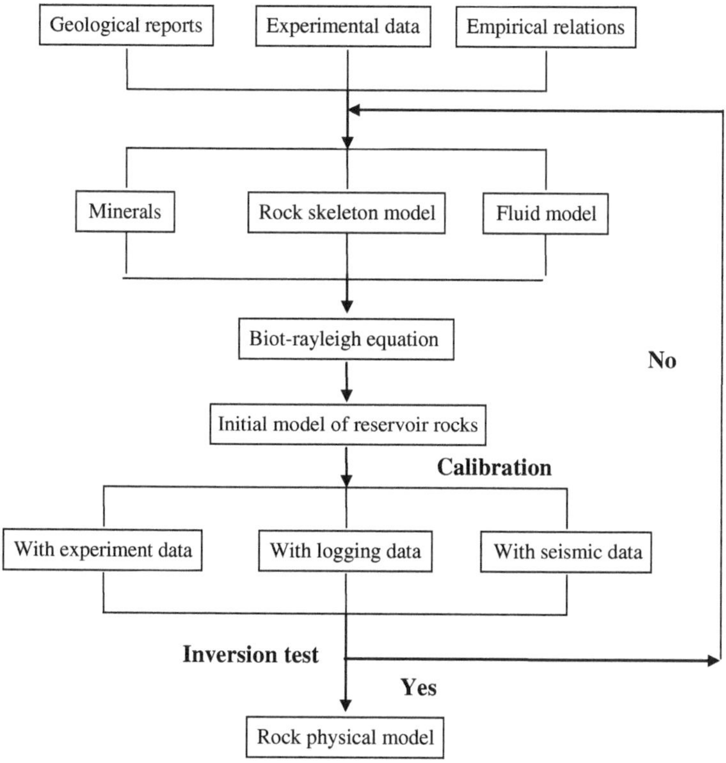

FIGURE 8.23 Multiscale rock physics modeling flowchart.

carbonate experiments, and the experimental data obtained below acoustic frequencies (including seismic frequencies) are especially rare. The experimental data of carbonates shown in Figure 8.24 are selected from Adam et al. (2006) and Cadoret et al. (1995) and they are mainly limestone. Adam et al. (2006) reports the wave observations of fully water-saturated and dry carbonate at the frequencies around 100 Hz and 800 KHz, and Cadoret et al. (1995) reports the wave observations of limestone at different water saturation degrees at the frequencies around 1 and 500 KHz. The former work includes only two data points of the same limestone respectively in full water saturation and dry status (close to gas saturation status), where the dry rock bulk modulus is obtained by a reverse substitution with Gassmann equation. Cadoret et al. (1995) includes the measurements at different saturation degrees. Rock properties for each sample are summarized in Table 8.1. Samples A, C, and G are selected from Cadoret et al. (1995) and samples B, D, E, and F are selected from Adam et al. (2006).

The comparison between template and experimental data suggests that the observation results of some samples (A, B, C, and D) are in good agreement with the template (for limestone of Metejan XVm layer) that are produced with the method in this chapter. The distribution of experimental data of these samples (each sample

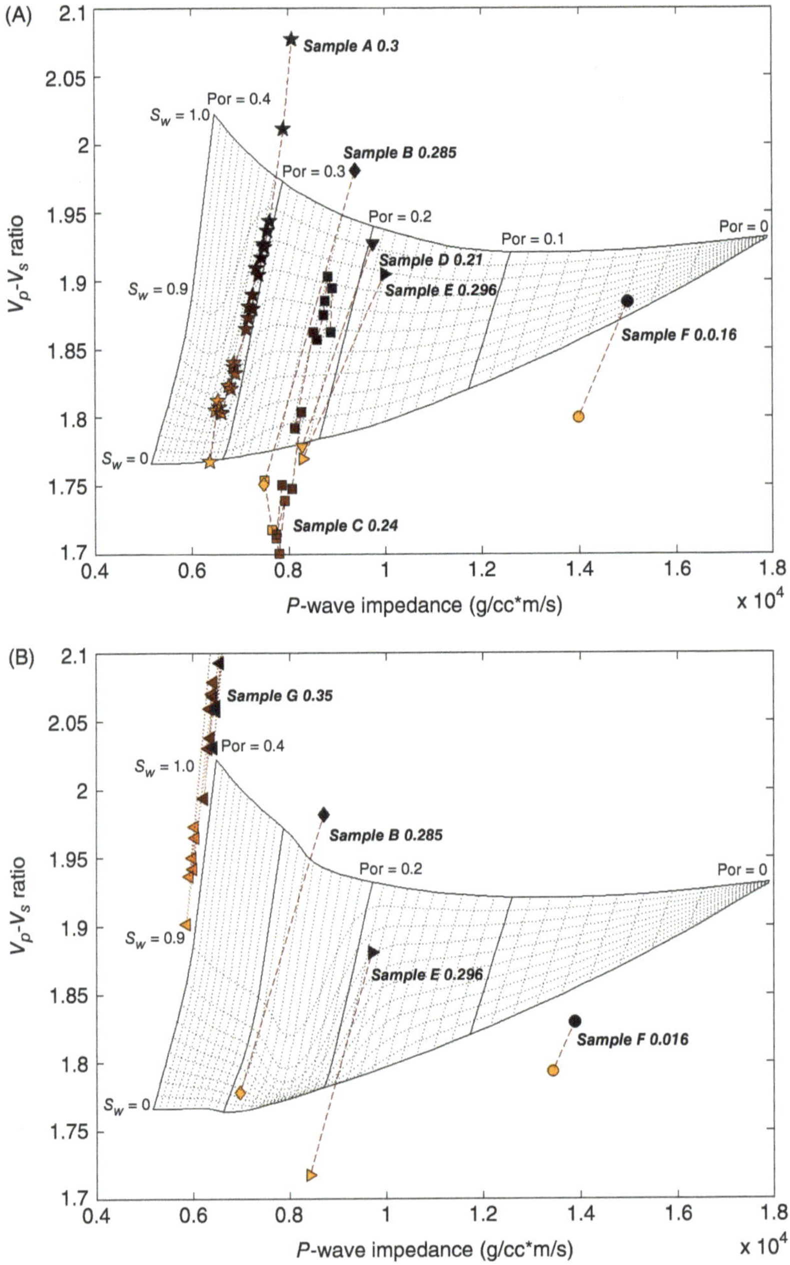

FIGURE 8.24 The comparison between rock physics templates and rock experimental data of carbonate in the two scales. (A) Ultrasonic frequencies (>500 KHz). (B) Below acoustic frequencies (<1 KHz). The changes of color of the symbols of experimental data (from dark to light) indicate the decreasing water saturation or increasing gas saturation.

TABLE 8.1 Properties of Rock Samples

	A	B	C	D	E	F	G
Porosity	0.3	0.285	0.24	0.21	0.296	0.016	0.35
Permeability/mD	255	0.43	46	5.5	103	0.03	4067
Lithology	Bioclastic limestone	–	Oolitic limestone	Dolomitic limestone	–	Dolomitic limestone	Bioclastic limestone

has a constant porosity) versus porosity are in good agreement with the longitudes of the rock physics template. In the experimental data, the wave response variation with saturation is the consistent with the parallel, and the wave response rangeability from full water saturation to full gas saturation also agrees with the template. According to the above results, although these samples are not collected from the target zone, the common characteristics of geology, structure, and pore fluid distribution of some limestone result in that similar rock physics templates can possibly be applicable to limestone from different areas. The mutual reference and calibration are realized for the rocks from different areas.

Samples (E, F, and G) are not in good agreement with the rock template. P-wave impedance observed on sample E is significantly higher than the theoretical template value at the same porosity. P-wave impedance observed on sample F is significantly lower than the theoretical template value at the same porosity. That is related to the specific mineral components and structural characteristics of those samples. The Vp/Vs ratio of sample F is also very low and there may contain cracks inside. Compared with rock template, P-wave impedance observed on sample G is lower while the Vp/Vs ratio is higher, which can be related to high permeability. Considering the diversity and complexity of pore structures of actual carbonate reservoir, it is impossible to describe all the observed phenomena with only one template. In the current stage, we have to select a specific type of reservoirs and perform modeling and inversion by use of the typical "rock template" with certain general statistical significance. Multisolution problem cannot be completely resolved.

Comparison between Figure 8.24A and B clearly shows that experimental data at ultrasonic frequencies are in better agreement with the template. There are fewer measured data at sonic frequencies with large error disturbance, which are in poor agreement with the template. At present, there is still potentials for the improvement of experimental observation precision on low-frequency elastic waves. In the seismic frequency band, there is still no work in literature on the studies of elastic wave measurement for carbonates with variable fluid saturation.

In Figure 8.25, the comparison between the calibrated rock physics template at acoustic frequencies and well-log interpretation results of some reservoir

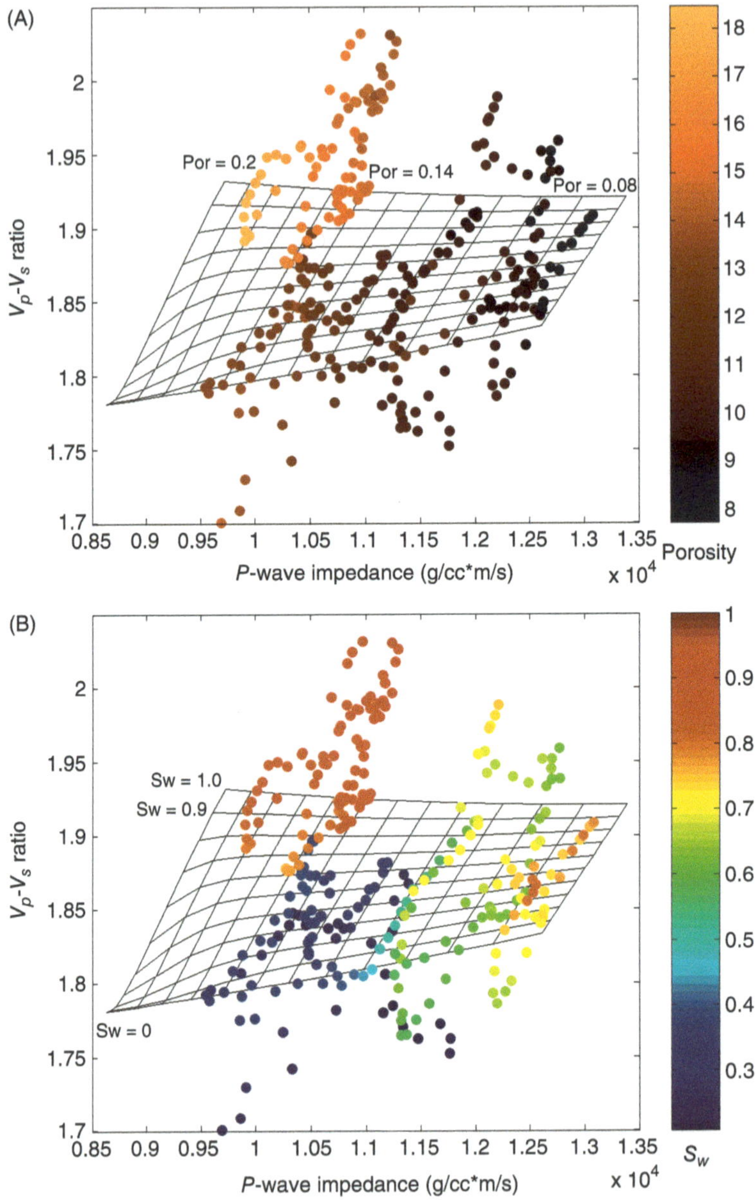

FIGURE 8.25 Comparison between rock physics template and well-log interpretation results of limestone. (A) Porosity (the color bar gives porosity scale). (B) Gas saturation (the color bar gives water saturation scale).

layers around Well Met22 is shown. The wave impedance and Vp/Vs ratio of the symbols are calculated on the basis of interval transit time of sonic waves in log data. In Figure 8.25A, the fine interpretation results of porosity from log data are in excellent agreement with the longitude distribution features on rock physics template. The log data show that the porosity of high-porosity reservoirs is nearly 0.2 and the lower limit of low-porosity reservoirs is around 0.08, respectively, consistent with the template porosity longitude curves for the porosities of 0.19 and 0.08. Both the log data and the template show that the porosities of reservoirs are mostly in range of 0.12–0.16. It is worth noting that, in the porosity results of an initial well-log interpretation, the porosity of high-porosity reservoirs is close to 0.24, and the interpreted values of porosity are generally higher. According to the analysis and comparison with rock physics template, initial interpretation results are found to be problematic. After an improved reinterpretation, the updated porosity is in agreement with template and consistent with the actual drilling reports and the geological understandings around the well.

In Figure 8.25B, some log data are distributed outside the range of rock physics template. That is caused by the complexity of internal pore structures and fluid distributions of carbonate. According to the P-wave impedance and Vp/Vs ratio variations versus water saturation, which is shown by the color bar, log data symbols of full water saturation are basically located close to the latitude curve of saturation 1 on template. Data points of nearly full water saturation are distributed along the latitude curve of saturation 0. Within the high porosity range (0.13–0.19), the interpreted saturation from log data has a polarization trend and is in better agreement with the template boundary. Within the low porosity range (0.08–0.13), the interpreted saturation has got obvious features of gradual variation and is in better agreement with the intermediate of the template.

Figure 8.26 shows the comparison between the calibrated rock physics template at seismic frequencies and the seismic data extracted from the data volume of poststack and prestack seismic inversion. The data is selected from the seismic traces around wells Met22 and Met23 in the block. According to the initial interpretation results, the reservoirs are classified into three categories, namely, water reservoirs, gas reservoirs, and low-porosity rocks. On the basis of log interpretation results and gas production reports of the two wells, after a well to seismic calibration, seismic data are sorted out from the corresponding water reservoirs, gas reservoirs, and low-porosity rocks near the top boundary of XVm layer. The possibility of misjudgements cannot be completely ruled out in the selection process of seismic data.

The yellow, blue, and pink colors in the figure denote gas reservoirs (including poor gas reservoirs), water reservoirs, and low-porosity rocks, which were the results of previous seismic interpretation.

As is shown in Figure 8.26, the seismic data of water reservoirs are generally distributed along the latitude curve of water saturation 1. The Vp/Vs ratios

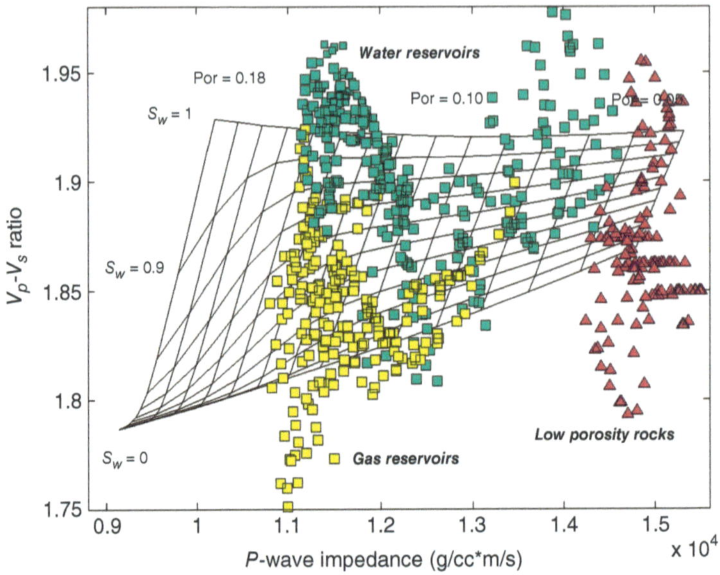

FIGURE 8.26 Comparison between rock physics template and seismic data of limestone.

of some symbols are higher than the upper bound of the template. Vp/Vs ratio in the seismic data of gas reservoirs show a wide distribution range, and the latitude curves on template corresponds to the gas reservoirs range from 30% to 100% for gas saturation. The Vp/Vs ratios of some symbols of gas reservoirs are lower than the lower bound of the template. In the porosity range of 0.12–0.15 (for the main reservoirs), the symbols of seismic data are in best agreement with the template. Therefore, it is feasible to detect gas in reservoirs on the basis of this template. The range of porosity of low-porosity rock symbols is 0.04–0.05. In the previous geological studies on the target stratum, the lower limit of porosity of reservoir was set as 0.05, which is slightly less than the value used in seismic interpretation and inversion. According to the figure, the rock physics template can efficiently distinguish between the defined reservoirs and low-porosity rocks.

8.3.4 Porosity and Gas Saturation Prediction

Based on the foregoing analysis, rock physics templates which are suitable for describing the seismic response characteristics of partially saturated limestone from target stratum is obtained. According to the template in seismic scale, the actual seismic responses can be related to rock porosity and saturation by using P-wave impedance and Vp/Vs ratio that are output from the conventional prestack seismic inversion.

Template in seismic scale is obtained on the basis of rock physics modeling. First, prestack and poststack inversions are performed for the target reservoir layers to get the data volume of wave impedances and Vp/Vs ratios. The wave impedance and Vp/Vs ratio are then extracted and projected on the template point by point. The grid point on template which is closest to the projection point is found. In the condition that the distance between a projection point and a grid point is less than a certain threshold, porosity and saturation of the grid point are the inversed values of porosity and saturation at the corresponding reservoir position. Quantitative prediction of reservoir properties and even a direct detection of hydrocarbons can be performed on the basis of this method.

In the target stratum, if the project point is located outside the range of template and has got large differences to the bound curves, it is generally taken as low-porosity rocks, and porosity and saturation are set to zero. In the inversion, only the main limestone reservoirs are concerned. Low-porosity tight limestone, mudstone, low-porosity marlite, and gypsum, which are located outside the template, are not the exploration targets; therefore, they can be directly treated as nonreservoirs.

Wave impedance, P-wave and S-wave velocities, and density data are obtained on the basis of prestack and poststack inversions. Rock physics template is applied between the top and bottom of XVm layer. Figure 8.27 shows the inversion results of porosity and saturation on a cross-well 2D seismic line. Well Met3 is an industrial gas-production well and has got sonic wave measurement. Wells Met1, 4, and 6 have no sonic log data. The inversion results around Met3 also exhibit strong features of gas saturation (indicated by the yellow color), followed by the wells Met1 and 4, with some slight features of gas-saturation. Well Met6 is the worst, around which the time profile shows only a small amount of water reservoirs (indicated by the blue color). According to the gas production test results, well Met3 is of high gas-production rate and wells Met1 and 4 are of low gas-production rate, while well Met6 only produces water in the target stratum. The results are consistent with that of the inversed porosity and saturation profiles.

The rock physics modeling technique that is proposed for partially saturated limestone gas reservoir is mainly applicable for the description of porous-type (Vuggy-type) reservoirs with simple mineral components, and the rock physics template established is mainly built for the Metejan limestone of XVm layer. For more complex reservoirs, rock physics models need to be further studied. If the various problems that are brought by different lithologies, structures, and fluids are expected to be solved, rock physics template must be reformed accordingly.

8.4 ROCK PHYSICS MODELING IN HETEROGENENOUS CARBONATE RESERVOIRS

8.4.1 Effect of Pore Shapes on Seismic Wave Velocity

The pore structures of carbonate reservoird are complex and have strong heterogeneity. Pore shapes may have significant effects on elastic wave velocity. Generally, elastic wave velocity in rocks decreases with the increasing porosity. With the

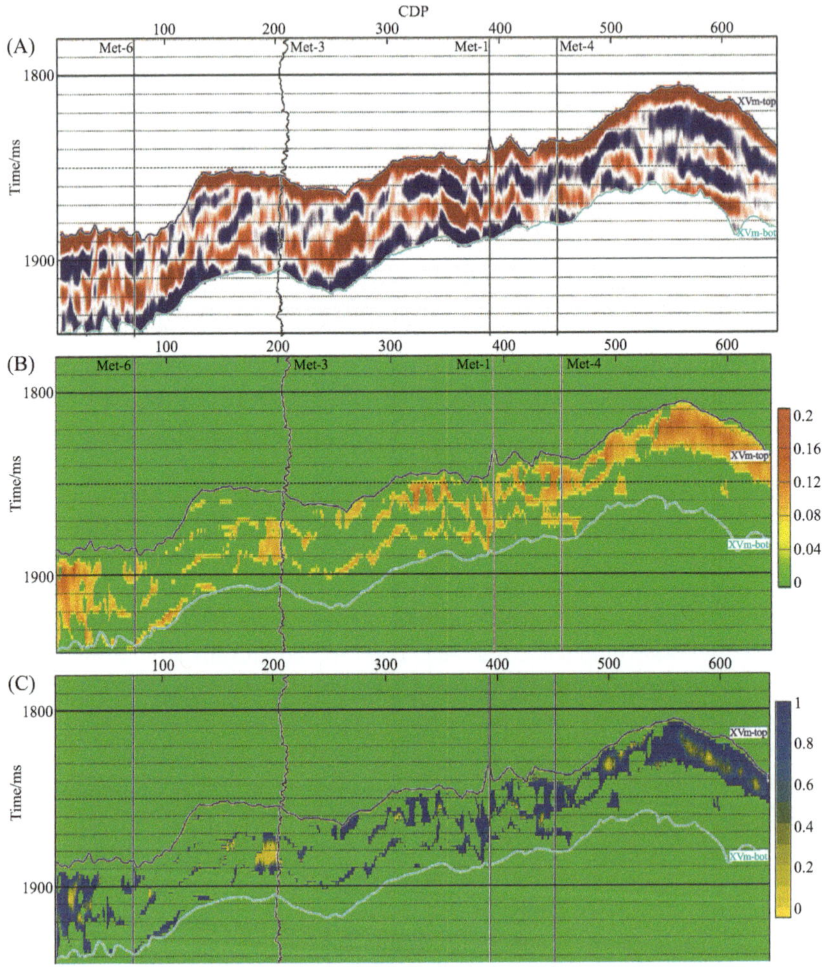

FIGURE 8.27 2D seismic data and inversion results from the profiles crossing wells Met6, Met3, Met1, and Met4. (A) Poststack seismic data. (B) Inversion results of reservoir porosity. (C) Inversion results of reservoir water saturation.

same porosity, however, the fluctuation of measured wave velocities in carbonate reservoir rocks may exceed 1000 m/s due to the influence of pore structures.

For the regions with the significant lateral variations in pore structure, rock physics modeling method based on a unique set of pore parameters will be problematic for applicability. In the actual applications, the variation patterns of pore shapes should be learned in reservoir rocks all over the block, then its effect could be considered in the modeling parameters and the precision of seismic prediction can be improved.

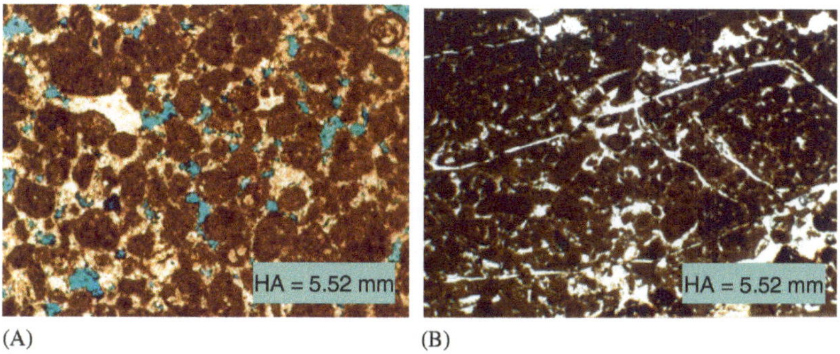

FIGURE 8.28 Pore shapes in Metejan limestone reservoirs. (A) Thin section from well Met22. (B) Thin section from well Met21.

Through the studies on lateral heterogeneities of actual carbonate reservoirs by means of thin section analysis, the different pore structure characteristics around well Met21 and well Met22 can be observed. The reservoir around well Met22 has got coarse rock grains and the dissolved pores are well developed, with plane porosity of about 7%, which will be advantageous for oil/gas accumulation (Figure 8.28A). The rocks around well Met21 have small calcite grains and cracks (flat pores) are developed, with plane porosity of about 2% (Figure 8.28B).

Rock physics experiments are performed on the core samples to investigate the effects of pore shapes on wave velocities in Metejan limestone reservoirs (Figure 8.29). The comparison shows that with the increase of porosity, the

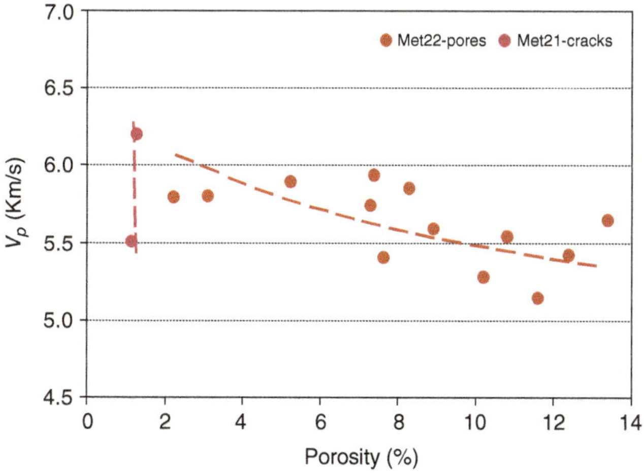

FIGURE 8.29 Effect of pore shapes of Metejan limestone reservoirs on measured wave velocity.

effect of dissolved pores on wave velocity exhibits a stable and gradual decline, while that of cracks is much more significant. In further applications, rock physics modeling needs to be performed according to the two types of pore shapes: dissolved pores and fractures.

8.4.2 Rock Physics Inversion Based on Pore Shape Analysis

8.4.2.1 Workflow for Rock Physics Modeling and Inversion

The related properties of in situ gas reservoirs include brine bulk modulus $K_f^{(1)} = 2.51$ GPa, brine viscosity $\eta^{(1)} = 0.001$ Pa · s and density $\rho_f^{(1)} = 1.04$ g/cm^3, gas bulk modulus $K_f^{(2)} = 0.081$ GPa, gas viscosity $\eta^{(2)} = 0.00015$ Pa · s and density $\rho_f^{(2)} = 0.17$ g/cm^3, and gas pocket radius $R_0 = 5$ mm. The input parameters include gas saturation S_g, rock porosity ϕ, tortuosity α (dependent on porosity), and permeability κ (for Metejan limestone, the empirical relation is $\kappa = 0.0012e^{63.34\phi}$).

Dry rock matrix's bulk modulus K_b and shear modulus μ_b, solid grain bulk modulus K_s and solid density ρ_s are calculated based on the initial input parameters. The initial input parameters are set according to the thin section analysis on rock samples from each well. However, the input parameters should be finally determined by seismic data calibration at the locations around each well.

Figure 8.30 illustrates the workflow to build the rock physics template, and the inversion of rock porosity and hydrocarbon saturation. Mineral components and pore structures are investigated by thin section analysis around each well so that the lateral heterogeneities are considered. The template is constructed iteratively by proper calibration with cross-well seismic data as well as available production tests. The accuracy of the template is improved by performing rock physics experiments on cores. Based on the rock physics template, the P-wave impedance and Vp/Vs ratio obtained from seismic inversion are projected to the template, rock porosity, and fluid saturation are estimated.

8.4.2.2 Rock Physics Templates for Heterogeneous Reservoirs

The templates corresponding to the seismic band are compared with the seismic inversion data in Figure 8.31 and the two templates are built according to the different pore structures of the two wells. The input parameters (Figure 8.30) and the calibrated template vary spatially because of the lateral heterogeneity in the target area. On each template, the porosity-related variations (3%–17%) correspond to longitude curves, whereas the saturation-related variations (0%–100%) correspond to latitude curves. The data (symbols in Figure 8.31) are extracted from the inverse data traces in the depth range of the target stratum and at the borehole locations. The XVm reservoirs were qualitatively classified into three categories, namely, water reservoirs, gas reservoirs, and low-porosity rocks, which are shown in blue, yellow, and purple, respectively. Comparisons show that the gas reservoirs are located close to the latitude curve of full gas saturation, whereas the water reservoir data are located closer

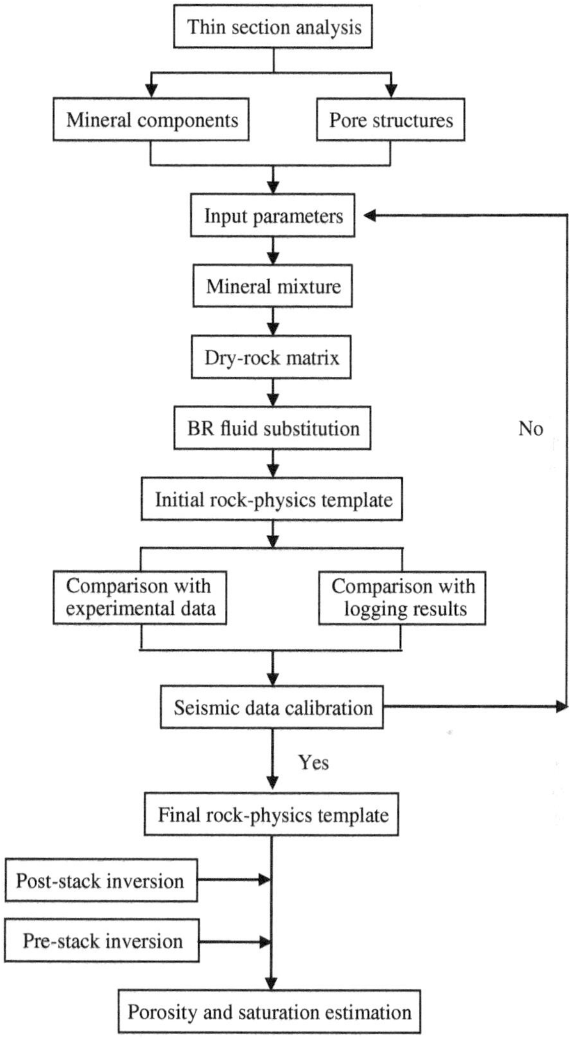

FIGURE 8.30 Workflow for the heterogeneous rock physics modeling and rock parameter inversion.

to the full water saturation curve. The purple triangles (low-porosity rocks), which are considered the least promising for gas production, are mainly located on the right-hand side of the longitude curve for porosity 0.05. According to the template, the rocks with relatively higher porosity (>0.12) and lower gas saturation (<50%) may also be considered gas reservoirs for potential production.

The seismic data (all symbols in Figure 8.31) and the template also suggest that most of the reservoir rocks have porosity in range of 0.05–0.13, which is in

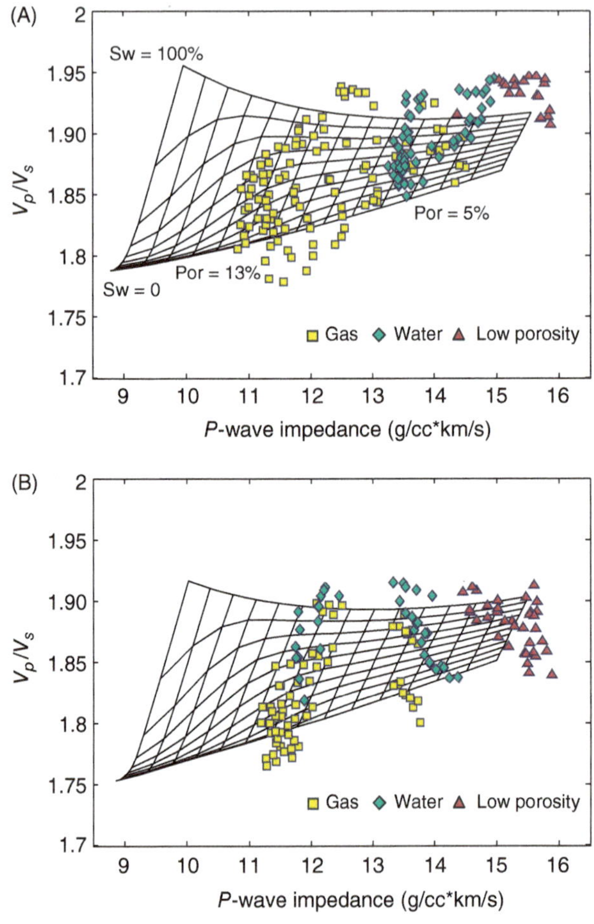

FIGURE 8.31 Rock physics templates of Met22 and Met21, corresponding to Figure 8.28A and B, respectively. (A) The calibrated rock physics template for Well Met22. (B) The calibrated rock physics template for Well Met21.

good agreement with the statistical results of core data from the target stratum (e.g., in Figure 8.32, almost 54% of the reservoir limestone has porosity in range of 5%–12.5%, which contributes to the major gas production of the whole stratum). Moreover, it is also evident in Figure 8.31 that the gas reservoirs with the best quality are those with porosity 8%–13%.

8.4.2.3 Porosity and Gas Saturation Estimate Based on Seismic Data

Based on the analysis of reservoir heterogeneity, we established the appropriate rock physics templates for the seismic properties of the partially saturated limestone. The P-wave impedance and Vp/Vs ratio obtained from the poststack and

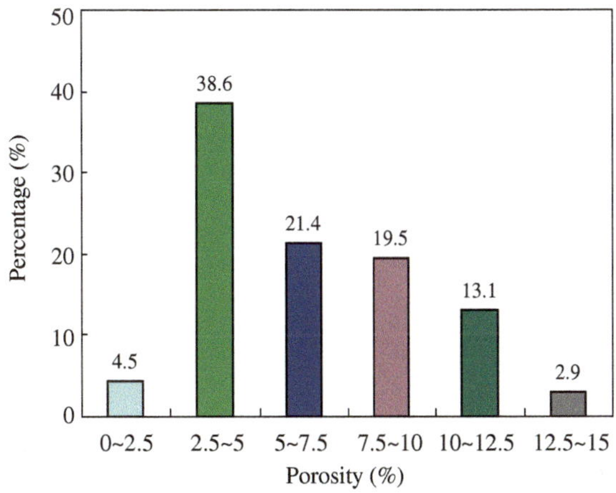

FIGURE 8.32 Statistical histogram of reservoir rock porosity versus percentage in the XVm layer of Metejan limestone.

prestack seismic inversion are used to predict the porosity and saturation. Only the reservoirs with high quality (porosity >8%) are considered.

Figure 8.33 shows the cross-sectional porosity (B) and saturation (C) profiles of wells Met22 and Met3 from the seismic data (A). As can be seen in Figure 8.33B, the reservoirs around wells Met22 and Met3 are well developed. The porosity around well Met22 is higher than that around well Met3. As can be seen in Figure 8.33C, wells Met22 and Met3 exhibit high gas saturation characteristics. According to the well test reports of gas production, wells Met22 and Met3 are high-production industrial gas wells (more than 6.7×10^5 m^3 and 7.6×10^5 m^3 gas per day, respectively, for the XVm stratum).

Figure 8.34 shows the cross-sectional porosity (B) and saturation (C) profiles crossing Well Met21, obtained from the seismic data (A). In the inversion saturation profile of Figure 8.34C, low gas saturation is observed around well Met21. According to the well test reports of Met21, nearly no gas is produced. The production rates validate the inversion results from the rock physics template.

Figure 8.35 shows the well-calibrated inversion results for 3D seismic data corresponding to the XVm stratum on an area of approximately 120 km^2 (the high part of the structure), where the poststack and prestack inversions provide the impedance and Vp/Vs ratio, respectively.

It is shown in Figure 8.35A that wells Met22 and Met23 are located in the high-porosity regions, which are the two high-production industrial gas wells. Wells Met1, Met4, and Met5, with medium to high porosities, have lower production rates. In addition, the three low-porosity wells Met6, Met8, and Met21 produce water or are dry wells with no gas present. Thus, porosity estimates can be used to identify gas reservoirs with some confidence. However, there may be exceptions. For instance, well Met3 with low-average porosity is a

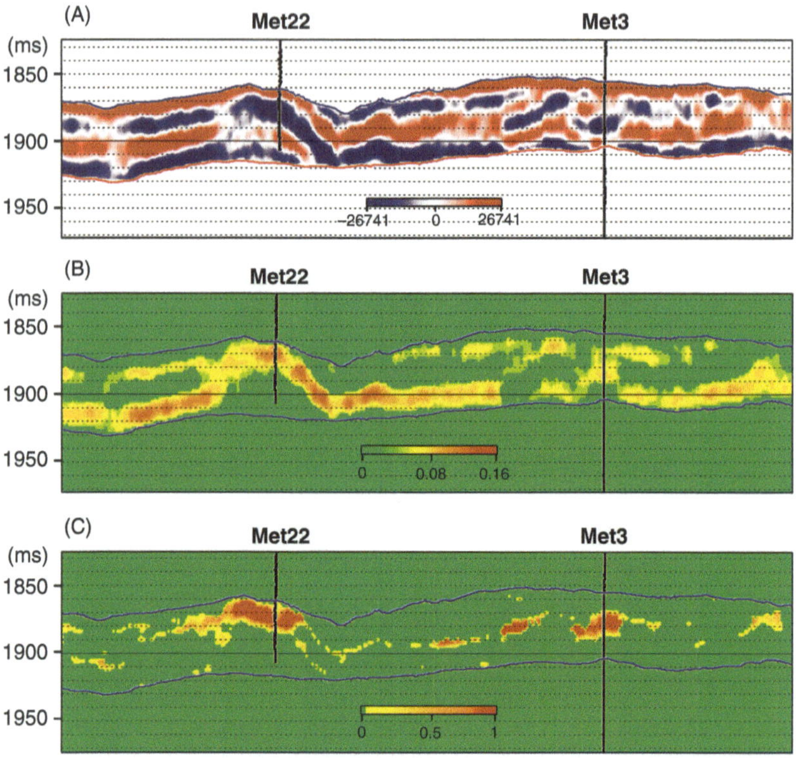

FIGURE 8.33 2D prediction results from cross-sectional profiles of wells Met22 and Met3. (A) Poststack seismic data. (B) Prediction results for porosity (the color bar shows the porosity scale). (C) Prediction results for gas saturation (the color bar shows the gas saturation scale).

high-production gas well, whereas well Met2 which is located in the high-porosity regions is a dry well.

Figure 8.35B shows the distribution of average gas saturation of the high-quality reservoirs (porosity >0.08). Wells Met3, Met22, and Met23 are all located in the high gas saturation regions, which can be clearly discriminated from the other wells. From well productions tests, wells Met3 and Met22 are highly gas saturated and well Met23 has medium-to-high gas saturation. The estimated results are consistent with the production rates.

8.5 CONCLUSIONS

In this chapter, based on the structure, reservoir, and distribution characteristics of the target carbonate stratum in the Metejan district of the Amu Darya Basin, the reservoir rocks of stratum are studied by the experimental methods including CT scanning, thin section analysis, and core ultrasonic measurement. Gas and water distribution features in rock is investigated by use of CT scanning.

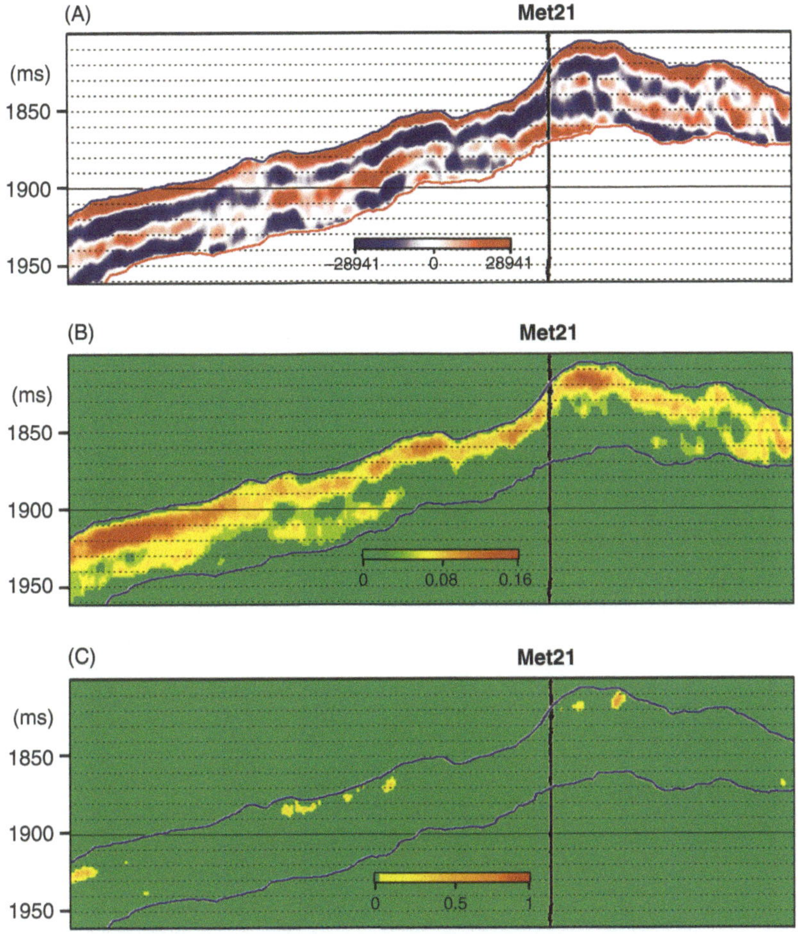

FIGURE 8.34 2D prediction results from cross-sectional profiles crossing wells Met21. (A) Poststack seismic data. (B) Prediction results for porosity (the color bar shows the porosity scale). (C) Prediction results for gas saturation (the color bar shows the gas saturation scale).

The correlation between mineral components, micropore structures, physical properties, and elastic characteristics of carbonate are established by means of thin section analysis. A joint use of a variety of experimental methods has provided a basis for the parameter setting in subsequent rock physics modeling.

For the limestone gas reservoirs in the Metejan district, rock physics templates are compared with experimental data, log data, and seismic data of different scales, respectively. The results show that the templates are basically consistent with the distribution of wave response data of all scales, and rock physics template can be used to quantitatively relate reservoir seismic response with rock porosity and fluid saturation. Based on poststack and prestack seismic

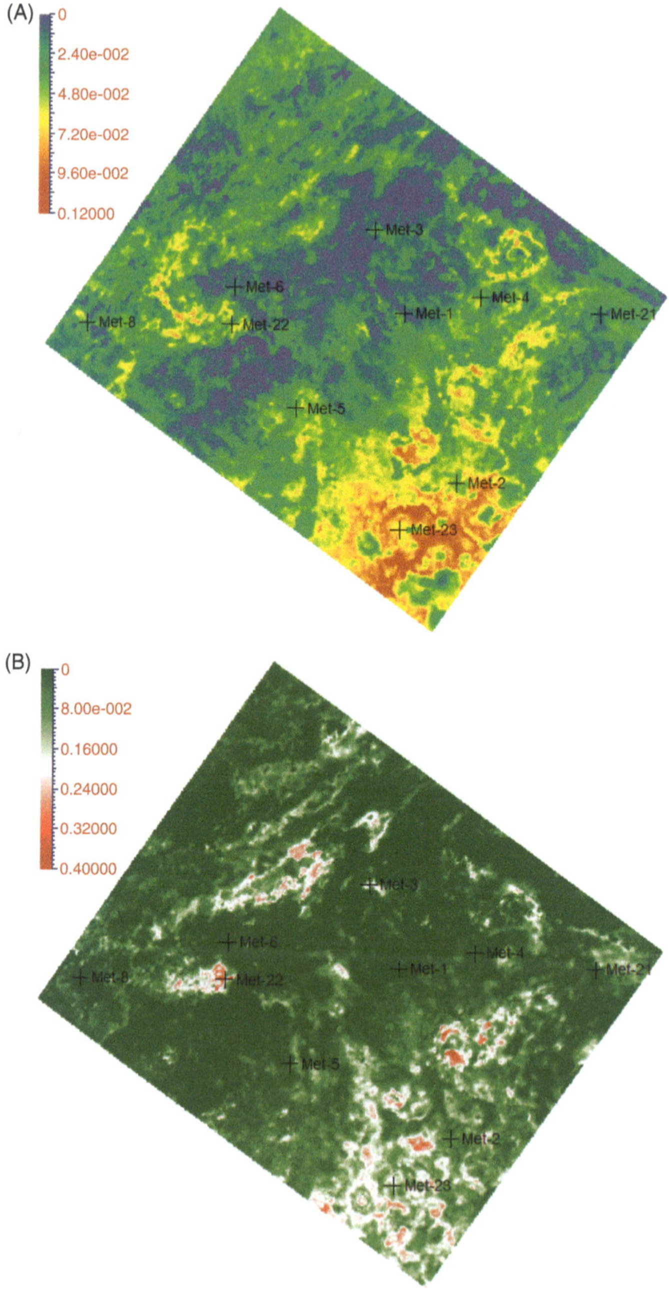

FIGURE 8.35 The predicted average porosity (A) and gas saturation (B) of 3D seismic data corresponding to the XVm stratum in the Metejan district. (A) Average porosity of the Metejan XVm stratum (the color bar is the average porosity scale). (B) Average gas saturation of the high-quality reservoirs of the Metejan XVm stratum (the color bar is the average gas saturation scale).

inversion, the porosity and saturation of actual reservoirs are estimated. The crossing-well inversion results are in good agreement with the actual gas production rates of wells.

As reservoir pore shapes have a significant effect on seismic wave velocities, the lateral heterogeneity characteristics of reservoirs are considered on the basis of the uniform-pore-structure modeling method. The method of rock physics modeling that is based on pore shapes analysis is proposed. Rock physics modeling is conducted respectively for the different pore structures that are developed in different geographical locations in the work area, so that the corresponding rock physics templates are made and gas saturation is estimated around each location. The prediction results agree with the production tests. The relevant techniques are extended from 2D to 3D seismic prediction and the prediction accuracy is improved.

REFERENCES

Adam, L., Batzle, M., Brevik, I., 2006. Gassmann's fluid substitution and shear modulus variability in carbonates at laboratory seismic and ultrasonic frequencies. Geophysics 71, F13–F183.

Ba, J., Nie, J.X., Cao, H., Yang, H.Z., 2008a. Mesoscopic fluid flow simulation in double-porosity rocks. Geophys. Res. Lett. 35, L04303, doi: 10.1029/2007GL032429.

Ba, J., Cao, H., Yao, F.C., Nie, J.X., 2008b. Double-porosity rock model and Squirt flow for laboratory frequency band. Appl. Geophys. 5, 261–276.

Ba, J., 2010. Wave propagation theory in double-porosity medium and experimental analysis on seismic responses. Scientia Sinica – Phys., Mech. Astron. 40, 1398–1409.

Ba, J., Yang, H.Z., Xie, G.Q., 2008c. AGILD Seismic Modelling for Double-Porosity Media. PIERS 2008 in Hangzhou, Hangzhou, China.

Ba, J., Carcione, J.M., Nie, J.X., 2011. Biot-Rayleigh theory of wave propagation in double-porosity media. J. Geophys. Res. 116, B06202, doi: 10.1029/2010JB008185.

Ba, J., Cao, H., Yao, F.C., 2010. Velocity Dispersion of P-Waves in Sandstone and Carbonate: Double-Porosity Theory and Local Fluid Flow Theory. SEG Annual Meeting 2010, Denver, US.

Ba, J., Carcione, J.M., Cao, H., Du, Q.Z., Yuan, Z.Y., Lu, M.H., 2012. Velocity dispersion and attenuation of P waves in partially-saturated rocks: wave propagation equations in double-porosity medium (in Chinese). Chin. J. Geophys. 55, 219–231.

Berryman, J.G., Wang, H.F., 1995. The elastic coefficients of double-porosity models for fluid transport in jointed rock. J. Geophys. Res. 100, 24611–24627.

Berryman, J.G., Wang, H.F., 2000. Elastic wave propagation and attenuation in a double-porosity dual-permeability medium. Int. J. Rock Mech. Min. Sci. 37, 63–78.

Biot, M.A., 1956a. Theory of propagation of elastic waves in a fluid-saturated porous solid: I. Low-frequency range. J. Acoust. Soc. Am. 28, 168–178.

Biot, M.A., 1956b. Theory of propagation of elastic waves in a fluid-saturated porous solid: II. Higher frequency range. J. Acoust. Soc. Am. 28, 179–191.

Cadoret, T., Marion, D., Zinszner, B., 1995. Influence of frequency and fluid distribution on elastic wave velocities in partially saturated limestones. J. Geophys. Res. 100, 9789–9803.

Che, Z.C., Luo, J.H., Liu, L., 1997. The basic structure classification and genetic analyses of oil and gas bearing basins in Central Asia and Northwestern China. Acta Geoscientica Sinica 18, 113–121.

Chen, X.L., Tang, X.M., 2012. Numerical study on the characteristics of acoustic logging response in the fluid-filled borehole embedded in crack-porous medium (in Chinese). Chin. J. Geophys. 55, 2129–2140.

Dutta, N.C., Seriff, A.J., 1979. On White's model of attenuation in rocks with partial gas saturation. Geophysics 44, 1806–1812.

Dutta, N.C., Odé, Hn., 1979. Attenuation and dispersion of compressional waves in fluid-filled porous rocks with partial gas saturation (White model), Part I-Biot theory. Geophysics 44, 1777–1788.

Dvorkin, J., Nur, A., 1993. Dynamic poroelasticity: a unified model with the Squirt and the Biot mechanisms. Geophysics 58, 524–533.

Gassmann, F., 1951. Über die Elastizität poröser Medien, Vier. der Natur. Gesellschaft in Zürich 96, 1–23.

Li, C.Y., 1982. Geotectonic Maps of Asia and Instructions. SinoMaps Press, Beijing, China.

Liu, L.F., Zhu, Y.X., 2007. Oil/Gas Geological Characteristics of the Pre-Caspian Basin and Central Asia Districts. China Petrochemical Press, Beijing, China.

Johnson, D.L., 2001. Theory of frequency dependent acoustics in patchy saturated porous media. J. Acoust. Soc. Am. 110, 682–694.

Johnson, D.L., Koplik, J., Dashen, R., 1987. Theory of dynamic permeability and tortuosity in fluid-saturated porous media. J. Fluid Mech. 176, 379–402.

Kuster, G.T., Toksöz, M.N., 1974. Velocity and attenuation of seismic waves in two-phase media: I. Theoretical formulations. Geophysics 39, 587–606.

Liu, J., Ma, J.W., Yang, H.Z., 2010. Research on P-wave's propagation in White's sphere model with patchy saturation (in Chinese). Chin. J. Geophys. 53, 954–962.

Liu, J., Ma, J.W., Yang, H.Z., 2009. Research on dispersion and attenuation of P wave in periodic layered-model with patchy saturation (in Chinese). Chin. J. Geophys. 52, 2879–2885.

Mavko, G., Mukerji, T., Dvorkin, J., 2009. The Rock Physics Handbook – Tools for Seismic Analysis of Porous Media, 2nd Edition. Cambridge University Press, New York.

Müller, T.M., Gurevich, B., Lebedev, M., 2010. Seismic wave attenuation and dispersion resulting from wave-induced flow in porous rocks: a review. Geophysics 75, 75A147- 75A164.

Nie, J.X., Yang, D.H., Yang, H.Z., 2004. Inversion of reservoir parameters based on the BISQ model in partially saturated porous media (in Chinese). Chin. J. Geophys. 47, 1101–1105.

Nie, J.X., Ba, J., Yang, D.H., Yan, X.F., Yuan, Z.Y., Qiao, H.P., 2012. BISQ model based on a Kelvin-Voigt viscoelastic frame in a partially saturated porous medium. Appl. Geophys. 9, 213–222.

Norris, A.N., 1985. A differential scheme for the effective moduli of composites. Mech. Mater. 4, 1–16.

Pride, S.R., Berryman, J.G., Harris, J.M., 2004. Seismic attenuation due to wave-induced flow. J. Geophys. Res. 109, B01201, doi: 10.1029/2003JB002639.

Rayleigh, F., 1917. On the pressure developed in a liquid during the collapse of a spherical cavity. Philos. Mag. 34, 94–98.

Sams, M.S., Neep, J.P., Worthington, M.H., King, M.S., 1997. The measurement of velocity dispersion and frequency-dependent intrinsic attenuation in sedimentary rocks. Geophysics 62, 1456–1464.

Santos, J.E., Corberó, J.M., Douglas, J., 1990a. Static and dynamic behavior of a porous solid saturated by a two-phase fluid. J. Acoust. Soc. Am. 87, 1428–1438.

Santos, J.E., Douglas, J., Corberó, J.M., Lovera, O.M., 1990b. A model for wave propagation in a porous medium saturated by a two-phase fluid. J. Acoust. Soc. Am. 87, 1439–1448.

Sun, W.T., Ba, J., Muller, T.M., Carcione, J.M., Cao, H., Du, Q.Z., Yan, X.F., 2012. P-Wave Dispersion and Attenuation in Patchy-Saturated Rocks. White, Dutta and Biot-Rayleigh Theories. 74th EAGE2012 in Copenhagen, Copenhagen, Denmark.

Tang, X.M., 2011. A unified theory for elastic wave propagation through porous media containing cracks: An extension of Biot's poroelastic wave theory (in Chinese). Scientia Sinica - Earth Sci 41, 784–785.

Toms, J., Müller, T.M., Ciz, R., Gurevich, B., 2006. Comparative review of theoretical models for elastic wave attenuation and dispersion in partially saturated rocks. Soil Dyn. Earthq. Eng. 26, 548–565.

White, J.E., 1975. Computed seismic speeds and attenuation in rocks with partial gas saturation. Geophysics 40, 224–232.

Xu, S.Y., White, R.E., 1995. A physical model for shear-wave velocity prediction. Geophys. Prosp. 44, 687–717.

Xu, S.Y., Payne, M.A., 2009. Modeling elastic properties in carbonate rocks. Lead. Edge 28, 66–74.

Yu, H., Li, J.S., Zhang, Y., Xu, G.C., 2013. Spectral decomposition in fault and reservoir identifications. OGP 48, 954–959.

Yu, H., Ba, J., Carcione, J., Li, J., Tang, G., Zhang, X., He, X., Ouyang, H., 2014. Rock physics modeling of heterogeneous carbonate reservoirs: porosity estimation and hydrocarbon detection. Appl. Geophys. 11, 9–22.

Zhao, H.B., 2007. Acoustic Wavefield in Porous Media Saturated With Two Immiscible Fluids (Ph.D. Dissertation). Institute of Acoustics, Chinese Academy of Sciences, Beijing.

Subject Index

A

B

C

G

H

N

O

P

Q

R

S

T

U

V

W

X

Z

Author Index

D

E

F

G

H

W

X

Y

Z

Kolibri 360 GmbH
Ferdinand-Jühlke-Straße 7
99095 Erfurt, Deutschland
produktsicherheit@kolibri360.de